Understanding Our Environment

An Introduction to Environmental Chemistry and Pollution

Third Edition

D0160941

Understanding Our Environment
An Introduction to Environmental Chemistry and Pollution

Third Edition

Edited by

Roy M. Harrison
The University of Birmingham, UK

RS•C
ROYAL SOCIETY OF CHEMISTRY

ISBN 0-85404-584-8

A catalogue record for this book is available from the British Library.

Published by The Royal Society of Chemistry,
Thomas Graham House, Science Park, Milton Road, Cambridge CB4 0WF, UK

For further information see our web site at www.rsc.org

Typeset by Paston PrePress Ltd, Beccles, Suffolk
Printed by Redwood Books Ltd, Trowbridge, Wiltshire

Preface

The field of environmental chemistry goes from strength to strength. Twenty-five years ago it existed in the UK in the form of a few isolated research groups in Universities, Polytechnics, and Research Institutes, but was very definitely a minority interest. It was not taught appreciably in academic institutions and few books dealt with any aspect of the subject. The awakening of environmental awareness, first in a few specialists and subsequently in the general public has led to massive changes. Environmental chemistry is now a component (optional or otherwise) of many chemistry degree courses, it is taught in environmental science courses as an element of increasing substance, and there are even a few degree courses in the subject. Research opportunities in environmental chemistry are a growth area as new programmes open up to tackle local, national, regional, or global problems of environmental chemistry at both fundamental and applied levels. Industry is facing ever tougher regulations regarding the safety and environmental acceptability of its products.

When invited to edit the second edition of 'Understanding Our Environment', I was delighted to take on the task. The first edition had sold well, but had never really met its original very difficult objective of providing an introduction to environmental science for the layman. It has, however, found widespread use as a textbook for both under-graduate and postgraduate-level courses and deserved further development with this in mind. I therefore endeavoured to produce a book giving a rounded introduction to environmental chemistry and pollution, accessible to any reader with some background in the chemical sciences. Most of the book was at a level comprehensible by others such as biologists and physicians who have a modest acquaintance with basic chemistry and physics. The book was intended for those requiring a grounding in the basic concepts of environmental chemistry and pollution. The third edition follows very much the same ethos as the second, but I have tried to encourage chapter authors to develop a more international approach through the use of case studies, and to make the book more easily useable for teaching in a wide range of contexts by the incorporation of worked examples where appropriate and of student questions. The book is a companion volume to 'Pollution: Causes, Effects and Control' (also published by the Royal Society of Chemistry) which is both more diverse in the subjects covered, and in some aspects appreciably more advanced.

Mindful of the quality and success of the second edition, it is fortunate that many of the original authors have contributed revised chapters to this book (A. G. Clarke, R. M. Harrison, B. J. Alloway, S. J. de Mora, C. N. Hewitt, R. Allott, and S. Smith). I am pleased also to welcome new authors who have produced a new view on topics covered in the earlier book (A. S. Tomlin, J. G. Farmer, M. C. Graham, and A. Skinner). The coverage is broadly the same, with some changes in emphasis and much updating. The authors have been chosen for their deep knowledge of the subject and ability to write at the level of a teaching text, and I must express my gratitude to all of them for their hard work and willingness to tolerate my editorial quibbles. The outcome of their work, I believe, is a book of great value as an introductory text which will prove of wide-spread appeal.

Roy M. Harrison
Birmingham

Contents

Contributors

R. Allott, *AEA Technology, Risley, Warrington, WA3 6AT, UK*

B. J. Alloway, *Department of Soil Science, University of Reading, Whiteknights, Reading, RG6 6DW, UK*

A. G. Clarke, *Department of Fuel and Energy, Leeds University, Leeds, LS2 9JT, UK*

S. J. de Mora, *Département d'Océanographie, Université du Québec à Rimouski, 300, allée des Ursulines, Rimouski, Québec, G5L 3A1, Canada*

J. G. Farmer, *Environmental Chemistry Unit, Department of Chemistry, The University of Edinburgh, King's Buildings, West Main Road, Edinburgh, EH9 3JJ, UK*

M. C. Graham, *Environmental Chemistry Unit, Department of Chemistry, The University of Edinburgh, King's Buildings, West Main Road, Edinburgh, EH9 3JJ, UK*

R. M. Harrison, *Institute of Public and Environmental Health, The University of Birmingham, Edgbaston, Birmingham, B15 2TT, UK*

C. N. Hewitt, *Institute of Environmental and Natural Sciences, Lancaster University, Lancaster, LA1 4YQ, UK*

A. Skinner, *Environment Agency, Olton Court, 10 Warwick Road, Solihull, B92 7HX, UK*

S. Smith, *Division of Biosphere Sciences, King's College, University of London, Campden Hill Road, London, W8 7AH, UK*

A. S. Tomlin, *Department of Fuel and Energy, Leeds University, Leeds, LS2 9JT, UK*

CHAPTER 1

Introduction

ROY M. HARRISON

1 THE ENVIRONMENTAL SCIENCES

It may surprise the student of today to learn that 'the environment' has
not always been topical and indeed that environmental issues have
become a matter of widespread public concern only over the past
twenty years or so. Nonetheless, basic environmental science has existed
as a facet of human scientific endeavour since the earliest days of
scientific investigation. In the physical sciences, disciplines such as
geology, geophysics, meteorology, oceanography, and hydrology, and
in the life sciences, ecology, have a long and proud scientific tradition.
These fundamental environmental sciences underpin our understanding
of the natural world, and its current-day counterpart perturbed by
human activity, in which we all live.

   The environmental physical sciences have traditionally been concerned with individual environmental compartments. Thus, geology is
centred primarily on the solid earth, meteorology on the atmosphere,
oceanography upon the salt water basins, and hydrology upon the
behaviour of freshwaters. In general (but not exclusively) it has been the
physical behaviour of these media which has been traditionally perceived
as important. Accordingly, dynamic meteorology is concerned primarily
with the physical processes responsible for atmospheric motion, and
climatology with temporal and spatial patterns in physical properties of
the atmosphere (temperature, rainfall, *etc.*). It is only more recently that
chemical behaviour has been perceived as being important in many of
these areas. Thus, while atmospheric chemical processes are at least as
important as physical processes in many environmental problems such as
stratospheric ozone depletion, the lack of chemical knowledge has been
extremely acute as atmospheric chemistry (beyond major component
ratios) only became a matter of serious scientific study in the 1950s.

There are two major reasons why environmental chemistry has flourished as a discipline only rather recently. Firstly, it was not previously perceived as important. If environmental chemical composition is relatively invariant in time, as it was believed to be, there is little obvious relevance to continuing research. Once, however, it is perceived that composition is changing (*e.g.* CO_2 in the atmosphere; ^{137}Cs in the Irish Sea) and that such changes may have consequences for humankind, the relevance becomes obvious. The idea that using an aerosol spray in your home might damage the stratosphere, although obvious to us today, would stretch the credulity of someone unaccustomed to the concept. Secondly, the rate of advance has in many instances been limited by the available technology. Thus, for example, it was only in the 1960s that sensitive reliable instrumentation became widely available for measurement of trace concentrations of metals in the environment. This led to a massive expansion in research in this field and a substantial *downward* revision of agreed typical concentration levels due to improved methodology in analysis. It was only as a result of James Lovelock's invention of the electron capture detector that CFCs were recognized as minor atmospheric constituents and it became possible to monitor increases in their concentrations (see Table 1). The table exemplifies the sensitivity of analysis required since concentrations are at the ppt level (1 ppt is one part in 10^{12} by volume in the atmosphere) as well as the substantial increasing trends in atmospheric halocarbon concentrations, as measured up to 1990. The implementation of the Montreal Protocol, which requires controls on production of CFCs and

Table 1 *Atmospheric halocarbon concentrations and trends**

Halocarbon	Concentration (ppt) Pre-industrial	1992	Annual Change (ppt) To 1990	Since 1992	Lifetime (years)
CCl_3F (CFC-11)	0	268	+ 9.5	0	50
CCl_2F_2 (CFC-12)	0	503	+ 16.5	+ 7	102
$CClF_3$ (CFC-13)	0	<10			400
$C_2Cl_3F_3$ (CFC-113)	0	82	+ 4–5	0	85
$C_2Cl_2F_4$ (CFC-114)	0	20			300
C_2ClF_5 (CFC-115)	0	<10			1700
CCl_4	0	132	+ 2.0	− 0.5	42
CH_3CCl_3	0	135	+ 6.0	− 10	4.9

*Data from: Intergovernmental Panel on Climate Change, 'Climate Change—The IPCC Scientific Assessment', ed. J. T. Houghton, G. J. Jenkins, and J. J. Ephraums, Cambridge University Press, Cambridge, 1990; and 'Climate Change 1995, The Science of Climate Change', ed. J. T. Houghton, L. G. Meira Filho, B. A. Callendar, N. Harris, A. Kattenberg, and K. Maskell, Cambridge University Press, Cambridge, 1996.

some other halocarbons, has led to a slowing and even a reversal of annual concentration trends since 1992 (see Table 1).

2 THE CHEMICALS OF INTEREST

A very wide range of chemical substances are considered in this book. They fall into three main categories:

(a) *Chemicals of concern because of their human toxicity.* Some metals such as lead, cadmium and mercury are well known for their adverse effects on human health at high levels of exposure. These metals have no known essential role in the human body and therefore exposures can be divided into two categories (see Figure 1). For these non-essential elements, at very low exposures the metals are tolerated with little, if any, adverse effect, but at higher exposures their toxicity is exerted and health consequences are seen. In the case of the so-called essential trace elements (see also Figure 1) the human body requires a certain level of the element, and if intakes are too low then deficiency syndrome diseases will result. These can have consequences as severe as the ones which result from excessive intakes. In between, there is an acceptable range of exposures within which the body is able to regulate an optimum level of the element.

Environmental exposure to chemical carcinogens is very topical despite the minuscule risks associated with many such exposures at typical environmental concentrations. Examples are benzene (largely from vehicle emissions) and polynuclear aromatic hydrocarbons (generated by combustion of fossil fuels). Figure 2 shows the structures of benzene, benzo(*a*)pyrene (the best known of the carcinogenic polycyclic aromatic hydrocarbons), and 2,3,7,8-tetrachlorodibenzodioxin (the most toxic of the chlorinated dioxin group of compounds). Despite great public concern over emissions of the last compound, the evidence for carcinogenicity in humans is quite limited.

(b) *Chemicals which cause damage to non-human biota but are not believed to harm humans at current levels of exposure.* Many elements and compounds come into this category. For example, copper and zinc are essential trace elements for humans and environmental exposures very rarely present a risk to health. These elements are, however, toxic to growing plants and there are regulations limiting their addition to soil in materials such as sewage sludge which is disposed of to the land. Another category of substance for which there is ample evidence of harm to biota,

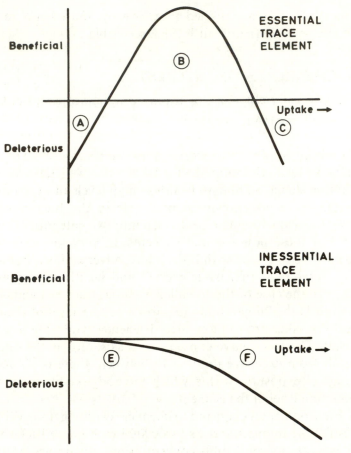

Figure 1   *Comparison of the consequences of exposure to essential and inessential trace elements. For the essential trace elements, Region A represents the deficiency syndrome when intakes are insufficient, Area B is the optimum exposure window and in Area C, excessive intake leads to toxic consequences. In the case of the inessential trace element at low exposures (Zone E) the element is tolerated and little if any adverse effect occurs. In Zone F toxic symptoms are developed*

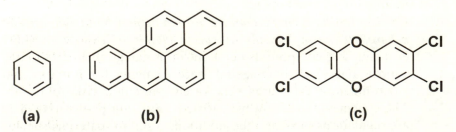

Figure 2   *Some molecules believed to have human carcinogenic potential: (a) benzene; (b) benzo[a]pyrene; (c) 2,3,7,8-tetrachlorodibenzodioxin*

but as yet little, if any, hard evidence of impacts on human populations, are the endocrine-disrupting chemicals (also termed oestrogenics). These synthetic chemicals mimic natural hormones and can disrupt the reproduction and growth of wildlife species. Thus, for example, bis-tributyl tin oxide (TBTO) interferes with the sexual development of oysters and its use as an anti-fouling paint for inshore vessels is now banned in most parts of the world. A wide range of other chemicals including polychlorinated biphenyls (PCBs), dioxins, and many chlorinated pesticides are also believed to have oestrogenic potential, although the level of evidence for adverse effects is variable.

(c) *Chemicals not directly toxic to humans or other biota at current environmental concentrations, but capable of causing environmental damage.* The prime example is the CFCs which found widespread use precisely because of their stability and low toxicity to humans, but which at parts per trillion levels of concentration are capable of causing major disruption to the chemistry of the stratosphere.

3 UNITS OF CONCENTRATION

The concentration units used in environmental chemistry are often confusing to the newcomer. Concentrations of pollutants in soils are most usually expressed in mass per unit mass, for example, milligrams of lead per kilogram of soil. Similarly, concentrations in vegetation are also expressed in $mg\,kg^{-1}$ or $\mu g\,kg^{-1}$. In the case of vegetation and soils, it is important to distinguish between wet weight and dry weight concentrations, in other words, whether the kilogram of vegetation or soil is determined before or after drying. Since the moisture content of vegetation can easily exceed 50%, the data can be very sensitive to this correction.

In aquatic systems, concentrations can also be expressed as mass per unit mass and in the oceans some trace constituents are present at concentrations of $ng\,kg^{-1}$ or $\mu g\,kg^{-1}$. More often, however, sample sizes are measured by volume and concentrations expressed as $ng\,l^{-1}$ or $\mu g\,l^{-1}$. In the case of freshwaters, especially, concentrations expressed as mass per litre will be almost identical to those expressed as mass per kilogram. As a kind of shorthand, however, water chemists sometimes refer to concentrations as if they were ratios by weight, thus, $mg\,l^{-1}$ are expressed as parts per million (ppm), $\mu g\,l^{-1}$ as parts per billion (ppb) and $ng\,l^{-1}$ as parts per trillion (ppt). This is unfortunate as it leads to confusion with the same units used in atmospheric chemistry with a quite different meaning.

Concentrations of trace gases and particles in the atmosphere can be expressed also as mass per unit volume, typically $\mu g\ m^{-3}$. The difficulty with this unit is that it is not independent of temperature and pressure. Thus, as an airmass becomes warmer or colder or changes in pressure so its volume will change, but the mass of the trace gas will not. Therefore, air containing 1 $\mu g\ m^{-3}$ of sulfur dioxide in air at 0 °C will contain less than 1 $\mu g\ m^{-3}$ of sulfur dioxide in air if heated to 25 °C. For gases (but not particles) this difficulty is overcome by expressing the concentration of a trace gas as a volume mixing ratio. Thus, 1 cm^3 of pure sulfur dioxide dispersed in 1 m^3 of polluted air would be described as a concentration of 1 part per million (ppm). Reference to the gas laws tells us that not only is this one part per 10^6 by volume, it is also one molecule in 10^6 molecules and one mole in 10^6 moles, as well as a partial pressure of 10^{-6} atmospheres. Additionally, if the temperature and pressure of the airmass change, this affects the trace gas in the same way as the air in which it is contained and the volume mixing ratio does not change. Thus, ozone in the stratosphere is present in the air at considerably higher mixing ratios than in the lower atmosphere (troposphere), but if the concentrations are expressed in $\mu g\ m^{-3}$ they are little different because of the much lower density of air at stratospheric attitudes. Chemical kineticists often express atmospheric concentrations in molecules per cubic centimetre (molec cm^{-3}), which has the same problem as the mass per unit volume units.

Worked Example

The concentration of nitrogen dioxide in polluted air is 85 ppb. Express this concentration in units of $\mu g\ m^{-3}$ and molec cm^{-3} if the air temperature is 20 °C and the pressure 1005 mb (1.005 $\times$ 10^5 Pa). Relative molecular mass of NO_2 is 46; Avogadro number is 6.022 $\times$ 10^{23}.

The concentration of NO_2 is 85 $\mu l\ m^{-3}$. At 20 °C and 1005 mb,

$$85\ \mu l\ NO_2\ \text{weigh}\ 46 \times \frac{85 \times 10^{-6}}{22.41} \times \frac{273}{293} \times \frac{1005}{1013}$$

$$= 161 \times 10^{-6}\ g$$

NO_2 concentration $= 161\ \mu g\ m^{-3}$

This is equivalent to 161 pg cm^{-3}, and

$$161\ \text{pg}\ NO_2\ \text{contain}\ 6.022 \times 10^{23} \times \frac{161 \times 10^{-12}}{46}$$

$$= 2.1 \times 10^{12}\ \text{molecules}$$

and NO_2 concentration $= 2.1 \times 10^{12}$ molec cm^{-3}.

4 THE ENVIRONMENT AS A WHOLE

A facet of the chemically centred study of the environment is a greater integration of the treatment of environmental media. Traditional boundaries between atmosphere and waters, for example, are not a deterrent to the transfer of chemicals (in either direction), and indeed many important and interesting processes occur at these phase boundaries.

In this book, the treatment first follows traditional compartments (Chapters 2, 3, 4, and 5) although some exchanges with other compartments are considered. Fundamental aspects of the science of the atmosphere, waters, and soils are described, together with current environmental questions, exemplified by case studies. Subsequently, quantitative aspects of transfer across phase boundaries are described and examples given of biogeochemical cycles (Chapter 6). Monitoring considerations are covered in Chapter 7, with the effects of chemical pollution in Chapter 8, and finally the regulatory aspects in Chapter 9.

5 BIBLIOGRAPHY

For readers requiring knowledge of basic chemical principles:
R.M. Harrison and S.J. de Mora, 'Introductory Chemistry for the Environmental Sciences', Cambridge University Press, Cambridge, 2nd Edn., 1996.

For more detailed information upon pollution phenomena:
'Pollution: Causes, Effects and Control', ed. R.M. Harrison, Royal Society of Chemistry, Cambridge, 3rd Edn., 1996.

CHAPTER 2

The Atmosphere

A. G. CLARKE AND A. S. TOMLIN

1 THE GLOBAL ATMOSPHERE

1.1 The Structure of the Atmosphere

1.1.1 Troposphere and Stratosphere. The vertical structure of the atmosphere, showing the features that are most relevant to the problems covered in this chapter, is illustrated in Figure 1. The figure shows the stratosphere, troposphere and boundary layer (that closest to the earth's surface). The difference between the layers is characterized by changes in temperature with height, and with changes in structure of the layers such as cloud cover and turbulence. The depth of the troposphere is 8–15 km, the lowest values occurring at the poles and the highest at the equator with some seasonal variations. Within this layer occurs most of the variability of conditions which leads to 'the weather' as the layman experiences it. The stratosphere is relatively cloud-free and considerably less turbulent—hence long distance passenger jets fly at stratospheric altitudes. Within the troposphere temperature decreases with height owing to the decreasing influence of radiation from the earth's surface, but as we enter the stratosphere the temperature starts to increase again. The turning point is called the tropopause. This situation of a layer of warmer, less dense air over a layer of cooler, denser air is quite stable. Consequently air is mixed across the tropopause very slowly unless special events such as tropospheric folding occur.

We normally think of 'air pollution' in terms of the troposphere, within which most pollutants have a fairly limited lifetime before they are washed out by rain, removed by reaction, or deposited to the ground. However, if pollutants are injected directly into the stratosphere they can remain there for long periods because of slow downward mixing, resulting in noticeable effects over the whole globe. Thus major volcanic

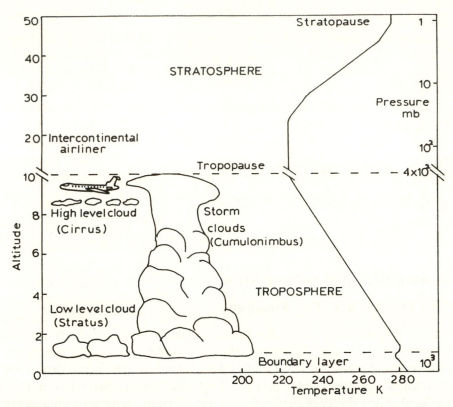

Figure 1   *The vertical structure of the atmosphere. The temperature profile would be
typical for latitude 60° N in summer. Note the change of scale used for the upper
half of the figure*

eruptions injecting fine dust into the stratosphere can lead to a reduction
in the amount of solar energy reaching the ground for more than a year
after the event. Other global problems relating to events in the stratosphere such as the possibility of damage to the ozone layer are discussed
later.

1.1.2   Atmospheric Circulation.   To understand both global and local
environmental problems we must first understand how pollutants
circulate throughout the atmosphere. The main driving forces for the
circulation of the atmosphere are the incident solar radiation and the
earth's rotation. Because of the sun's angle, the amount of solar energy
falling on a given area varies with latitude so that the poles are cold and
the equatorial regions warm. Warm air rises at the equator and cold air
flows inwards from both North and South. A similar situation occurs at
the poles where warm air flows towards them and falls in the cold regions
there. The rotation of the earth affects the circulation patterns in a

fundamental way due to an effect called the Coriolis force. For example, air moving south towards the equator gives the impression of being influenced by a force in the westerly direction. The net result is the tendency for air to circulate in large-scale eddies around the 'low' and 'high' pressure regions on synoptic weather charts.

The proportions of incident energy reflected back to space, absorbed by the land or sea, and re-radiated at a longer wavelength all vary from place to place and affect the temperature distribution and circulation patterns. This energy balance is crucial to the determination of the global climate and is considered in more detail in Section 1.2. The processes of evaporation of water, cloud formation, and precipitation also affect the energy balance and circulation patterns. The presence of the ground has only a small effect on the overall pattern of atmospheric circulation and at most altitudes air movements approximate to those of a non-viscous fluid. The theoretical wind speed can be calculated from the pressure gradient and the rotational velocity of the earth—the so-called geostrophic wind speed. The pressure gradient is reflected on a weather chart by the closeness of the isobars, lines of constant pressure. If the isobars are close together the wind speed will be high.

1.1.3   The Boundary Layer.   Near to the ground the situation is more complicated due to the effects of frictional and buoyancy forces. Turbulence is generated by mechanical forces as air flows over uneven ground features such as hills, buildings, or trees. The ground may also warm or cool the air next to it resulting in upcurrents and downcurrents. In the language of fluid mechanics we have turbulent transport of momentum and energy with corresponding velocity and temperature gradients in the vertical direction. Consider the variation of wind speed with height over the lowest few hundred metres of the atmosphere. This variation is greatest over rough surfaces (*e.g.* a city) where the effect could be a reduction of 40% of the wind speed aloft, that is, the geostrophic wind. Over smooth surfaces (*e.g.* sea, ice sheets) the effect is less and the reduction may be only 20%. The changing effect of friction with height also causes a variation of wind direction as we move away from the earth's surface, *i.e.* 'wind shear'. A plume from a tall chimney may therefore appear to be travelling at an angle to the ground level wind.

Within the troposphere we therefore define a boundary layer within which surface effects are important. This is of the order of 1 km in depth but varies significantly with meteorological conditions (Figure 1). Vertical mixing of pollutants within the boundary layer is largely determined by the atmospheric stability which relates to the intensity of the buoyancy effects previously mentioned. This is the subject of a later section.

Table 1   *Time and distance scales for atmospheric dispersion of emissions*

Time of travel	Typical distances	Area affected
Hours	Tens of km	Throughout the boundary layer
Days	Thousands of km	Pollutant escaping from boundary layer into free troposphere
Weeks	Round the earth	The whole troposphere in one hemisphere Transport to other hemisphere beginning
Months	Round the earth	Whole global troposphere. Some penetration into lower stratosphere

As a generalization, mixing within the boundary layer is relatively rapid whereas mixing through the remainder of the troposphere is slower. This gives rise to the idea of a mixing depth within which pollutants are retained and may be transported long distances. So, for example, models of pollutant transport from the UK to the rest of Europe involve a distance scale of about 1000 km and often assume vertical mixing depth of perhaps 1 km with the pollutants uniformly distributed within this layer. Table 1 indicates the time and distance scales involved in the dispersion of pollutants emitted from the ground. No account is taken in this table of the rates of removal of any pollutant by reaction, deposition to the ground, *etc.*

1.2   Greenhouse Gases and the Global Climate[1–6]

1.2.1   The Global Energy Balance.   The energy that reaches the earth comes from the sun, and the absorption and loss of radiation from the earth and its atmosphere determine our climate. If the earth had no atmosphere, the mean surface temperature would be 255 K, well below the freezing point of water. The atmosphere serves to retain heat near the surface and the earth is thereby made habitable. This accounting for incoming and outgoing energy is called the global energy balance and

[1] 'The Greenhouse Effect, Climate Change and Ecosystems' (SCOPE 29), ed. B. Bolin, B. R. Doos, J. Jager, and R. A. Warwick, John Wiley & Sons, Chichester, 1986.
[2] 'Climate Change 1995: The Science of Climate Change', ed. J. T. Houghton, L. G. Meira Filho, B. A. Callander, N. Harris, A. Kattenberg, and K. Maskell, Cambridge University Press, Cambridge, 1996.
[3] 'The Greenhouse Effect and Terrestrial Ecosystems of the UK', ed. M. G. R. Cannell and M. D. Hooper, ITTE Research Publication No. 4, HMSO, London, 1990.
[4] 'Climate Change, The IPCC Scientific Assessment', ed. F. T. Houghton, G. J. Jenkins, and J. J. Ephraums, Cambridge University Press, Cambridge, 1990.
[5] http://www.ipcc.ch/ 'The Intergovernmental Panel on Climate Change (IPCC) home page', December 1997.
[6] http://www.bna.com/prodhome/ens/text4.htm, 'The Kyoto Protocol', December 1997.

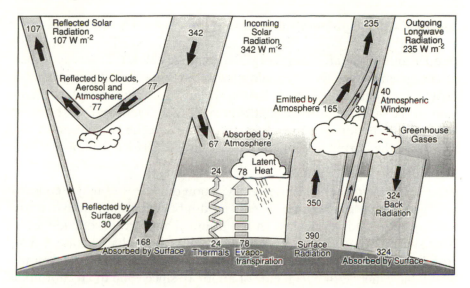

Figure 2 *The earth's radiation and energy balance for a net incoming solar radiation of 342 W m^{-2}*
(Reproduced with permission from the Intergovernmental Panel on Climate Change[2])

could potentially be upset by any significant changes to the earth's atmosphere.

Most of the radiant energy from the sun lies in or near the visible region of the spectrum (*i.e.* at short wavelength *ca.* 0.6 μm) with some in the UV region. The stratosphere absorbs UV radiation primarily due to the ozone present and this results in the warming shown in Figure 1. The lower atmosphere is transparent to visible light so it gains relatively little energy from incoming radiation. Some of the transmitted radiant energy in the visible region penetrates to the ground and is absorbed. Some light is reflected unchanged from clouds or from the ground (especially by snow or ice). The fraction of reflected light is termed the albedo and is over 0.5 for clouds but below 0.1 for the oceans. The global average albedo is about 0.3. Figure 2 shows the amounts of radiation for different components of the overall energy balance. The radiation emitted from the ground lies in the infrared region of the spectrum (long wavelength, *ca.* 10–15 μm) and several atmospheric constituents absorb in this region. Carbon dioxide, water vapour, and ozone are the most important of these. Methane, nitrous oxide, and chlorofluorocarbons (CFCs) are also significant. Some of the absorbed energy will still be re-radiated back to space but a part will be returned to the ground or retained in the atmosphere. The final factors that result in surface to atmosphere transfer of energy are direct warming of the air nearest the ground together with evaporation/condensation processes. The net effect is that more energy is retained near the surface of the earth and the mean temperature is therefore higher (global average 288 K). This is described as the 'greenhouse effect' by analogy with the properties of glass. Glass is largely transparent to solar radiation while absorbing completely radiation in the infrared at wavelengths greater than 3 μm. In fact the most important function of a greenhouse is to prevent the circulation of air, inhibiting the normal cooling processes, but the term 'greenhouse effect' has none the less been retained.

1.2.2   The Carbon Dioxide Cycle.   Carbon dioxide is of major concern as a greenhouse gas because there is no doubt that human activities are leading to a gradual increase in the atmospheric carbon dioxide level. This suggests that we may eventually modify the global climate. Fossil fuel burning is the main contributor to the global annual emissions, which have increased by a factor of about 10 since 1900 to an enormous 6.1×10^9 tonnes in 1994. Deforestation adds about another 1.6×10^9 tonnes per annum.[2] This must be considered in relation to the total atmospheric content of CO_2 which is about 750×10^9 tonnes, corresponding to a concentration of around 358 ppmv in 1994 as opposed to 280 ppmv in pre-industrial times. The various components of the overall global balance of carbon dioxide are generally understood but not easily quantified. Figure 3 shows the global carbon cycle and carbon reservoirs. CO_2 is removed from the atmosphere by photosynthesis in plants thus fixing CO_2 into a biomass reservoir. CO_2 is released in the processes of respiration and decay and these processes are naturally in balance unless we destroy the biomass reservoir or burn fixed forms of carbon. The oceans contain vast amounts of CO_2 in inorganic form as well as in association with living organisms such as plankton. Exchange of gas between the atmosphere and the upper layers of the ocean is rapid and subsequent transfer to deep ocean regions slow. In some areas there may be net release of CO_2 and in other areas net removal but overall the oceans represent a net sink for CO_2 although on a slow time-scale. It is estimated that the time taken for the atmosphere to adjust to changes in sources and sinks of CO_2 is between 50 and 200 years although this is difficult to quantify because each part of the carbon cycle has its own time-scale.

1.2.3   Global Warming.   The rate of concentration change of CO_2 and other greenhouse gases is shown in Table 2. What is important however is not just the rate of increase but the effect each species could have on global warming. Molecule for molecule changes in CH_4, N_2O and the CFCs have more effect than changes in CO_2 although their overall concentrations are lower. The Global Warming Potential (GWP) is a

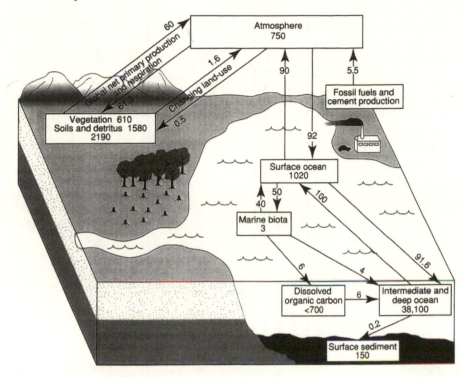

Figure 3 *The carbon dioxide cycle, showing the reservoirs (in GtC) and fluxes (GtC/yr) relevant to the anthropogenic perturbation averages over the period 1980 to 1989* (Reproduced with permission from the Intergovernmental Panel on Climate Change[2])

quantified measure of the relative effect of each species on radiative forcing of the atmosphere, including both direct and indirect effects. The index is defined as a cumulative radiative forcing between the present time and some specified time in the future caused by a unit mass of gas emitted at present, relative to CO_2.[2] The total global warming effect of each gas is then determined by multiplying by the amount of gas emitted. A typical uncertainty in the figures is about 35%. Table 2 demonstrates the GWPs for a number of greenhouse gases. Although CO_2 is the most important contributor, the other gases taken together contribute about half the overall radiative forcing. Some of the species represent CFCs and their replacements following the Montreal Protocol. Although these species have high GWPs their concentrations are small and their total impact less than 3%. We will return to these species in Section 1.3. The situation with ozone is quite complex and depends on its vertical distribution with particular sensitivity around the tropopause. The reduction in lower stratospheric ozone over the last 15–20 years has been shown to have a slight negative effect on global warming. In the troposphere there appears to be a gradual increase in the level of ozone

Table 2 Concentration changes, lifetimes and global warming potentials of greenhouse gases[2]

Species	Pre-industrial concentration	Concentration in 1994	Rate of concentration change	Atmospheric lifetime (years)	GWP (time horizon 20 years)	GWP (time horizon 100 years)
CO_2	280 ppmv	358 ppmv	0.4% yr^{-1}	50–200	1	1
CH_4	700 ppbv	1720 ppbv	0.6% yr^{-1}	12	56	21
N_2O	275 ppbv	312 ppbv	0.25% yr^{-1}	120	280	310
CFC-11	0	268 pptv	0% yr^{-1}	50	4500	3500
CF_4	0	72 pptv	12% yr^{-1}	50 000	4400	6500
HCFC-22 (a CFC substitute)	0	110 pptv	5% yr^{-1}	12	1500	510

due to emissions of NO_x and hydrocarbons, and this will have a positive effect on global warming. The net effect of changing ozone levels is predicted to be positive although small and may vary from region to region. Aerosols (including particles, small droplets and soot) may also affect global warming by either scattering and absorbing radiation or through their effects on clouds. Although the effect of aerosols shows a complex dependency on their size and distribution there have been significant advances in quantifying their contribution to global warming, and current models predict they have a negative (*i.e.* a cooling) overall effect. Aerosols are very short-lived species and hence their radiative forcing will respond quickly to changes in emissions. An example of this is the cooling effect caused by the 1991 volcanic eruption of Mount Pinatubo.

1.2.4 Climate Change. There now seems to be some consensus that mean surface temperatures have been increasing since the late 19th century at a rate over and above natural variability. However, the climatological consequences of global warming are still not well understood. Modelling the effect of increased greenhouse gas levels on the global climate is an enormously complex problem requiring high performance computers. Three-dimensional global circulation models describe the vertical, latitudinal, and longitudinal variations in conditions and attempt to compare the present situation with various scenarios for future emissions based on predicted population and economic growth, energy availability, *etc.*[2] The feedback processes described above are being better and better represented in the models as is the coupling between atmosphere and ocean. Predictions show that changes in surface and atmospheric temperatures, cloud cover, evaporation and precipitation, *etc.* are all affected by the changed radiation balance but the effects are not equally distributed over the globe. For a medium emissions scenario, predicting CO_2 doubling by 2100, most models now suggest an increase of about 2 °C in mean global temperatures relative to 1900,[2,5] with the largest increases of 8–10 °C over high Northern latitudes as shown in Plate 1*, 2–3 °C in Europe and N. Asia ($>$50 N) in winter, and increases of up to 4 °C in Antarctica. Although the global precipitation might increase by 5–10%, the studies suggest that the tropics and areas bordering the eastern coasts of continents would become generally wetter and the subtropical regions become drier. This might be critical for some central African regions which already suffer severe drought conditions. A rise in sea level resulting from the melting of ice-caps and thermal expansion of the oceans is predicted at 15–50 cm by 2100. A significant thinning of the Arctic ice seems to have occurred already and is currently being studied.

* Plates are between pages 48 and 49.

1.2.5 International Response. International discussions have been taking place for some years with a view to limiting the emissions of greenhouse gases. The second World Climate Conference met in Geneva in 1990. It had as its technical basis a report[4] from the UN Intergovernmental Panel on Climate Change, an international body of 300 scientists. A follow-up meeting in Berlin (1995) agreed that targets suggested by the Rio summit were inadequate and industrialized nations should act within a shorter time. The third session of the Conference of the Parties (COP)[5] to the Climate Change Convention took place in Kyoto in December 1997 where agreements were finally made and a Protocol established.[6] The Parties to the Convention, shown in Table 3, have agreed individually or jointly to ensure that their aggregate anthropogenic carbon dioxide equivalent emissions of the greenhouse gases (CO_2, CH_4, N_2O, hydrofluorocarbons, perfluorocarbons, and sulfur hexafluoride) do not exceed their assigned amounts by the year 2010. The countries who have signed the Protocol are listed below along with their allowed emissions as a percentage of their emissions in the base year, 1990. Some countries (marked with an *) are agreed to be undergoing the transition to a market economy and therefore their base year is different.

Reductions are expected to be achieved in a number of ways including: the enhancement of energy efficiency; sustainable forest management and agricultural practices; the development and increased use of new and renewable forms of energy and carbon dioxide sequestration technologies; the progressive reduction or phasing out of tax exemptions and subsidies that are counter to the objective of the Convention; measures to limit and/or reduce emissions of greenhouse gases in the transport sector; and the limitation and/or reduction of methane emissions through

Table 3 *Annex B of the Kyoto Protocol—Agreed reductions or limitations in emissions of greenhouse gases not covered by the Montreal Protocol*[6]

Party	*Agreed emission limitation or reduction (% base year)*
Iceland	110
Australia	108
Norway	101
New Zealand, Russian Federation*, Ukraine*	100
Croatia*	95
Canada, Hungary*, Japan, Poland*	94
United States of America	93
Bulgaria*, Czech Republic*, Estonia*, European Community, Monaco, Switzerland, Latvia*, Liechtenstein, Lithuania*, Romania*, Slovakia*, Slovenia*	92

recovery and use in waste management as well as in the production, transport and distribution of energy.

Developing countries remain exempt for the present although there is a 'clean development mechanism' aimed at promoting sustainable development through technology transfer. Without this it is likely that emissions from developing nations will rise steeply and may account for 60% of emissions over the next two decades.

It is probable that the wider use of natural gas as a substitute for coal will result in some benefit since the mass of CO_2 emitted per unit of heat released is less: gas 0.43, oil 0.62, coal 0.75 ktonne $(MW\ yr)^{-1}$. This trend will occur in the UK as natural gas is increasingly used for power production and the amount of coal used is reduced. However, many developing nations such as India and China have large coal reserves. The other alternatives are the use of renewable energy sources such as wind, solar, wave, and tidal power or the further development of nuclear energy. This seems unlikely in the short term for both economic and environmental reasons. The use of biomass is also a possibility on a local scale since the replanting of biomass fuels makes it a sustainable energy source. National emissions of CO_2 for several countries are listed below. Figures for China and former USSR are calculated estimates.

UK (1994)   551 Tg[7]   Germany (1994)   874 Tg[7]   France (1994) 308 Tg[7]
USA (1995) 4786 Tg[8]   Former USSR (1995) 3804 Tg[9]   China (1995) 2389 Tg[9]

1.3 Depletion of Stratospheric Ozone[10-20]

1.3.1 The Ozone Layer.   Although ozone occurs in the troposphere and plays an important role in air pollution chemistry, about 90% of the

[7] http://www.aeat.co.uk/netcen/corinair/94/corin94.html 'European topic centre on air emissions', December 1997.
[8] http://www.epa.gov/globalwarming/inventory/ 'US EPA global warming emissions inventories'.
[9] http://cesimo.ing.ula.ve/GAIA/Countries/co2_total.html
[10] UK Review Group on Stratospheric Ozone, 'Stratospheric Ozone', HMSO, London, 1987.
[11] UK Review Group on Stratospheric Ozone, 'Stratospheric Ozone 1988', HMSO, London, 1988.
[12] UK Review Group on Stratospheric Ozone, 'Stratospheric Ozone 1990', HMSO, London, 1990.
[13] http://www.unep.ch/ozone/home.htm 'The Ozone Secretariat WWW Home Page: UNEP', December 1997.
[14] http://www.al.noaa.gov/WWWHD/pubdocs/WMOUNEP94.html 'Executive summary of the WMO/UNEP Scientific Assessment of Ozone Depletion: 1994 document', December 1997
[15] 'Scientific Assessment of Ozone Depletion: 1994', World Meteorological Organization, Global Ozone research and Monitoring Project, Report No. 37, WMO, Geneva, 1995.
[16] J. C. Farman, B. J. Gardiner, and J. D. Shanklin, *Nature*, 1985, **315**, 207.
[17] http://www.epa.gov/ozone/mbr/mbrqa.html 'The US EPA Methyl Bromide Phase Out Web Site', December 1997.
[18] http://jwocky.gsfc.nasa.gov/ 'The TOMS home page', December 1997.
[19] S. Solomon and D. L. Albritton, *Nature*, 1992, **357**, 33.
[20] http://www.he.net/~afeas/index.html 'AFEAS, Alternative Fluorocarbons Environmental Acceptability Study', December 1997.

total ozone content of the atmosphere occurs in the stratosphere at altitudes between 15 and 50 km. The ozone layer acts as a filter for ultraviolet radiation from the sun, removing most of the radiation below 300 nm. This serves to protect humans from the adverse effects of UV which become significant below 320nm since decreasing wavelength corresponds to higher energy photons which can cause sunburn and types of skin cancer. Any depletion of stratospheric ozone would therefore lead to a larger amount of UV radiation incident on the earth's surface and an increased risk of the induction of cancers.

Concern was first expressed about this risk in the early 1970s in connection with emissions of nitrogen oxides from supersonic aircraft such as Concorde, which fly in the lower stratosphere. Nitrogen oxides are potential catalysts for the destruction of ozone. This particular effect is now thought to be relatively minor and attention switched in the 1980s to halogen compounds, especially CFCs or freons. Freons are a group of chlorofluorocarbons which have been used as aerosol propellants, refrigerants and as gases for the production of foamed plastics. Their attraction lies in the fact that they are non-toxic, non-flammable and chemically inert. Global production of the two commonest gases, CFC 11 ($CFCl_3$) and CFC 12 (CF_2Cl_2), rose rapidly from below 50,000 tonnes per annum in 1950 to 725,000 tonnes per annum by 1976 decreasing slightly to 650,000 tonnes in 1985. About 90% was released directly to the atmosphere while the remainder, representing refrigerant use, will be released when the equipment is eventually discarded. The actual concentration of CFCs in the atmosphere is extremely small, (less than 1ppb, see Table 2), but has risen dramatically this century at a rate which correlates well with known emissions. This rise in CFCs in the 1980s has clearly affected stratospheric ozone levels via the processes described below and has been the subject of control measures in the 1990s as will be discussed later.

Because they are chemically inert CFCs are resistant to attack by molecules, radicals, or the UV radiation present in the troposphere and are not subject to significant dry deposition or rain-out. The higher energy UV radiation in the stratosphere can, however, lead to photo-dissociation forming chlorine atoms which can in turn lead to the destruction of ozone. Despite the slow exchange of air between the troposphere and the stratosphere this effect is now known to be significant.

1.3.2 Ozone Depletion.   The chemistry of ozone depletion is complex but a basic outline of the important processes is as follows. Ozone is formed from the dissociation of molecular oxygen by short wavelength UV radiation in the upper stratosphere:

$$O_2 \overset{UV}{\rightarrow} O^\bullet + O^\bullet \qquad (1)$$

$$O^\bullet + O_2 + M \rightarrow O_3 + M \quad (M = \text{inert third body}) \qquad (2)$$

However, ozone itself is rapidly photodissociated:

$$O_3 \overset{UV}{\rightarrow} O_2 + O^\bullet \qquad (3)$$

and the so-called 'odd oxygen' species and O_3 may interconvert many times before they destroy one another by:

$$O^\bullet + O_3 \rightarrow O_2 + O_2 \qquad (4)$$

In fact, measurements of the ozone profile in the atmosphere suggest that ozone destruction must be considerably faster than could be achieved by reaction (4) alone and that other reactions must be involved. These other mechanisms can be represented by

$$X + O_3 \rightarrow XO + O_2 \qquad (5)$$

$$XO + O^\bullet \rightarrow X + O_2 \qquad (6)$$

Net effect $\qquad\qquad O^\bullet + O_3 \rightarrow O_2 + O_2$

X may represent a range of species including $Cl^\bullet$, $Br^\bullet$, NO, $OH^\bullet$ and $H^\bullet$ and is not consumed in the overall ozone destruction process. If X = NO, the reactions form and destroy NO_2; if X = $Cl^\bullet$, the reactions form and destroy ClO, but because this is a catalytic cycle small concentrations of X can have a significant effect on ozone levels. Other sets of reactions involving NO and $Cl^\bullet$ simply achieve the interconversion of O_3 and $O^\bullet$ and therefore have no effect on the net ozone levels. The reactive NO_x and $Cl^\bullet$ species can be removed by the formation of the relatively stable 'reservoir' molecules HNO_3 and HCl or the somewhat shorter lived chlorine nitrate $ClONO_2$. About half the stratospheric content of NO_x is stored as HNO_3 and about 70% of the chlorine as HCl. Although these may be reactivated by conversion back to NO_x and $Cl^\bullet$, they may eventually be transferred back to the troposphere and removed to the ground by rain-out.

1.3.3 The Antarctic Ozone 'Hole'.[10–16] The above was the general picture (although a highly simplified account) of the homogeneous stratospheric chemistry as understood before 1985. In that year Farman *et al.*[16] published the results of ground-based measurements in Antarctica showing very significant depletions, of the order of 50%, in the total column ozone content of the atmosphere. The Antarctic ozone 'holes' of

1992 and 1993 were the most severe on record, with ozone being locally depleted by more than 99% between about 14 and 19 km in October, 1992 and 1993. Subsequent aerial surveys and analysis of satelite data confirmed this phenomenon and led to a complete reappraisal of the chemistry and meteorology involved.

During the dark, cold Antarctic winter upper stratospheric air moving from low to high latitudes subsides and as it does so develops a strong westerly circulation pattern. This produces a vortex which effectively isolates the air in the lower stratosphere over the Antarctic continent from the air at lower latitudes. Within the vortex the temperature falls progressively until below about $-80\,°C$ polar stratospheric clouds (PSCs) may form. These are composed of very small particles (1 μm) of nitric acid trihydrate ($HNO_3.3H_2O$). A further drop in temperature of about $5\,°C$ may result in water ice crystals being formed. These are rather larger (10 μm). It is the heterogeneous reactions involving these cloud crystals which dramatically alter the chemistry of the stratosphere. Basically these reactions convert chlorine from its inactive, reservoir forms (HCl, $ClONO_2$) into forms which are active ozone depletors ($Cl^•$, ClO). HCl is readily incorporated into ice crystals and can undergo reaction with $ClONO_2$:

$$\underset{\text{ice}}{HCl} + \underset{\text{gas}}{ClONO_2} \rightarrow \underset{\text{gas}}{Cl_2} + \underset{\text{ice}}{HNO_3} \tag{7}$$

and

$$H_2O + ClONO_2 \rightarrow HOCl + HNO_3 \tag{8}$$

The nitric acid is left in the ice phase. The chlorine remains in the gas phase until the polar spring when the sun reappears and photodissociates it to chlorine atoms:

$$Cl_2 + h\nu \rightarrow Cl^• + Cl^• \tag{9}$$

The $Cl^•$ atoms rapidly react with ozone generating ClO:

$$Cl^• + O_3 \rightarrow ClO^• + O_2 \tag{10}$$

In the winter the stratosphere is thus chemically 'preconditioned' by heterogeneous reactions so that in the spring very rapid ozone depletion occurs. In addition to the ozone destruction cycle represented by reactions (5) and (6) with $X = Cl^•$ it is now recognized that chlorine monoxide dimers are also important. This was realized because the oxygen atom concentrations in the lower stratosphere are too low to account for the observed ozone destruction rates.

$$ClO + ClO + M \rightarrow Cl_2O_2 + M \tag{11}$$

$$Cl_2O_2 + hv \rightarrow Cl^\bullet + ClOO^\bullet \tag{12}$$

$$ClOO^\bullet + M \rightarrow Cl^\bullet + O_2 + M \tag{13}$$

These reactions by-pass the $ClO + O^\bullet$ reaction as a route for reconversion of ClO back to $Cl^\bullet$.

The importance of ClO within the Antarctic stratosphere is illustrated in Figure 4. Reactions of bromine atoms in addition to those of chlorine atoms are now thought to account for about 20% of the ozone depletion. Bromine emissions occur in the form of methyl bromide which has natural and synthetic sources such as soil fumigation, biomass burning, and the exhaust of automobiles using leaded gasoline, plus another family of halocarbons the Halons—Halon 1301 is CF_3Br, Halon 1211 is CF_2BrCl (Table 4). Methyl bromide is currently being targeted for phase out and is covered under amendments to the Montreal Protocol.[17] Current models of ozone depletion also consider the effects of CO_2. Increased CO_2 levels would lead to a lowering of temperatures in the stratosphere which may serve to slow down the destruction reactions: $O + O_3$ and $NO + O_3$. The role of sulphate aerosols also has to be considered in the lower stratosphere.[15]

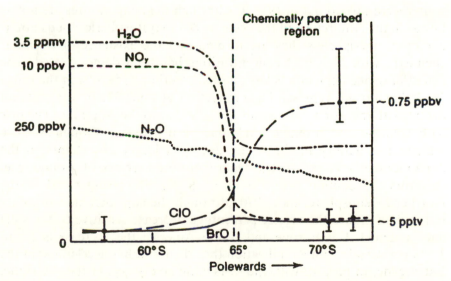

Figure 4 *Profiles of ClO and other species in the Antarctic stratosphere, 18 km altitude, near the boundary of the chemically perturbed region. The decreases in water vapour and NO_x are due to condensation of water and nitric acid in polar stratospheric clouds followed by gravitational settling*
(Reproduced with permission from UK Review Group on Stratospheric Ozone[11])

Table 4   *Atmospheric lifetimes and ozone depletion potentials for halogenated compounds*

Compound name	Chemical formula	Atmospheric lifetime[2,12] (yr)	Ozone depletion potential[13]
CFC 11	$CFCl_3$	50	1.0
CFC 12	CF_2Cl_2	102	1.0
CFC 113	$C_2F_3Cl_3$	85	1.1
Carbon tetrachloride	CCl_4	42	1.08
Methyl chloroform	CH_3CCl_3	4.9	0.12
HCFC 22	CH_3CCl_3	12.1	0.05
HFC 134a	$C_2H_2F_4$	14.6	0
Halon 1301	$CFBr$	65	12.5
Halon 1211	CF_2BrCl	20	3.0
Methyl bromide	CH_3Br	1.2	0.6

But is the 'hole' in the Antarctic ozone layer significant for the rest of the globe? Several facts suggest that it is. As the Antarctic spring progresses the vortex breaks up and the ozone depleted air can then be transported to lower latitudes. Such an event has been observed in Australia. Large increases of surface UV are observed in Antarctica and the southern part of South America during the period of the seasonal ozone 'hole'. In the Northern Hemisphere airborne studies of the Arctic winter were carried out in 1988/9. Although the temperatures are not as low as in the Antarctic and the occurrence of stratospheric clouds not as common, nevertheless they are formed and the existence of a similar chemistry with high ClO concentrations has been demonstrated. The extent of ozone depletion is less marked—perhaps 15–20% in the range 20–25 km altitude representing a reduction of some 3% in total column ozone, with the worst years again 1992/1993 as in the Antarctic. The link with chlorine and bromine levels has been proven although is more difficult to quantify in the Arctic due to a larger uncertainty in the dynamics of the Arctic vortex. Data from the network of ground level monitors for column ozone show a clearly decreasing trend in the Northern Hemisphere since 1970, although the considerable variability in the data makes the precise percentage decrease sensitive to the start date assumed. For Europe and North America the decrease is 2.5 to 3.5% per decade with an indication that the trend has accelerated in the last decade, in parallel with the worsening conditions in the Antarctic. Data from the Total Ozone Mapping Spectrometer (TOMS) shown in Plate 2 illustrates the ozone depletion in the Northern Hemisphere.[18]

1.3.4   Effects of International Control Measures.   The UN Convention on the Protection of the Ozone Layer (the 'Vienna Convention') was agreed in 1985 and subsequently measures to reduce the emissions of various halocarbons were incorporated into the 'Montreal Protocol' in September 1987. Meetings in London in 1990 and Copenhagen in 1992 further tightened the restrictions under the Protocol which has as its final objective the elimination of ozone depleting substances. More than 160 countries are now Parties to the Convention and the Protocol[13] with some special agreements for developing countries. The use of both CFCs and Halons will be phased out by the year 2000 as will the use of carbon tetrachloride CCl_4. Methyl chloroform will be phased out by 2005. Replacement chemicals such as the HCFCs are now being used and the change in reported emissions[20] is illustrated in Figure 5. Because these contain hydrogen atoms as well as halogens they are more reactive in the troposphere and have shorter lifetimes as illustrated in Table 4. Their potential impact on the stratosphere is therefore much reduced.

Evidence is now emerging which indicates that the atmospheric growth rates of the main ozone depleting substances are slowing, showing that the Montreal Protocol is having an effect. Total tropospheric organic chlorine increased by only about 60 ppt/year (1.6%) in 1992, compared to 110 ppt/year (2.9%) in 1989.[14] Even so, the expectation is that the total stratospheric chlorine loading will peak after 2000 at about 4 ppb and only decline slowly through the remainder of the 21st century.[12] Figure 6 shows the predicted fall in stratospheric chlorine in the 21st century but also demonstrates the sources of this chlorine.[20] CFCs are predicted to be a major source well into the next century because of their long lifetimes. How long it takes for the chlorine loading to fall to the pre-war level of below 1 ppb depends on the extent of global participation in implementing restrictions and the extent of future use of alternative chlorine containing compounds such as the HCFCs (Table 4). Global ozone losses and the Antarctic ozone 'hole' are predicted to recover in about the year 2045 as long as the Montreal Protocol and amendments are closely adhered to.

2 ATMOSPHERIC TRANSPORT AND DISPERSION OF POLLUTANTS

2.1 Wind Speed and Direction

In the previous section we dealt with global pollution problems. We now move on to local pollution problems which we think of as directly affecting the air quality around us. Local air quality is significantly influenced by the rate of mixing of pollutants which are emitted and therefore we need to have an understanding of issues such as wind speed

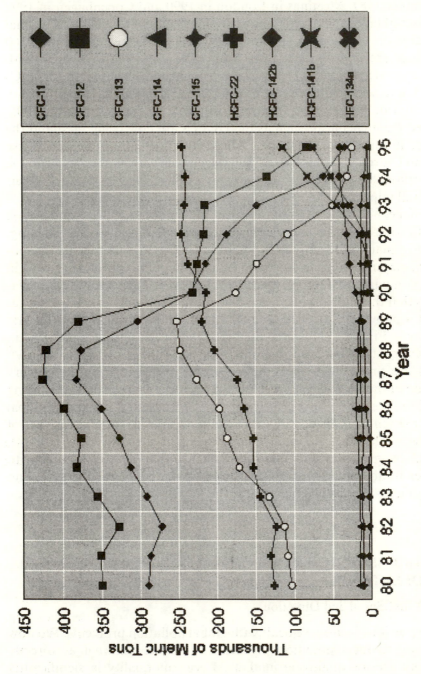

Figure 5 *Annual Production of Fluorocarbons Reported to AFEAS 1980–1995*
(Reproduced with permission from the Alternative Fluorocarbons Environmental Acceptability Study[20])

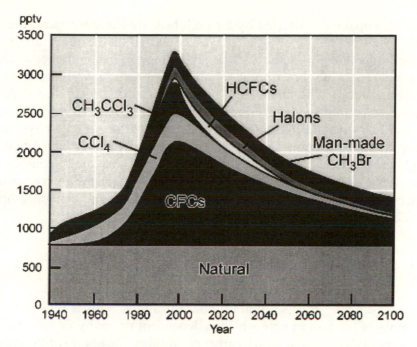

Figure 6 *Predicted chlorine loading for the next century by source. Bromine from all sources is shown as its equivalent in chlorine*
(Reproduced with permission from the Alternative Fluorocarbons Environmental Acceptability Study[20])

and atmospheric stability. In general, low wind speeds result in high pollutant concentrations and vice versa. If we imagine the wind blowing across the top of a chimney emitting smoke at a constant rate, the volume of air into which the smoke is emitted is directly proportional to the wind speed. The concentration of smoke in the air is thus inversely proportional to the wind speed. A similar description can be applied to a source distributed uniformly over a wide area, *e.g.* domestic emissions from a city. The concentration of pollutants in the hypothetical box of air into which the pollutants are mixed is proportional to the emissions rate and inversely proportional to the wind speed. In practice this picture is grossly oversimplified and the concentration of pollutants measured in urban areas rarely decreases with wind speed as rapidly as predicted.

For reasons which will be discussed later, the wind speed at ground level tends to drop overnight and rise again during the morning, especially during cloud-free conditions. Of course, emissions also tend to drop overnight—fewer fires, boilers and furnaces are alight, fewer cars are on the roads. So some of the highest pollution levels occur in the morning when emissions increase rapidly before the wind speed picks up and dispersion conditions improve.

The people most affected by air pollution are those who are situated downwind of the major sources. A knowledge of the prevailing wind direction is therefore important in predicting the likely impact of these sources. The wind direction at a point is not sufficient to identify high pollution levels with the effect of distant sources. Over the scale of hundreds to thousands of kilometres the overall path of the particular mass of air must be computed for periods of perhaps 1 to 3 days. Such trajectories can be quite curved as illustrated in Plate 3, which relates to high smog episodes over the UK. The origin of the smog precursors is clearly the pollutant emissions from the southern UK and/or continental Europe. Corresponding trajectories from the Atlantic generally bring much less polluted air.

2.2   Atmospheric Stability

In addition to wind direction, the extent of vertical mixing in the atmosphere also affects pollutant concentrations. This is related to the stability of the atmosphere which is dependent on many factors such as the time of day, the synoptic weather conditions, the nature of the earth's surface, *etc.*

2.2.1   The Lapse Rate.   The roughness of the ground produces a certain amount of turbulence in the lowest layer (boundary layer) of the atmosphere which promotes the mixing and dispersion of pollutants. This effect increases with the scale of the surface roughness and is greater for a city with large buildings than for open ground with few obstructions. However the major factor affecting atmospheric stability and turbulence is thermal buoyancy. The pressure in the atmosphere decreases exponentially with height. Ascending air expands as the pressure decreases and as it expands it cools. A simple calculation based on the properties of gases leads to the conclusion that we should expect a decrease in temperature with height of $9.86\,°C\,km^{-1}$ or about $1\,°C$ for every 100 m of dry air. The figure is somewhat lower for moist air. The variation of temperature with height is called the lapse rate and the calculation for the ideal case leads to what is known as the adiabatic lapse rate (a.l.r.). In the real atmosphere the lapse rate can be greater than, smaller than, or close to the adiabatic lapse rate. This fundamentally affects the extent of vertical mixing of air as the following two examples show:

   Case (a). Temperature decreases with height more rapidly than the a.l.r., Figure 7a. Air that is slightly warmer than its surroundings starts to rise and to cool at the a.l.r. The

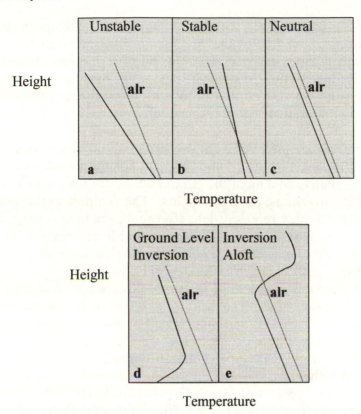

Figure 7 *Schematic illustration of the atmospheric lapse rate for various stability categories. Full lines: actual temperature profile; dashed lines: adiabatic lapse rate (a.l.r.)*

temperature difference between the rising parcel of air and its surroundings increases with height, thus the upward movement due to thermal buoyancy continues and is magnified. The atmosphere is unstable. Upcurrents in one location are balanced by downcurrents elsewhere and rapid vertical mixing of air occurs, promoting rapid dispersion of pollutants.

Case (b). Temperature decreases with height less rapidly than the a.l.r. or actually increases with height, Figure 7b. Air that is slightly warmer than its surroundings starts to rise and cool at the a.l.r. and the temperature difference between the rising parcel of air and its surroundings soon decreases to zero. Upward movement due to thermal buoyancy ceases. The atmosphere is stable since any vertical movement of air tends to be damped out. Lower polluted layers stay near the ground and pollutant concentrations will be high.

Unstable situations occur with bright sunlight warming the ground to a temperature above that of the air. The air adjacent to the ground is subsequently warmed and rises due to its buoyancy. Such situations are common during daytime in summer, especially when the wind speed is low. High wind speeds tend to lead to neutral conditions with the lapse rate close to the adiabatic value (Figure 7c).

2.2.2   Temperature Inversions.   Stable situations occur when the lowest layer of air is cooled by the ground beneath. The most common cause is overnight radiative cooling of the ground which often leads to low level temperature inversions on clear nights. The temperature profile, in Figure 7c which may represent late afternoon, gradually changes into profile 7d overnight. The effect is that ground level emissions become trapped in the stable inversion layer which may not be more than 100–200 m deep. Emissions from tall industrial chimneys however may be above the inversion layer and be vertically dispersed by relatively good mixing conditions aloft. The following day the inversion layer is gradually eroded by the warming effect of the sun until, by mid-morning, the temperature profile has returned to that of neutral conditions (Figure 7c) and the trapped pollutants are effectively released to be dispersed to higher levels.

Another factor contributing to high pollution levels during inversion conditions is the lowered wind speed. Since high level, faster moving air does not mix with low level air, there is no mechanism for the downward transport of momentum. The lowest level air therefore becomes stagnant. In the low temperature conditions, dew, frost, or fog formation may occur. Fog adds to the problem by slowing down the break-up of the overnight inversion layer because the sun's energy is reflected by its upper surface and does not reach the ground. The ground therefore stays cool rather than warming. In extreme cases fog may persist for several days as happened in the London 1952 smog. Polluted fogs are more persistent than clean fogs because the chemicals dissolved in the water droplets prevent complete evaporation even when the relative humidity drops well below 100%.

The other main type of inversion occurs during anticyclonic conditions and is described as a subsidence inversion. Within an anticyclone, air is diverging from a high pressure region and at the centre subsides from a high level to lower levels of the atmosphere. As it subsides it progressively warms with decreasing height resulting in the development of an elevated inversion layer as illustrated in Figure 7e. Below the inversion layer the air may be neutral or even unstable so that good mixing occurs but only up to the inversion height. Local urban pollution levels are rarely as great under such conditions as during ground level inversions. However, subsidence inversions are often associated with warm, dry weather and they provide the ideal conditions for a long range transport of pollution. Summer haze conditions, in which the UK receives already polluted air from Europe before the addition of our own emissions, can lead to particulate pollution levels as high as those on the most polluted winter days. At the same time the levels of photo-chemical oxidants such as ozone are also high as shown in Plate 3.

In the discussion so far, no reference has been made to the geographical situation in which air pollution levels are being considered. Towns situated in valleys are particularly susceptible to pollution problems. Cool air will tend to flow downhill into the valley so aggravating the problem of low level inversions. Mixing between the air in the valley and the air above is reduced. Fogs will persist longer. Often in winter, a layer of polluted air over the town with cleaner air above can be clearly seen from the higher ground. Towns situated by the sea may be subject to sea breezes. The proximity of relatively warm ground to the cool water surface results in a circulation of air from sea to land. The sea breeze will be cooler than the overlying air, another example of inversion conditions, so in polluted areas conditions will be worse than for a corresponding inland site. Sea mists may blow inland aggravating the general discomfort. Los Angeles is one example of a city whose geographical location exagerates pollution problems caused by high emissions. It is situated in a basin area surrounded by large hills which inhibit mixing out of the city. Inversions are also common due to the presence of subsidence conditions over the Pacific. In fact, five of the worst polluted cities of the world are situated in the Pacific basin region.[21]

2.3 Dispersion from Chimneys[22]

2.3.1 Ground-level Concentrations. The effluent gases leaving a chimney gradually entrain more and more air, and the plume width both in the vertical and in the horizontal directions increases with downwind distance. In the cross wind direction there is a rapid fall-off in concentration as we move away from the plume centre line which is defined as being along the average wind direction. The ground-level concentration very close to the chimney will be zero because it takes some time for vertical mixing to drive the plume to the ground. Some distance downwind the dispersing plume reaches ground level and the concentration

---

[21] D. Mage , G. Ozolins, P. Peterson, A. Webster, R. Orthofer, V. Vandeweerd, and M. Gwynne. *Atmos. Environ.*, 1996, **30**, 681. This is a follow up report to the WHO/UNEP report 'Urban Air Pollution in Megacities of the World' published in 1992 in which the detailed data are included.
[22] 'Recommended Guide for the Prediction of the Dispersion of Airborne Effluents', ed. J. R. Martin, American Society of Mechanical Engineers, New York, 3rd Edn., 1979.

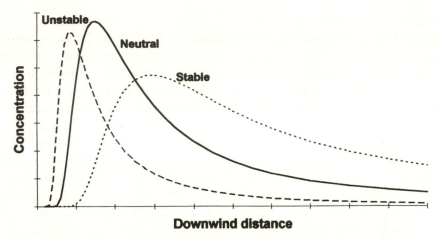

Figure 8   *The variation in ground-level concentration along the plume centre line*
downwind of a chimney for various stability classes

rises rapidly to a maximum value. For example, for Eggborough Power
Station in northern England this occurs up to about 8 km from the
200 m chimney. Thereafter the concentration falls off with distance as
the plume becomes more and more dilute, Figure 8. The distance at
which maximum concentrations occur depends on stability conditions.
Unstable conditions (significant vertical mixing) bring the plume down
to ground rapidly, that is closer to the chimney, but the rate of dilution is
large so the concentration falls rapidly from its maximum value as we go
to greater distances. Stable conditions result in the plume dispersing very
slowly. It may remain visible for a considerable downwind distance. The
point of maximum ground level concentration is a long way from the
chimney. Above the chimney top, the vertical dispersion of the plume
can be hindered by an inversion layer. The pollutants are then trapped
between the inversion height and the ground. At sufficient distance
downwind, the concentration may be virtually constant at all heights as
the plume becomes well dispersed within the mixing layer.

2.3.2   Plume Rise.   The basic models of plume dispersion suggest that
the maximum ground level concentration depends inversely on the
square of the chimney height. A 40% increase in chimney height should
roughly halve the ground level impact. However, the height of the plume
depends on both the chimney height and the plume rise caused by the exit
conditions of the release. The major factor is the thermal buoyancy of the
plume. The hot flue gases will rise relative to the cooler surrounding air
and the hotter the release the greater the plume rise. A lesser factor is the
vertical momentum of the gases due to their efflux velocity out of the

chimney top. Plume rise can mean that the effective chimney height is double the physical chimney height under low wind speed conditions and so is a considerable asset in achieving effective dispersion. Conversely, any control technology for pollutant reduction which also reduces the flue gas temperature (such as a scrubber) results in poorer dispersion of the plume. If we take plume rise into account then the effect of raising the stack height to reduce ground level concentrations is less than would otherwise be predicted.

2.3.3 Time Dependence of Average Concentrations. The above description of plumes is really a simplification since atmospheric turbulence is random, making the dispersion process irregular. Visible plumes will often look ragged and, from the point of view of an observer making spot measurements on the ground, one minute the concentration may be close to zero and the next very high. The longer averaging time taken, the less the variability of the results. This can be illustrated by the results of measurements made by the Central Electricity Generating Board near Eggborough, a 2000 MW power station.[23] At the radius of maximum effect, which is about 8 km from the chimney, the peak 3 minute concentrations of sulfur dioxide due to the power station were approaching 1000 μg m^{-3}. Such peaks were extremely rare and 95% of the 3 minute averages were below 10 μg m^{-3}. The highest daily averages were around 100 μg m^{-3} and the highest monthly averages around 10 μg m^{-3}. The overall annual average contribution of the power station to the ambient SO_2 concentration was 2–3 μg m^{-3} in an area where the prevailing concentration from other sources was about 40 μg m^{-3}.

2.4 Mathematical Modelling of Dispersion[24–37]

There are a number of techniques for modelling the dispersion and reaction of atmospheric pollutants. Space does not allow us a full study

[23] A. J. Clarke, 'Environmental Effects of Utilising More Coal', ed. F. A. Robinson, Royal Society of Chemistry, London, 1980, p. 55.
[24] John H. Seinfeld and Spyros N. Pandi, 'Atmospheric Chemistry and Physics, from Air Pollution to Climate Change', John Wiley & Sons, New York, 1998, pp. 916–919, 1193–1285.
[25] D. G. Atkinson, D. T. Bailey, J. S. Irwin, and J. S. Touma, *J. Appl. Meteorol.*, 1997, **36**, 1088.
[26] D. J. Carruthers, H. A. Edmunds, K. L. Ellis, C. A. McHugh, B. M. Davies, and D. J. Thomson, *Int. J. Environ. Poll.*, 1995, **5**, 382.
[27] Z. Zlatev, J. Christensen, and A. Eliassen, *Atmos. Environ.*, 1993, **27**, 845.
[28] C. Pilinis and J. H. Seinfeld, *Atmos. Environ.*, 1988, **22**, 1985.
[29] J. S. Chang, R. A. Brost, I. S. A. Isaksen, S. Madronich, P. Middleton, W. R. Stockwell, and C. J. Walcek, *J. Geophys. Res.—Atmospheres*, 1987, **92**, 14681.
[30] W. J. Collins, D. S. Stevenson, C. E. Johnson, and R. G. Derwent, *J. Atmos. Chem.*, 1997, **26**, 223.
[31] R. I. Sykes and R. S. Gabruk, *J. Appl. Meteorol.*, 1997, **36**, 1038.
[32] J. C. Weil, L. A. Corio, and R. P. Brower, *J. Appl. Meteorol.*, 1997, **36**, 982.
[33] D. M. Lewis and P. C. Chatwin, *J. Appl. Meteorol.*, 1997, **36**, 1064.

here but the main methods will be introduced. Choosing a particular type of model depends on many factors such as the distance scale in which we are interested, *e.g.* urban roadside, plume dispersion, regional scale smog modelling, global circulation modelling, *etc*. We may also need to take into account whether we need to include chemical reactions and deposition, and the computer resources available. Available methods include Gaussian formulations,[22,24-26] Eulerian grid modelling,[27-29] Lagrangian trajectory modelling[30] and turbulence or statistical models which try to account for random fluctuations.[31-34]

The simplest approach is the Gaussian model which uses empirical formulae to describe the distribution of pollutants downwind of a single source or a group of sources. Fast screening type calculations of the dispersion of a pollutant from a point source such as a large chimney are usually based on the Gaussian Plume Model[22] which is discussed in Chapter 7. This treats steady state emissions and situations which can be described as a superposition of a number of different steady states. For example long term average distributions around a source can be calculated by using statistically averaged meteorological data. The model can be adapted to treat area sources such as wind-blown dust from a stockpile or odours from a sewage works. Urban areas can be modelled as a sum of area sources representing domestic and commercial emissions plus larger point sources such as factories or power stations. Another application is dispersion from sources distributed along a line such as a motorway. Some current urban models use Gaussian formulations to predict the dispersion from a range of source types but at present the validation of such models for such a complex situation is limited.

Because it is a steady-state model the Gaussian plume approach has limitations in that it cannot represent diurnal variations of pollutants. On its own it also has limitations in that it can only represent first-order reaction or deposition processes although it can be coupled with chemical box models. This is particularly important for secondary pollutants such as ozone whose concentrations (as we shall see in a later section) depend on a series of non-linear chemical reactions involving NO_x and hydrocarbon species, and levels of sunlight. Also because it is steady state the Gaussian model cannot represent the random fluctuations present in a real plume since for this a turbulence model would be required. The Gaussian model is best suited to averaged concentrations

[34] F. Pasquill and F. B. Smith, 'Atmospheric Diffusion', 3rd Edn., Ellis Horwood, Chichester, 1983.
[35] J. Saltbone and H. Dovland, 'Emissions of sulphur dioxide in Europe in 1980 and 1983'. EMEP/ CCC Report 1/86. Norwegian Institute for Air Research, Lillestrom, Norway, 1986.
[36] H. S. Eggleston and G. McInnes, 'Methods for the compilation of UK air pollutant emission inventories', Report LR 634(AP), Warren Spring Laboratory, Stevenage, 1987.
[37] http://www.aeat.co.uk/netcen/airqual/emissions/den-naei.html 'National Atmospheric Emissions Inventory for the UK', December 1997.

of non-reactive pollutants but has the advantage that it uses a small amount of computer resources and is therefore commonly used as an air quality management tool.

The more complex processes involved in long range transport and the reactions producing smog and acid rain, which also involve a number of source types, require numerical simulation on large computers. Here we may choose to use either Lagrangian or Eulerian models. Lagrangian models follow the trajectories of single or multiple air parcels as they are transported by the winds over long distances. The Eulerian approach uses a grid which covers the whole of the domain and therefore equations have to be solved at each grid point. There are advantages and disadvantages to both methods. The Lagrangian approach, because it can simulate only a few air parcels, can have lower computer simulation times and can therefore accommodate large chemical reaction mechanisms, describing the formation of smog for example. However, if limited trajectories are used we may not get a full picture and will only simulate smog concentrations at a few points in the region. Models of regional transport of SO_2 and O_3 are often based on large numbers of back-trajectories (see Plate 3) for specified receptor points taking account of emissions and loss processes along the path. The Eulerian approach gives much better coverage of the domain but at the expense of computer simulation time. Often large parallel computers are used for Eulerian simulations,[27-29] *e.g.* smog models, global climate models, *etc.*

We can see that the choice of model type depends on the required outcome and accuracy needed. The accuracy of all types of model, however, depends heavily on the input data. Detailed knowledge of emissions data for a range of sources is needed along with an accurate representation of meteorological data such as wind speed and direction, stability class, and precipitation. The resolution of the input data significantly affects the results. Detailed modelling of urban pollution requires an emissions inventory on a grid basis, for example 1 km squares, along with point source information. Emissions inventories for the most common pollutants are now available for the whole of Europe on a national basis,[7] on the 150 km × 150 km grid used by the European Monitoring and Evaluation Programme (EMEP)[35] and on grid sizes down to 10 km × 10 km for the UK.[36,37]

3 EMISSIONS TO ATMOSPHERE AND AIR QUALITY

3.1 Natural Emissions

3.1.1 Introduction. The primary components of pure dry air are nitrogen N_2 (78.1%), oxygen O_2 (20.9%), argon Ar (0.9%), and carbon

Table 5   *Natural emissions of S^{38} and N^{39} compounds*

Source	$Tg\ S\ yr^{-1}$	Source	$Tg\ N\ yr^{-1}$
Volcanoes	9.3	Lightning	8
Biomass burning	2.2	Biomass burning	5
Marine biosphere	15.4	Soil—biogenic	7
Terrestrial biosphere	0.35	NH_3 oxidation	0.9
		From stratosphere	0.6
Natural total	27.25	Natural total	22
Anthropogenic	77	Anthropogenic	22

dioxide CO_2 (0.035% or 350 ppm). Water vapour is present in amounts which typically range from 0.5 to 3% at ground level, depending on temperature and relative humidity. Analysis of air samples reveals the presence of hundreds of other substances in trace amounts. Some of these can be explained directly in terms of their emissions either from natural sources or human activity. Others must have arisen indirectly by chemical processes in the atmosphere. We distinguish these as *primary* and *secondary* pollutants. The secondary pollutants, including gases like ozone and particulate compounds such as sulfates, are dealt with in Section 4. Here we concentrate on the primary pollutants.

Within densely populated areas of land, the levels of most pollutants are dominated by the contributions for which humans themselves are responsible. However, nature is also a generator of what we call 'pollutants' and on a global scale the natural emissions may be comparable to human emissions. This is illustrated in Table 5.

3.1.2 Sulfur Species.   Sulfur in the form of SO_2 and some H_2S is emitted most dramatically from volcanoes (Table 5). Significant amounts are emitted from biological processes. In the absence of air, biological decay results in emissions of hydrogen sulfide, H_2S, and organic compounds such as dimethyl sulfide (DMS). Carbonyl sulfide (OCS) emissions also occur together with small amounts of carbon disulfide (CS_2) and dimethyl disulfide (DMDS). From the oceans the emissions related to phytoplankton are primarily of DMS. The grouping together of all sulfur compounds is appropriate since H_2S and organic sulfur compounds are converted to SO_2 in the atmosphere. Various estimates place the total emissions at 20–30 Tg sulfur per year (1 Tg = 1 million tonne). Since the global SO_2 from combustion emissions is 70–100 Tg S yr^{-1}, the natural sulfur emissions are of the order of one third or one quarter of the anthropogenic emissions. Anthropogenic emissions are greatest in the northern hemisphere and for the southern hemisphere natural and anthropogenic emissions are of a similar order.

This assessment omits the largest single component of the atmospheric sulfur budget which is represented by the sulfate content of sea-salt aerosols. There is obviously a sharp vertical gradient of these aerosols and a large proportion return to the sea quite quickly. Emission figures of the order of 200 Tg yr^{-1} are quoted for aerosols in the 10–20 m height range but the particle number concentrations in the free troposphere (>2 km) may be 3 orders of magnitude lower than in the oceanic boundary layer.[38]

3.1.3 Nitrogen Species. Biological processes in soil lead to the release of all of the common nitrogen oxides, NO, NO_2, and N_2O. The amounts involved are very uncertain but for NO and NO_2 are of the order of 5–10 Tg N yr^{-1}. Lightning and biomass burning are other major sources. Oxidation of NH_3 to NO occurs in the troposphere and some nitrogen in the form of HNO_3 is transferred to the troposphere from the stratosphere. These sources total 20–30 Tg N yr^{-1} (Table 5). In comparison the anthropogenic emissions of NO + NO_2 from combustion are about 20 Tg N yr^{-1}. Lee *et al.*[39] estimate the total global NO_x emissions to be 44 Tg N yr^{-1} with an uncertainty range of 23–81 Tg N yr^{-1}.

The major source of nitrous oxide, N_2O, is the release from soil especially in situations where fertiliser has been added. Smaller releases occur from the oceans and from combustion processes. The sources are not as well understood as the main loss mechanism which is decomposition in the stratosphere (6–10 Tg N_2O yr^{-1}). Industrial emissions occur during the production of nitric acid and adipic acid (an intermediate in the production of nylon). Total UK emissions were estimated at 94 000 tonnes in 1994[40] as compared with 470 000 tonnes in the US in 1995.[8]

The other important nitrogen species is ammonia, NH_3. Its sources may be considered partly natural and partly anthropogenic since they are dominated by animal excreta. But there are also significant releases from biomass burning, crops, and the oceans. Several estimates[41] of global emissions place the total at around 50 Tg N yr^{-1}. The sources of ammonia are difficult to quantify since the earth's surface can act as a source and a sink and the overall emission rate depends on soil and

[38] T. S. Bates, B. K. Lamb, A. Guenther, J. Dignon, and R. E. Stoiber, *J. Atmos. Chem.*, 1992, **14**, 315.

[39] D. S. Lee, I. Kohler, E. Grobler, F. Rohrer, R. Sausen, O. Gallard, L. Klenner, J. G. J. Oliver, and F. J. Dentener, *Atmos. Environ.*, 1997, **31**, 1735.

[40] Digest of Environmental Statistics No. 18, HMSO, London, 1996.

[41] A. F. Bouwman, D. S. Lee, *et al.*, A global high resolution emissions inventory for ammonia, Global Biogeochemical Cycles, 1997, **11**, 561.

climatic conditions. The official figures for the UK for 1993 are, in ktonnes yr^{-1}:

Cattle	130
Pigs	25
Sheep	15
Poultry	25
Fertilizers and crops	30
Agricultural Total	225
Non-Agricultural sources	35
Overall total	260

Currently ammonia emissions in Europe are estimated at 2.7 Mtonnes yr^{-1} for the countries of the European Union and 6.4 Mtonnes yr^{-1} for all the European countries covered by the EMEP area (European Monitoring and Evaluation Programme).[42] Currently there are no international targets for ammonia emissions but future proposals by the EU are likely to cover ammonia.

3.1.4 Hydrocarbons. The importance of methane as a greenhouse gas has been discussed earlier. The largest natural sources of methane are anaerobic fermentation of organic material in rice paddies and in northern wetlands and tundra, plus enteric fermentation in the digestive systems of ruminants (*e.g.* cows). Methane is also released from insects, from coal mining, gas extraction, and biomass burning. Total emissions are 300–550 Tg yr^{-1} and appear to be increasing at a rate of 50 Tg yr^{-1}. Within the UK, animals (29%), mining (8%), landfill gas (46%), and gas leakage (9%) account for most of the 3.9 Tg yr^{-1} emissions (1994 estimate[40]).

Heavier hydrocarbons, such as isoprene, α and β-pinene and other terpenes, are released directly to the atmosphere from trees and can contribute to the formation of photochemical smog in high emission areas (see Section 4). Global emissions of these Biogenic Volatile Organic Compounds (BVOCs) are estimated to be 1150 Tg C yr^{-1} of which 88% is from trees and shrubs and 10% from crops.[43]

3.2 Emissions of Primary Pollutants

3.2.1 Carbon Monoxide and Hydrocarbons. Efficient combustion can be achieved in most stationary combustion appliances providing they are

[42] 'Acid Deposition in the United Kingdom 1992–94', Fourth Report of the Review Group on Acid Rain, Department of the Environment, Transport and the Regions, London, 1997.
[43] A. Guenther, C. N. Hewitt, *et al., J. Geophys. Res.* 1995, **100**, D5, 8873.

Table 6 *Typical urban emissions from current European technology cars (g km^{-1}) assessed in on-road tests*[44]

Vehicle type	CO	Hydrocarbons	NO$_x$	PM
Standard petrol without catalyst	27	12.8	1.7	–
Petrol with catalyst	2.0	0.2	0.4	–
Diesel	0.9	0.3	0.8	0.4

Table 7 *Present European Standards for passenger cars applicable to all models from 1.1.97 (Directive 94/12/EC)*

Vehicle type	CO g km^{-1}	HC + NO$_x$ g km^{-1}	PM g km^{-1}	Evaporative losses g/test
Petrol	2.2	0.5	–	2.0
IDI Diesel	1.0	0.7	0.08	–
DI Diesel*	1.0	0.9	0.10	–

*DI diesel limits to be equal to those for IDI engines from 1.10.99.

properly adjusted, and CO and hydrocarbon concentrations do not give serious cause for concern. Faulty or improperly adjusted appliances can produce dangerous amounts of CO (several per cent in the flue gas) usually due to some abnormal limitations of the air supply.

CO emissions from internal combustion engines are more of a problem. In petrol engines lacking any control devices, incomplete burnout of the fuel leads to high carbon monoxide and significant hydrocarbon emissions, especially during idling and deceleration. Installation of catalytic converters results in reductions of 90% for CO and hydrocarbons although significant emissions may take place under cold start conditions before the catalyst 'lights off'. Diesel engines have much lower carbon monoxide and hydrocarbon emissions (Table 6).[44]

Permitted emissions from European vehicles have been progressively reduced and the current standards are indicated in Table 7. Further reductions have been proposed for the years 2000 and 2005.

The term Volatile Organic Compounds (VOCs) is used to describe organic material in the vapour phase excluding methane. There are many non-combustion sources of VOC emission of which the most important is the use of solvents, including those released from paints. Evaporative losses of gasoline during storage and distribution are also significant (see Table 8).

[44] Diesel Vehicle Emissions and Urban Air Quality. Second Report of the Quality of Urban Air Review Group, Department of the Environment, London, 1993.

Table 8   *Estimated UK emissions of primary pollutants by UNECE source category for 1994* (ktonne)[40]

	CO	NO_x[a]	SO_2	VOC[b]	Black smoke	$PM10$[c]
Power stations	21	526	1759	6	19	40
Domestic	272	69	90	37	93	37
Commercial/public service	5	34	74	2	3	5
Refineries	6	45	138	2	3	5
Iron and steel	26	48	89	4	2	20
Other industrial combustion	36	128	422	7	13	18
Solvent use				658		
Non-combustion processes	–	5	12	411	–	63
Extraction/distillation of fossil fuels[d]	47	109	2	243	–	
Road transport—petrol	4117	653	19	545	15	21[e]
Road transport—diesel	203	442	44	76	232	45[e]
Other transport	40	151	57	25	3	7
Waste treatment and disposal	48	4	4	21	41	–
Agriculture	11	3	9	–	1	1
Forests	–	–	–	80[f]	–	–
Total	4833	2218	2718	2117	426	263

[a] NO_x is expressed as NO_2 equivalent.
[b] Volatile organic compounds excluding methane.
[c] 1993 data. Secondary particulate pollutants such as sulfates and nitrates are not accounted for in this total.
[d] Includes offshore contribution from gas flaring and use of gas.
[e] Authors' estimates from the road transport total of 66 ktonne.
[f] Estimate of natural emissions.

VOCs are important in atmospheric chemistry for the formation of photochemical smog which will be discussed in Section 4. Various control measures are therefore being implemented so as to enable the European Member States to meet their agreed obligation under a 1991 UNECE Protocol of reducing VOCs by 30% relative to 1988 by 2000. Between 1990 and 1994 a 9% reduction in VOC emissions across Europe was reported.[45]

3.2.2 Nitrogen Oxides. Although there are small emissions of NO_x from industrial processes such as nitric acid production, the main emissions are from combustion. In stationary combustion plant, factors such as the fuel type, fuel nitrogen content, burner design, the intensity of combustion, the overall shape and size of the furnace, and the amount of excess air all influence NO formation and can be modified to achieve a certain measure of control. However this falls considerably short of eliminating the emissions. Typical flue gas concentrations for industrial

[45] http://tiger.eea.eu.int/, 'The European Environment Agency', December 1997.

coal-burning are about 550 ppm but with new designs of low-NO_x burners this is reduced by about 50%. Nitrogen dioxide, NO_2, forms only a small fraction of the waste gases from combustion (usually less than 10%) so the description 'NO_x emissions' actually refers mainly to NO emissions. Once in the atmosphere, oxidation of NO to NO_2 occurs, as described later, and the relative proportions of the two oxides may then be comparable.

There is negligible nitrogen in gasoline or diesel fuels so the NO_x arises from the N_2 and O_2 in the air. Nitrogen oxide emissions are high due to the high temperatures and pressures and are at a maximum during acceleration and minimum during idling. Diesel engines have comparable NO_x emissions to petrol engines as shown in Tables 6 and 7.

3.2.3 Sulfur Dioxide. Sulfur dioxide, SO_2, arises from the sulfur present in most fuels amounting to 1–2% wt in coal, 2–3% in heavy fuel oils and decreasing amounts in lighter oil fractions. There are also non-combustion emissions from sulfuric acid plant and from the roasting of sulfide ores during non-ferrous smelting.

The limit for the sulfur content of diesel fuels has been progressively reduced and now stands at 0.05% in the USA and much of Europe. Typical gasoline sulfur content is about half this. During combustion, conversion to sulfur dioxide is virtually complete although about 5–10% may be retained by coal ash. For the highest sulfur fuel oils (3% S) the flue gas concentrations can reach 2000 ppm while for a typical power station coal of 1.6% S the concentration is about 1200 ppm. No control over SO_2 emissions can be achieved by modification of combustion conditions, and reductions must be sought by pre-treatment of the fuel or desulfurisation of the flue gases after combustion.

3.2.4 Particulate Matter. Dust can arise by the disturbing action of outdoor industrial activity on the ground or on raw materials. Quarrying, open cast mining, tipping, building activities, the action of heavy lorries, or simply a strong wind acting on stock piles can lead to grit and dust blowing beyond the site and causing a nuisance to others in the area. These are described as 'fugitive' emissions and the particles tend to be quite coarse (≥ 5 μm diameter).

During combustion, formation of *soot* commonly accompanies carbon monoxide formation and is generally due to inadequate air supply. Soot particles produced from gas phase fuel/air reactions are commonly sub-micrometre (< 1 μm) and because they are of comparable size to the wavelength of light they are effective both at light absorption and light scattering. A relatively small mass concentration will therefore render the exhaust of flue gases opaque and give a dark stain on a filter paper. This is the basis of the reflectance technique used for measurement of 'smoke'.

In the combustion of coal, volatile material is released and burns as a gas, the remaining char burns more slowly, and a final residue of ash is left. Smoke can be produced from incomplete combustion of the volatile matter. Anthracite, coke, and the other manufactured solid fuels (such as the formed briquettes widely used for cooking stoves in Asia) contain low levels of volatile material and so reduce the smoke problem. Incomplete combustion of volatile matter is also the major problem with wood smoke and smoke from bonfires or biomass burning. This can lead to serious problems in developing countries where the use of biomass is still common for domestic cooking and heating in cities causing high emissions at low level.

Smoke formation is not a significant problem with spark ignition engines because the petrol and air are well mixed before entering the cylinders. Diesel engines suffer from the disadvantage of producing more smoke, especially under heavy load or acceleration conditions, because of the relatively poor mixing of air with the fuel injected as a spray into the cylinder. This produces regions which are too rich in fuel for complete combustion, leading to soot formation.

Four important terms relating to particulate matter in the atmosphere are:

Smoke—the particulate material assessed in terms of its *blackness* or *reflectance* when collected on a filter, as opposed to its *mass*. This is the historical method of measurement of particulate pollution in the UK. The size of particles collected into the sampler is below 10–15 μm.
TSP—Total Suspended Particulate matter. Mass concentration determined by filter weighing, usually using a 'Hi-Vol' sampler which collects all particles up to about 20 μm depending on wind speed.
PM10—Particulate matter below 10 μm aerodynamic diameter. This corresponds to the particles *inhalable* into the human respiratory system, and its measurement uses a size selective inlet.
PM2.5—Particulate matter below 2.5 μm. This is closer to, but slightly finer than, the definitions of *respirable* dust that have been used for many years in industrial hygiene to identify dusts which will penetrate the lungs.

Particulate vehicle emissions, for example diesel smoke, are predominantly in the PM2.5 range. The *total number* of particles emitted is dominated by ultrafine particles in the 30–80 nm (0.03–0.08 μm) range. However, a high proportion of the *total mass* is contributed by larger particles—one 0.5 μm particle has the same mass as 1000 particles of

50 nm. The health significance of these ultrafine particles is an area of current research.

3.2.5 Emissions Limits. Emissions from industrial processes are controlled via emissions limits set by international bodies such as the European Union or by national organisations such as the UK Environment Agency or US EPA. Processes subject to control receive an authorization which includes the permitted levels of emission. For example, large European municipal incinerators burning more than 3 tonne h^{-1} of waste must comply with the following emissions standards in mg m^{-3} of dry gas.

Particulates	30	CO	100
SO_2	300	HCl	50
HF	2	Volatile organics	20 (as total C)
Pb + Cr + Cu + Mn	5	Ni + As	1
Cd + Hg	0.2		

3.2.6 Emissions Inventories. The use of *spatially* disaggregated emissions inventories in modelling has been mentioned above. Knowledge of particular emission sources directs attention to the best targets for control and emissions reductions. Table 8 shows the UK national breakdown of emissions by *sector* for various pollutants while Table 9 illustrates *national* emissions for several European countries. These tables will be referred to in following discussions.

Table 9 *Emissions inventory data for European Countries from the CORINAIR Project* (ktonne per annum)[7]

Country	NMVOC		NO_x (as NO_2)		SO_2	
	1990	1994	1990	1994	1990	1994
Germany (all)	3323	2531	2980	2266	5255	2998
Germany West	2484	–	2424	–	912	–
Former GDR	839	–	556	–	4343	–
France	2866	2307	1590	1682	1300	1013
UK	2682	2430	2773	2387	3787	2696
Italy	2549	2785	2053	2157	2253	3671
Spain	1894	1970	1257	1227	2206	2060
Poland	1295	–	1446	–	3273	–
Former Czechoslovakia	531	–	1010	–	2409	–

3.3   Air Quality

3.3.1   Air Quality Standards.   Air quality is assessed by reference to air quality standards.[46] Those applicable in Europe and the UK are indicated in Table 10. The existing European standards are subject to revision in the near future following a new Framework Directive on Air Quality Assessment and Management (Directive 96/62/EC). The values included in the first 'daughter' directive agreed in June 1998 are indicated in the Table. Standards proposed by the UK Expert Panel on Air Quality Standards (EPAQS) are also indicated. These were promulgated to support the new responsibilities for air quality management of the Local Authorities following the 1995 Environment Act and are also used to define objectives within the government's National Air Quality Strategy.[47]

3.3.2   Air Quality Monitoring.   A critical activity in air quality management is the monitoring of the pollutant concentrations. Continuous monitoring data for common pollutants such as CO, NO_x, O_3, SO_2, and particulate matter (TSP or PM10) are recorded at many sites around the world. Information from an increasing number of these sites is available on the World Wide Web.[48] Sample data from Leeds in the UK is illustrated in Figure 9 and shows the diurnal cycles of the main urban pollutants.

A global monitoring network, the GEMS/AIR programme, was initiated by the WHO in 1973, and in 1975 became part of the Global Environment Monitoring System GEMS,[49] now jointly managed by WHO and UNEP. GEMS/AIR was implemented to strengthen urban air pollution monitoring and assessment capabilities, to improve validity and comparability of data among cities, and to provide global assessments on levels and trends of urban air pollutants, and their effects on human health. Approximately 170 monitoring stations in 80 cities in 47 countries participate in the programme measuring SO_2 and suspended particulate matter (55 cities in 33 countries) with other stations measuring NO_2, CO, O_3, and Pb. An indication of the air quality in major cities of the world from this programme is given in Table 11.

---

[46] Quality of Urban Air Review Group, 'Urban Air Quality in the United Kingdom', Department of the Environment, London, 1993.
[47] 'The United Kingdom National Air Quality Strategy', Department of the Environment, The Stationery Office Ltd, London, 1997.
[48] UK Air quality data is available at http://www.aeat.co/uk/netcen/ A review of other sites is given in J. Perks, *Clean Air*, 1997, **27**, 2, 44.
[49] http://www.gsf.de/UNEP/gemsair.html

Table 10   *Air quality standards for European Community and UK*

Pollutant	Measure	Concentration[a]	Source[b]
Smoke	Daily means—98%	$250\ \mu g\ m^{-3}$	EC(1)
	Winter median	$130\ \mu g\ m^{-3}$	
	Annual median	$80\ \mu g\ m^{-3}$	
PM10	Running 24 h means—99%	$50\ \mu g\ m^{-3}$	EPAQS
	Stage 1:		EC(5)
	24 h—35 times/yr	$50\ \mu g\ m^{-3}$	
	Annual mean	$40\ \mu g\ m^{-3}$	
	Stage 2:		EC(5)
	24 h—7 times/yr	$50\ \mu g\ m^{-3}$	
	Annual mean	$20\ \mu g\ m^{-3}$	
SO_2	Daily means—98%	$250\ \mu g\ m^{-3}$ (smoke <150)	EC(1)
		$350\ \mu g\ m^{-3}$ (smoke <150)	
	Winter median	$130\ \mu g\ m^{-3}$ (smoke >60)	
		$180\ \mu g\ m^{-3}$ (smoke <60)	
	Annual median	$80\ \mu g\ m^{-3}$ (smoke >40)	
		$180\ \mu g\ m^{-3}$ (smoke <40)	
	15 min means—99.9%	100 ppb	EPAQS
	1 h—24 times/yr	$350\ \mu g\ m^{-3}$	EC(5)
	24 h—3 times/yr	$125\ \mu g\ m^{-3}$	
	1 yr, ecosystem protection	$20\ \mu g\ m^{-3}$	
NO_2	Hourly means—98%	$200\ \mu g\ m^{-3}$	EC(2)
	Hourly means—18/yr	$200\ \mu g\ m^{-3}$	EC(5)
	Annual mean	$40\ \mu g\ m^{-3}$	
$NO + NO_2$	1 yr, vegetation protection	$30\ \mu g\ m^{-3}$ (as NO_2)	EC(5)
CO	Running 8 h mean	10 ppm	EPAQS
Ozone	A. 8 h mean	$110\ \mu g\ m^{-3}$	EC(3)
	B. 24 h mean	$65\ \mu g\ m^{-3}$	
	1 h mean	$200\ \mu g\ m^{-3}$	
	C. 1 h mean	$180\ \mu g\ m^{-3}$	
	D. 1 h mean	$360\ \mu g\ m^{-3}$	
	Running 8 h mean—97%	50 ppb	EPAQS
Lead	Annual mean	$2\ \mu g\ m^{-3}$	EC(4)
	Annual mean	$0.5\ \mu g\ m^{-3}$	EC(5)

[a] Units: 100 ppb corresponds to $266.1\ \mu g\ m^{-3}$ for SO_2, $191.2\ \mu g\ m^{-3}$ for NO_2, $116.4\ \mu g\ m^{-3}$ for CO, and $199.5\ \mu g\ m^{-3}$ for O_3 at a reference temperature of 20 °C.

[b] Sources: EC(1) Directive 80/779/EEC; EC(2) Directive 85/203/EEC; EC(3) Directive 92/72/EEC. Levels for ozone are reference levels not quality standards: A. Health protection standard, B. Vegetation protection standards, C. Population information, D. Population warning; EC(4) Directive 82/884/EEC; EC(5) Directive on limit values for sulfur oxide, oxides of nitrogen, particulate matter and lead in ambient air, 1998; EPAQS—UK Expert Panel on Air Quality Standards.

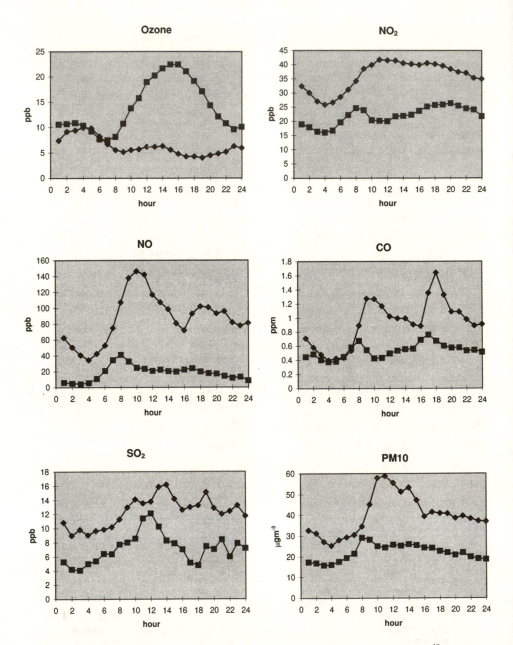

Figure 9   *Diurnal variation of pollutant concentrations for the Leeds AUN site.*[48] *Each point represents the monthly average for that hour of the day.* ◆ *January 1997,* ■ *July 1997*

Table 11 *Air quality data for megacities of the world. Annual average data for 1989 or 1990* [21,b]

City	Population 1990 (10^6)	Motor vehicle registrations (10^6)	SO_2 $\mu g\ m^{-3}$	NO_2 $\mu g\ m^{-3}$	TSP $\mu g\ m^{-3}$
Tokyo	20.5	4.40	20–30	60	50–60
Mexico City	19.4	2.50	100–200	–	200–500
New York	15.6	1.78	30–50	–	50–70
London	10.6	2.70	30–40	60–80	20[a]
Beijing	9.7	0.308	20–120	–	250–400
Moscow	9.4	0.665	–	50–100	100
Delhi	8.6	1.66	20–40	40–50	350–500

[a] Smoke level. This will be significantly below the corresponding TSP level.
[b] The data in the table represent the approximate ranges across several monitoring sites in each city and are to be regarded as indicative.

3.3.3 Air Quality Trends. The most significant trends in air quality on a national basis relate to a rather small number of factors such as:

- changes in the patterns of fossil fuel usage
- growth in vehicular transport
- improved technology for vehicles giving lower emissions
- tighter control of industrial emissions, especially those from large scale combustion for power generation.

There is however a significant difference between industrialized and developing nations with respect to the major contributors to air quality problems. A UNEP report has shown that cities in developing countries seem to be following the same historical trends as in other countries. Before industrialization, pollution is mainly from domestic sources and light industry. Concentrations are generally low but increase as the population increases. As industrial development and energy usage increase air pollution levels begin to rise dramatically and emission controls have to be introduced.[21] Emissions and air quality trends will be illustrated in the next few sections by reference to several of the commoner pollutants. Ozone, as a secondary pollutant, is dealt with later.

3.3.4 Vehicular Emissions—CO and Hydrocarbons. A major feature of the development of most countries in the last three decades has been the growth in vehicular traffic. The number of cars in the UK doubled from 10 to 20 million between 1970 and 1994, an average growth rate of 3% per annum. But in other countries the increase has been even more rapid—China, for example, has seen growth of 13% per annum representing a trebling of the vehicle numbers in 10 years, reaching 9.5 million by 1994.

The importance of vehicular emissions relative to other sources for the UK is shown in Table 8. Carbon monoxide emissions are predominantly from road vehicles whilst about 30% of VOCs also come from vehicles. Historically, CO emissions increased as traffic density increased through the 1980s but are now on a downward trend arising from the progressive introduction of catalytic converters (Figure 10). The significance of vehicular emissions is shown in the diurnal variation of the CO levels in major cities as illustrated in Figure 9. The double humped curve corresponding to the morning and evening rush hours is clearly seen. Carbon monoxide levels beside busy roads can, under very adverse conditions, reach approximately half the occupational exposure limit of 50 ppm but general urban levels in much of Europe and the US are usually below 10 ppm, and annual averages reach only 1–2 ppm in the UK and 5 ppm in the US.[50] In major cities where traffic is heavy but catalytic converters are not common, CO levels can be high; for example in 1992 Mexico City exceeded WHO guidelines for CO by a factor of more than 2.[21]

3.3.5 Nitrogen Oxides. Emissions of nitrogen oxides in the UK gradually increased until the early 1990s due to increased traffic density (Figure 10) and are only now coming down slowly as new vehicular emissions regulations begin to take effect and the NO_x emissions from power stations also decrease. Countries which have high coal consumption and/or high vehicle populations tend to have high NO_x emissions. Figure 11 and Table 9 illustrate this for several European countries using data from the CORINAIR inventory.[7] Allowing for the populations, the per capita emissions do not vary to the same extent as the total national emissions. At the moment it is considered unlikely by the European Environment Agency that targets for a 30% reduction by 2000 across the EU will be met.[45]

Since nitrogen dioxide is the main health hazard, most attention has been paid to monitoring urban levels of NO_2 as opposed to total NO_x. Annual average concentrations in urban areas of the UK range from 25 to 45 ppb. A significant wintertime episode of NO_2 pollution was experienced in London in December 1991 when the levels exceeded 300 ppb for 8 hours over one night and exceeded 100 ppb continuously for 3 days. The current EU standard for hourly averages is 104.6 ppb (200 $\mu g\, m^{-3}$). Throughout Europe there are a number of cities with long

[50] http://www.epa.gov/oar/aqtrnd96/toc.html 'National Air Quality and Emissions Trends Report 1996'.

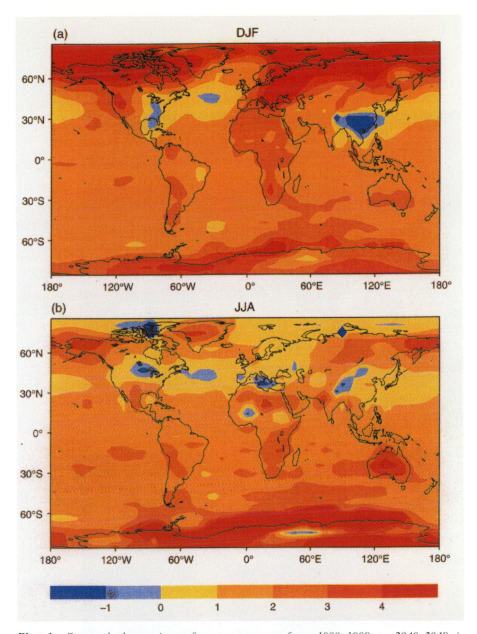

Plate 1 *Seasonal change in surface temperature from 1880–1889 to 2040–2049 in simulations with aerosol effects included. Contours are every 1° C. (a) December to February (b) June to August*
(Reproduced with permission from the Intergovernmental Panel on Climate Change[2])

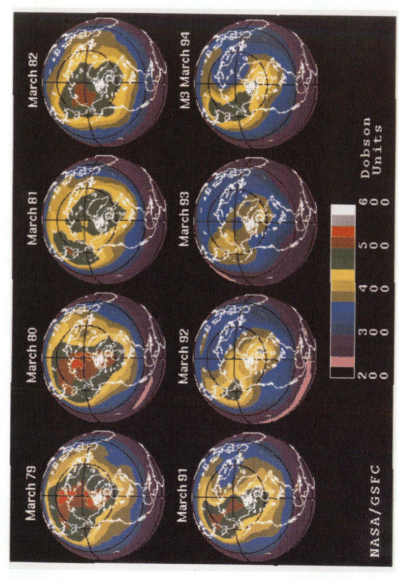

Plate 2 *TOMS total ozone mapping showing ozone depletion in the Northern Hemisphere since 1979* (Reproduced with permission from the Ozone Processing Team at NASA Goddard Space Flight Center[18])

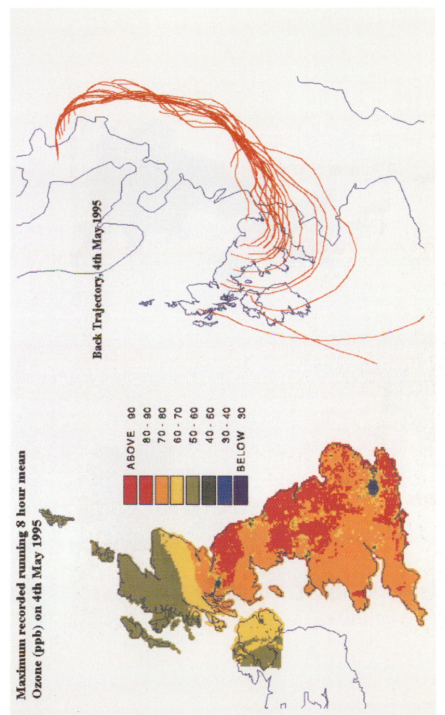

Plate 3 *Trajectory modelling study showing a smog episode across the UK* (Reproduced with permission of NETCEN)

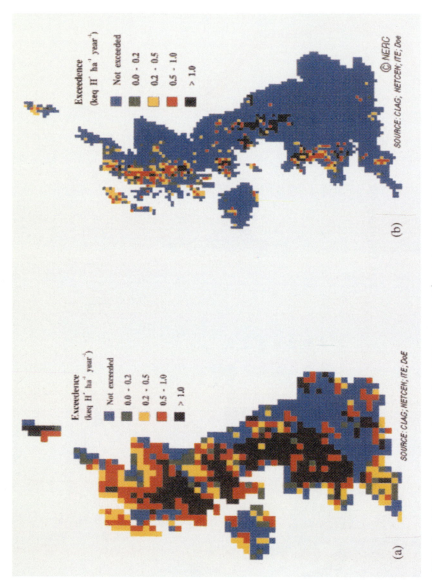

Plate 4 *Critical load exceedences for acidity in k eq H^+ per ha per annum. (a) soils, (b) fresh waters (Reproduced with permission from NETCEN[64])*

Exceedence
(keq H⁺ ha⁻¹ year⁻¹)

Not exceeded
0.0 - 0.2
0.2 - 0.5
0.5 - 1.0
> 1.0

SOURCE: CLAG; NETCEN; ITE; DoE

(b)

Exceedence
(keq H⁺ ha⁻¹ year⁻¹)

Not exceeded
0.0 - 0.2
0.2 - 0.5
0.5 - 1.0
> 1.0

SOURCE: CLAG; NETCEN; ITE, DoE

(a)

© NERC

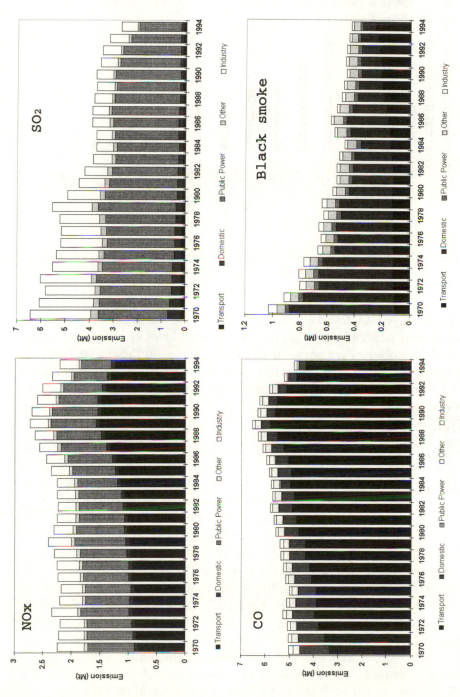

Figure 10 *UK annual emissions of pollutants by category of emissions source*[48]

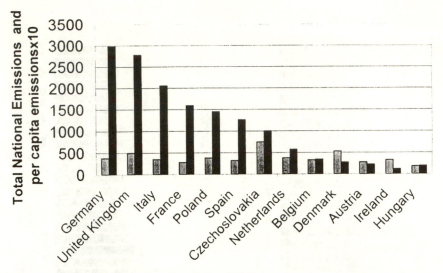

Figure 11   *National NO$_x$ emissions (ktonne per annum) for various European countries from the CORINAIR inventory[7] for 1990 and the corresponding emissions per capita*

term average NO$_2$ levels above the proposed EU guideline of 40 μg m^{-3}. Maximum concentrations do not show exceedences so often, however, and NO$_2$ episodes are not seen to cause a threat to those living outside urban areas.[45] Figure 9 shows that NO shows a similar diurnal variation to CO because of the traffic contribution but that NO$_2$, which can be considered as a secondary pollutant, does not show as marked a trend. This is discussed later.

3.3.6   Sulfur Oxides.   Following the outcry over the London smog of 1952 and the Clean Air Act of 1956, a concerted effort was made in the UK to reduce levels of smoke and sulfur dioxide in the air, which arose mainly from coal combustion. The components of this effort were control of visible smoke emissions from industrial chimneys, control of chimney heights to ensure adequate dispersion of SO$_2$, and control of domestic smoke emissions by local authorities through the promotion of the use of alternative fuels and the introduction of 'smoke control zones'. In practice, because of the widespread availability of natural gas from the North Sea in the years following 1970, gas has captured most of the domestic heating market in the UK and shares the commercial market with oil. Domestic coal use is limited and the major users are now the electricity supply industry and steel making. During the 1990s even this pattern changed with the generating companies switching to gas for new power stations for both economic and environmental reasons.

The urban concentrations of smoke declined dramatically from 1960 to 1980 but for sulfur dioxide the overall upward trend in emissions was not reversed until much later—1970 was the peak year. This pattern is mirrored in some but not all of the other European countries. Exceedences of the EU air quality standards for smoke and SO_2 are now very rare—none of the UK monitoring sites showed exceedences of the daily standards in the period 1993–95. One of the cities with a remaining smoke and SO_2 problem is Belfast in Northern Ireland, where coal burning in homes is still common since natural gas is only just becoming available. Many cities in Eastern Europe where coal is still in heavy use, have serious problems reminiscent of England in the 1960s. Concentrations reach 100–150 μg m^{-3} in some areas of central and north-western Europe during episodic 'winter smog' situations.[45]

A comparison of sulfur dioxide emissions across the countries of Europe shows wide variations depending on the fuel mix in each country. Table 9 shows data for 1990 and 1994 from the CORINAIR emissions inventory. France has relatively low emissions owing to the dominance of nuclear power for electricity generation. Many of the Eastern European countries have high emissions due to dependence on relatively poor quality coal. The former German Democratic Republic had especially high emissions (over 80% of the 1990 figure for 'Germany' shown in the table). These have been markedly reduced since reunification as shown by the 1994 figure.

3.3.7 Vehicular Particulates.[51] Black smoke emissions were originally dominated by coal smoke and as control efforts were targeted on this source rapid improvement of air quality was achieved in many cities. However, the more recent decline has not been as rapid because of the rise of smoke derived from vehicles, particularly diesels. Diesel engine soot is three times blacker than coal smoke and this is incorporated into the estimation of the 'black smoke' emissions shown in Table 8. In contrast, the small amount of smoke from a petrol engine is much lighter than coal smoke because it consists largely of unburnt hydrocarbons rather than elemental carbon. Figure 10 and Table 8 show how UK smoke emissions, once dominated by coal, are now dominated by vehicular emissions.

In terms of the *mass* of fine PM10 particles the picture is not so one-sided (Table 8) and both petrol and diesel engines make a significant contribution. Although for a given mass diesel smoke is much blacker than petrol-derived smoke and gives a much greater influence on the reflectance (and hence the 'smoke' measurement) this is irrelevant if a direct mass measurement technique is used. Analysis of UK PM10 measurements suggest that on average 40–50% of urban PM10 is

[51] 'Airborne particulate matter in the United Kingdom', Third Report of the Quality of Urban Air Review Group, Department of the Environment, available from The University of Birmingham, Institute of Public and Environmental Health, 1996.

contributed by vehicle exhausts in winter, with higher proportions when the PM10 concentration itself is high and lower proportions in summer.[48] In addition to the primary emissions indicated in Table 8 account must also be taken of the secondary particulate pollutants such as sulfates and nitrates. These are discussed in Section 5.1.

On a global scale, particulate matter represents the biggest problem for the urban environment in terms of non-attainment of air quality standards. Many of the world's largest cities persistently exceed WHO limits, exposing large populations to a health risk. Although PM10 levels are falling in developed areas such as the US and Europe, it is currently estimated that in nearly three-quarters of major EU cities WHO Air Quality Guidelines for SO_2 and particulate matter (PM) are exceeded at least once in a typical year.[45]

3.3.8 Heavy Metals. Emissions of lead, Pb, from automobile exhausts arise from the use of lead tetra-alkyl anti-knock additives to improve the octane rating of the petrol. About 70% of the lead is emitted from the tail pipe, mainly as mixed halides. Increasing concern over the potentially harmful effects of lead on health, particularly the health of children, resulted in a gradual decrease in the maximum permitted amount which can be added to petrol in many countries. This is currently 0.15 g Pb l^{-1} in Europe. All new cars must now be able to run on unleaded gasoline. The emissions of lead in 1995 were about 14% of what they were in 1975 and 70% of total gasoline sales in the UK are now of unleaded petrol.

The decrease in the use of lead has led to a marked reduction in the atmospheric concentrations in many countries, with Europe following some years behind the USA. For UK urban sites, long term averages have come down from a typical 500 ng m^{-3} to around 100 ng m^{-3} in 1998, with similar decreasing trends being mirrored in the US where averages were around 50 ng m^{-3} in 1995. These figures may be compared with the current EU air quality standard of 2 or the new standard of 0.5 μg m^{-3} annual average. Although declining in Europe, airborne lead remains an important air pollutant in cities where leaded petrol is still a major transport fuel. UNEP reported that several of the world's megacities still exceed WHO limits for lead, with Cairo and Karachi showing the highest exceedences in 1992.[21]

3.3.9 Toxic Organic Micropollutants (TOMPS). This phrase is used to describe several groups of organic pollutants which are present in the atmosphere in very small concentrations but which, by inhalation or ingestion of contaminated food products, could result in human health effects. TOMPS may be present in the vapour phase or adsorbed onto particles, the distribution between vapour and particulate phases depending on temperature. They include polynuclear aromatic compounds (described variously as PNA, PAH or PAC)—high molecular weight hydrocarbons often found in association with soot, some of which are carcinogenic. Benzo[a]pyrene is one example. Following the US EPA, analyses are commonly made on a group of 16 compounds with 2–6 rings.

There has been a long-standing interest in the presence of such carcinogens in the general environment and in the levels emitted from combustion processes. Attention has been switched from coal smoke to diesel engine exhaust smoke as a source of these compounds but it remains to be established whether there is a significant health hazard at the very low concentrations to which the general public are exposed.

The other main group of TOMPS are polychlorinated aromatic compounds. These include polychlorinated biphenyls (PCBs) which have entered the environment following wide scale use as insulating fluids in electrical transformers, and polychlorinated dibenzodioxins and dibenzofurans ('dioxins', PCDDs; 'furans', PCDFs), which have combustion origins, especially from incinerators. There are numerous isomers of each of these classes of compounds depending on the number and locations of the chlorine atoms in the molecules.

Extensive monitoring of air, water and soil samples for TOMPS has been undertaken in recent years. The average air concentrations in London between 1991 and 1995, for example were as follows:[40]

PAHs 50–150 ng m^{-3}; PCBs 1.0–1.5 ng m^{-3}; Dioxins 100–200 fg TEQ m^{-3}.

The US EPA carried out a thorough review of dioxins in the environment and their health effects, published in 1994.[52] Airborne concentration data reported in the USA and Europe also indicated average air concentrations of the order of 100 fg TEQ m^{-3}.

Note: 1 ng = 10^{-9} g, 1 fg = 10^{-15} g. TEQ = Toxic Equivalent—in summing the masses of different dioxins each is weighted by its relative toxicity with 2,3,7,8 tetrachlorodibenzo-p-dioxin taken as the reference.

4   GAS PHASE REACTIONS AND PHOTOCHEMICAL OZONE

4.1   Gas Phase Chemistry in the Troposphere

4.1.1   Atmospheric Photochemistry and Oxidation.   Although the emissions patterns and dispersion discussed in the previous sections give a part of the picture of where high levels of pollutants can be found, we

[52] 'Estimating Exposure to Dioxin-like Compounds' (3 Vols) and 'Health Assessment Document for 2,3,7,8-Tetrachlorodibenzo-*p*-dioxin (TCDD) and Related Compounds' (3 Vols), US Environmental Protection Agency, Office of Research and Development, Washington, 1994.

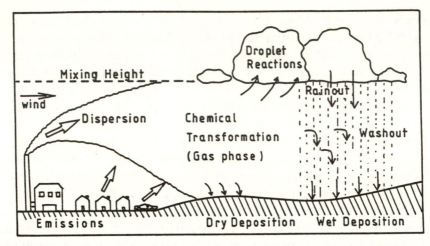

Figure 12   *Processes which may be involved between the emission of an air pollutant and its ultimate deposition to the ground*

must combine this with an understanding of other atmospheric processes to see how secondary pollutants such as ozone and acid rain are formed. These processes including dispersion, chemical reaction, and deposition are illustrated in Figure 12. The chemical processes involved are complex and changes both in the gas phase and in the aqueous droplet phase are important. Not only do we have to consider the transformation of the primary pollutants but also the formation of secondary pollutants such as ozone which can have adverse effects on the environment and on human health.

Of basic importance to an understanding of the gas phase chemistry is the effect of the sun. Photons of ultraviolet light provide a means of initiating chemical reactions which would otherwise not take place. In addition to stable molecules, photochemical reactions involve free radicals such as hydroxyl $OH^{\bullet}$, hydroperoxy $HO_2^{\bullet}$ and methyl $CH_3^{\bullet}$. Free radicals are extremely reactive and have very short lifetimes. Their concentrations in the atmosphere are small but none the less significant. For example, $OH^{\bullet}$ concentrations in polluted atmospheres may be in the range 10^6–10^7 radicals cm^{-3}, *i.e.* one radical for every 10^{13} nitrogen molecules.

One of the most important overall processes we have to describe is oxidation, that is, the combination of atmospheric oxygen with the primary pollutants. For the three commonest inorganic pollutants the overall results are

carbon monoxide	$CO \rightarrow CO_2$	carbon dioxide
nitrogen oxides	$NO, NO_2 \rightarrow HNO_3$	nitric acid
sulphur dioxide	$SO_2 \rightarrow H_2SO_4$	sulphuric acid.

Later we shall consider the involvement of the two acids in particles and droplets but as far as the gas phase chemistry is concerned these species mark the end point of the process. For organic hydrocarbon species there may be a number of intermediate stable molecules formed, but the overall process is rather like combustion with the end product being carbon monoxide; for example:

methane $CH_4 \rightarrow$ formaldehyde HCHO $\rightarrow$ carbon monoxide

although the time taken to complete this process can be very long. In all these cases the main species initiating the sequence of reactions is the hydroxyl radical:

$$CO + OH^{\bullet} \rightarrow CO_2 + H^{\bullet} \qquad (14)$$

$$NO_2 + OH^{\bullet} \rightarrow HNO_3 \qquad (15)$$

$$SO_2 + OH^{\bullet} \rightarrow HSO_3^{\bullet} \qquad (16)$$

$$CH_4 + OH^{\bullet} \rightarrow CH_3^{\bullet} + H_2O \qquad (17)$$

In the case of nitric acid formation, there are no remaining free radicals to continue the chain of reactions. In the other cases the hydroxyl radical is eventually regenerated but only after several further steps which interlink the chemistries of the various species (Table 12). Carbon monoxide oxidation is a slow process and the lifetime of CO in the atmosphere is several years. The oxidation rate of SO_2 can be around $1\% \ h^{-1}$ resulting in an overall lifetime of a few days. Radicals other than

Table 12 *Chemical reactions for the atmospheric oxidation of CO, SO_2, CH_4*

Carbon monoxide
$$CO + OH^{\bullet} \rightarrow CO_2 + H^{\bullet}$$
$$H^{\bullet} + O_2 + M \rightarrow HO_2^{\bullet} + M \quad (M = \text{an inert third body})$$

Sulfur dioxide
$$SO_2 + OH^{\bullet} \rightarrow HSO_3^{\bullet}$$
$$HSO_3^{\bullet} + O_2 \rightarrow SO_3 + HO_2^{\bullet}$$
$$SO_3 + H_2O \rightarrow H_2SO_4$$

Methane
$$CH_4 + OH^{\bullet} \rightarrow CH_3^{\bullet} + H_2O$$
$$CH_3^{\bullet} + O_2 + M \rightarrow CH_3O_2^{\bullet} + M$$
$$CH_3O_2^{\bullet} + NO \rightarrow CH_3O^{\bullet} + NO_2$$
$$CH_3O_2^{\bullet} + O_2 \rightarrow HCHO + HO_2^{\bullet}$$

$HO_2^{\bullet}$ to $OH^{\bullet}$ conversion
$$HO_2^{\bullet} + NO \rightarrow OH^{\bullet} + NO_2$$

OH$^{\bullet}$ such as HO$_2^{\bullet}$, CH$_3$O$_2^{\bullet}$, or other hydrocarbon peroxy radicals will also attack SO$_2$ but at a slower rate and their contribution to the overall oxidation is thought to be relatively small.

4.1.2 Ozone. We must now address the question of where the hydroxyl radicals come from in the first place. One source, present even in non-polluted atmospheres at a background level of 20–40 ppb, is ozone. Ozone is a secondary pollutant formed through photochemical reactions and can have a harmful effect on human health causing respiratory problems, and on crop yields. It is the primary constituent of photochemical smog. Ultraviolet light of wavelength below 310 nm can dissociate ozone producing electronically excited oxygen atoms O* which rapidly split molecules of water vapour:

$$O_3 \xrightarrow{\text{UV light}} O_2 + O* \tag{18}$$

$$H_2O + O* \longrightarrow 2OH^{\bullet} \tag{19}$$

Aldehydes (R.CHO, including formaldehyde H.CHO) can also be photolysed producing hydrogen atoms which eventually result in OH$^{\bullet}$ radicals via reactions already shown in Table 12.

Of basic importance to the understanding of polluted urban atmospheres is the photolysis of NO$_2$ and the subsequent formation of ozone above background levels. It was noted earlier that only a small proportion of NO$_x$ emissions are in the form of NO$_2$, the rest being NO. NO emitted into the atmosphere can be slowly oxidized to NO$_2$ by the reaction with molecular oxygen

$$O_2 + 2NO \rightarrow 2NO_2 \tag{20}$$

Photolysis of NO$_2$ by UV light below 395 nm produces oxygen atoms and subsequently ozone:

$$NO_2 \xrightarrow{\text{UV light}} NO + O^{\bullet} \tag{21}$$

$$O_2 + O^{\bullet} \longrightarrow O_3 \tag{22}$$

However the process is reversed by the reaction

$$O_3 + NO \rightarrow O_2 + NO_2 \tag{23}$$

so the net result is an ozone level in equilibrium with NO and NO$_2$ and dependent on the intensity of the solar radiation. The observed levels of ozone are higher than would be predicted on the basis of this limited

Table 13 *Species lifetimes, k_{OH} and POCP weighted emissions for selected hydrocarbon species. POCP values are calculated relative to a POCP of 1 for ethene based on 1990 data*[53]

Species	$k_{OH} \times 10^{12}$ cm^3 molecule^{-1} s^{-1}	Lifetime (h)	POCP weighted emission (tonnes ethene yr^{-1})
methyl benzene (toluene)	5.96	58.3	85423
ethene	8.52	40.7	78925
1,3-dimethyl benzene	23.6	14.7	66086
butane	2.54	136.7	58649
propene	26.3	13.2	36946
benzene	1.23	282.3	9418
ethane	0.268	1295	3017

scheme. High ozone levels imply a high NO_2/NO ratio or significant $NO \rightarrow NO_2$ conversion which cannot be achieved by the molecular reaction (20). The types of reactions responsible have already been shown in Table 12; they are the transfer of an O-atom from $HO_2^{\bullet}$, $CH_3O_2^{\bullet}$ and other peroxy radicals. Crudely we can say that the photochemical reaction mixture catalyses the NO to NO_2 conversion resulting in the build-up of ozone. Hydrocarbon molecules differ in their photochemical ozone creation potential (POCP),[53–55] largely related to how quickly they react with the $OH^{\bullet}$ radical. Methane has a very low potential but other species, including substituted benzenes such as toluene and the xylene isomers, and light unsaturated hydrocarbons such as ethene, propene, and but-2-ene have high POCPs. POCPs represent the increase in ozone concentration caused by an increase in the concentration of the hydrocarbon species relative to ethene which has a value 1. The aromatics are present in high concentration in gasoline and the unsaturated compounds are typical products of engine combustion. Table 13 gives some examples of the rate of reaction with $OH^{\bullet}$ for selected hydrocarbon species, along with associated POCP weighted emissions. These are calculated by multiplying the POCP by the annual emission and therefore indicate which VOCs are best targeted for control strategies.

As they are driven by the chemistry described above, we can see that the concentrations of O_3 and NO_2 depend on sunlight and emissions and will therefore vary diurnally as well as with the time of year. The diurnal variations of NO, NO_2, and O_3 typically detected are illustrated in

[53] UK Photo-chemical Oxidants Review Group, 'Ozone in the United Kingdom', Department of the Environment, London, 1993.
[54] R. G. Derwent and M. E. Jenkin, 'Hydrocarbon involvement in photochemical ozone formation in Europe', AERE–R13736, UK Atomic Energy Authority, Harwell.
[55] UK Photo-chemical Oxidants Review Group, Fourth Report, 'Ozone in the United Kingdom', Department of the Environment, Transport and the Regions, London, 1997.

Figure 9. The morning rush hour peak in the NO and hydrocarbon emissions is followed by the gradual conversion to NO_2 and subsequent rise of O_3, which decays as the sun goes down in late afternoon.

The production of ozone is much greater in the trace representing summer conditions than for winter due to higher photolysis rates. For NO and NO_2 the winter hourly averages are higher. In an air mass moving downwind from a city, the ozone peak may be worse in the surrounding countryside than in the city centre because of less destruction of O_3 by NO. During the night most of the reactions that have been described die down but there is an additional route for conversion of NO_2 to nitric acid via the nitrate radical $NO_3^{\bullet}$ which is formed from reaction of NO_2 with O_3. $NO_3^{\bullet}$ is photolytically unstable in daylight. Dry deposition is also an important loss process for O_3 and can account for up to 30% of its loss.

Space does not permit a detailed discussion of the hydrocarbon chemistry in the atmosphere, which is extremely complex. In addition to reactions with $OH^{\bullet}$ radicals and molecular oxygen similar to those for methane shown in Table 12, hydrocarbon species are attacked by the oxygen atoms released in reaction (21) and by ozone. One result of the interaction of the hydrocarbon and NO_x chemistries is the formation of a group of lachrymatory substances including peroxyacetyl nitrate (PAN) and peroxybenzoyl nitrate (PBzN). The thresholds for eye irritation for these compounds are only 700 and 5 ppb respectively. They are formed through the reaction of NO_2 with oxidation products of aldehydes:

$$RCHO \xrightarrow{OH^{\bullet}} RCO^{\bullet} \xrightarrow{O_2} RCO\!-\!O\!-\!O^{\bullet} \xrightarrow{NO_2} RCO\!-\!O\!-\!O\!-\!NO_2$$

$R = CH_3$ gives peroxy acetyl nitrate; $R = C_6H_5$ gives peroxy benzoyl nitrate.

4.2   Trends in Ozone Levels

In the UK, elevated ozone levels generally occur in summer, anticyclonic conditions such as those illustrated in Plate 3 when photolysis rates are high and wind speeds low. Concentrations tend to be higher in southern England where polluted air masses from continental Europe have a significant effect. The highest recorded concentration in the UK was 258 ppb at Harwell in 1976 as compared with a high of 450 ppb in the Los Angeles basin in 1979. Los Angeles provides an ideal environment for ozone formation due to high traffic emissions and its meteorological conditions. Ozone levels in LA are currently falling due to emissions legislation, although the number of days a year on which the State limit of 90 ppb is exceeded is still of the order of 100.

In Europe there is evidence that ozone is slightly increasing throughout the troposphere. Over 1000 monitoring stations are now reporting ozone levels in Europe. In the summer of 1996 the EEC threshold for warning the public of 1 h $> 360\ \mu\text{g m}^{-3}$ was exceeded at three stations in Europe, two in Athens and one in Firenze in Italy. The threshold for information of the public of 1 h $> 180\ \mu\text{g m}^{-3}$ was exceeded in all EU states except for Ireland. In terms of episodes of elevated ozone levels, the UK and Scandinavia are the least affected out of all the European countries with southern Europe being most affected. Urban areas show less exceedence because of high NO conditions.

The formation of photochemical smog is governed by emissions of NO_x and volatile organic carbons (VOCs). Because the chemistry is complex and non-linear, abatement strategies for ozone and smog are difficult to devise. Within Europe some regions will respond better to reductions in NO_x emissions and some better to a reduction in VOCs. In general whether a region is NO_x or VOC limited depends on the NO_x:VOC ratio, with regions of high NO_x often responding better to reductions in VOCs. The UNECE Protocol now require a 30% reduction in VOCs from 1988 to 1999 and an EC directive on ozone requires national monitoring networks to assess conditions according to the reference levels given in Table 10.

5 PARTICLES AND ACID DEPOSITION

5.1 Particle Formation and Properties

5.1.1 Particle Formation. The nitric and sulfuric acids formed in the gas reactions described earlier generally undergo further changes. They are both water soluble and will be rapidly absorbed into water droplets if these are present. They may react with solid particles forming sulfates and nitrates. For example, limestone particles (calcium carbonate, $CaCO_3$) can be converted to calcium sulfate, and salt particles (NaCl), of marine origin, can be converted to sodium sulfate, Na_2SO_4, or sodium nitrate, $NaNO_3$, with the displacement of hydrogen chloride gas, HCl. However, the most common reactions are those involving ammonia:

$$NH_3 + HCl \rightleftharpoons NH_4Cl \tag{24}$$

$$NH_3 + HNO_3 \rightleftharpoons NH_4NO_3 \tag{25}$$

$$NH_3 + H_2SO_4 \rightarrow NH_4HSO_4 \tag{26}$$

$$NH_3 + NH_4HSO_4 \rightarrow (NH_4)_2SO_4 \tag{27}$$

Reactions (24) and (25) are actually reversible and the position of equilibrium depends on the concentrations and the temperature. Significant dissociation of NH_4Cl and NH_4NO_3 occurs during warm summer weather. One end product of the oxidation of SO_2 in the atmosphere is ammonium sulfate, a substance better known as a fertilizer. The average total mass of these secondary sulfates and nitrates in the UK varies from about 10–12 $\mu g\ m^{-3}$ in the south east to 6–7 $\mu g\ m^{-3}$ in Scotland[56] and, being fine particles, they contribute to PM10. These figures should be compared to the typical PM10 levels in the atmosphere discussed earlier.

Particles formed by gaseous reactions or condensation are initially very small ($< 0.1\ \mu m$) but grow rapidly either by surface accumulation of material from the gas or by particle–particle coagulation. Once in the size range 0.1 to 2.0 μm, they become relatively stable towards further growth and can remain airborne for periods of days. Most smoke emissions are in this category along with the sulfates and nitrates. At the other end of the size spectrum (2–50 μm) is coarse dust, either emitted from industrial processes or raised by the wind from the ground, which has a shorter lifetime.

5.1.2 Particle Composition. Figure 13 presents a typical breakdown of the components of the total suspended particulate matter for an urban area based on results from Brisbane, Australia.[57] The fine particles ($< 2.5\ \mu m$) are dominated by the ammonium sulfate and nitrates plus carbonaceous material. About one-third of this is elemental carbon and the other two-thirds organic carbon (*i.e.* high molecular weight hydrocarbons). The coarser particles are dominated by wind-blown dusts (crustal materials such as clays, silica, limestone, *etc.*) and include the sea salt component but have smaller amounts of ammonium sulfate and carbon. These results are fairly similar to those reported in the UK[51] although the sea salt component is higher.

5.1.3 Deliquescent Behaviour. The particles of water soluble compounds such as sulfates, nitrates, and chlorides will exist in the atmosphere either as solid particles or as droplets depending on the relative humidity. For pure compounds the transition from solid to liquid is a sharp one. For example, it takes place with salt, NaCl, at 75% relative humidity and with ammonium sulfate at 86%. Particles of mixed composition may continuously grow in size by water absorption from 60–70% up to 100% relative humidity. Insoluble carbonaceous particles are, of course, not subject to this phenomenon. Particles are

[56] A. G. Clarke, M. J. Willison, and E. M. Zeki, *Atmos. Environ.*, 1985, **19**, 1081.
[57] T. C. Chan, R. W. Simpson, G. H. McTainsh and P. D. Vowles, *Atmos. Environ.*, 1997, **31**, 3773.

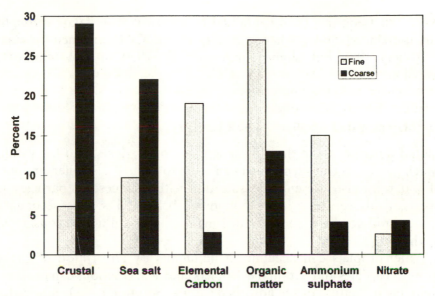

Figure 13 *Composition of coarse (> 2.5 μm) and fine atmospheric particles in Brisbane, Australia*[57]

important in cloud and fog formation since they act as condensation nuclei on which the much larger droplets may begin to be formed from water vapour.

5.1.4 Optical Properties.[58] Fine particles in the 0.1–2 μm size range are effective at scattering light, and some, especially soot, are also effective at absorbing light. These effects contribute to a reduction in visibility through the atmosphere. Quantitatively, the visibility is the maximum distance at which a large dark object can be seen against the horizon sky (sometimes called the visual range). In clean air the distance can be over 50 km but in air polluted by particles this can be severely reduced. A mass concentration (TSP) of 200–300 $\mu g\,m^{-3}$ will reduce the visibility to below 5 km. The phenomenon of a mid-summer haze on a hot day reflects high particulate pollutant levels. It is a totally different phenomenon to an early morning mist which is caused by water droplets at a relative humidity near 100% in Europe. Such highly polluted summer days are usually associated anticyclonic conditions (*cf.* Plate 3). Wintertime visibility has been improved in many cities and the frequency of fogs reduced in urban areas by the measures taken to control smoke and SO_2 emissions. Any additional measures to reduce pollutant emissions, especially of NO_x and SO_2, will bring a corresponding benefit in terms of improved visibility throughout the

[58] A. P. Waggoner, R. E. Weiss, *et al.*, *Atmos. Environ.*,1981, **15**, 1891.

year. This trend has been confirmed by measurements taken throughout the United States where visibility in the East worsened between 1970 and 1980 but slightly improved by 1990 in line with SO_2 emissions.[50]

5.2   Droplets and Aqueous Phase Chemistry

Liquid water occurs in the atmosphere as clouds, mists, and fog within which the concentration of water can be up to 1 g m^{-3}. Smaller amounts of water are also present in association with deliquescent particles as discussed above. At relative humidities below 95%, this secondary amount will normally be less than 1 mg m^{-3} and will not be considered further.

Water droplets can accumulate pollutants by adsorption of either gases or particles and within the droplets chemical reactions can proceed, changing the nature of the adsorbed species. Solution of SO_2 into water results in a mixture of the species SO_3^{2-} (sulfite), HSO_3^- (bisulfite) and H_2SO_3 (undissociated sulfurous acid) depending on the pH which is defined as $-\log_{10}$ (H^+ ion concentration). At typical cloudwater pH values the bisulfite ion is the dominant ion formed from:

$$SO_2 + H_2O \rightleftharpoons H^+ + HSO_3^- \tag{28}$$

There are several mechanisms for the oxidation of bisulfite to sulfuric acid in the aqueous phase. In a cloud the most important oxidants appear to be ozone and hydrogen peroxide, which is formed from the reaction of two $HO_2^{\bullet}$ radicals. The oxidants must first be adsorbed into water from the gas phase and then take part in the reactions:

$$O_3 + HSO_3^- \rightarrow H^+ + SO_4^{2-} + O_2 \tag{29}$$

$$H_2O_2 + HSO_3^- \rightarrow H^+ + SO_4^{2-} + H_2O \tag{30}$$

The sulfuric acid formed will be completely dissociated to ions H^+ and SO_4^{2-}. The ozone reaction is inhibited by low pH whereas the H_2O_2 reaction is not. In acidic droplets the oxidation by H_2O_2 is therefore dominant whereas at high pH the O_3 becomes more significant. In a cloud or fog where there has been a significant input of particulate pollution, the oxidation of SO_2 by atmospheric oxygen catalysed by metal ions (iron, Fe^{3+}, or manganese, Mn^{2+}) or by soot can also be important. Although not indicated by the simple reactions 29 and 30, free radical reactions are as important in solution chemistry as they are in the gas phase, and many different species are involved. The results of a recent European Project in this area have been published.[59]

It is difficult to distinguish experimentally between the photochemical reaction mechanism leading to H_2SO_4 with subsequent absorption of the acid into water and the aqueous phase oxidation of SO_2. However, it appears that both routes are important.

In the case of nitrogen oxides the routes for nitrate formation in clouds are

- dissolution of gaseous nitric acid vapour formed by reactions previously described in Section 4
- dissolution of nitrate-containing particles into droplets
- absorption of nitrogen oxides or nitrous acid, HONO, into droplets followed by oxidation of nitrite ions NO_2^- by oxidants such as H_2O_2.

Because of the low solubility of NO_2 and especially NO in water it is likely that the dominant processes are the first two.

Despite partial neutralization of dissolved acids by ammonia, the water in polluted fogs and clouds can be much more acidic than in collected rainfall. pH values down to 2 have been measured in urban fogs—more acid than vinegar. Similarly low values have been measured in the plumes of large power stations.

5.3 Deposition Mechanisms

5.3.1 Dry Deposition of Gases. As illustrated in Figure 12, the life cycle of an air pollutant normally involves emission, dispersion and transport, chemical transformation and finally deposition to the ground. Understanding the rates and mechanisms of deposition is important to the assessment of the environmental impact of many pollutants. Experimentally we can measure the concentration of the pollutant ($\mu g\ m^{-3}$) and the total rate of deposition ($\mu g\ m^{-2}\ s^{-1}$). The higher the ground level concentration, the more rapid the deposition but the ratio of these two quantities gives a useful measure of the efficiency of the deposition process. It is called the *deposition velocity*.

$$\text{Deposition velocity} = \frac{\text{deposition rate}}{\text{air concentration}} \quad (\text{units: m s}^{-1})$$

[59] 'Heterogeneous and Liquid-Phase Processes', ed. P. Warneck, Springer, Berlin, 1996.

Different surfaces (water, soil, ice, *etc.*) will have correspondingly different deposition velocities. The description 'dry deposition' is used in all cases even if the removal is to a wet surface.

Deposition to vegetation is rather more complex than deposition to a plane surface. The pollutant's progress may be retarded either by the slowness of diffusion through the air within the canopy of vegetation (*i.e.* between the leaves, *etc.*) or by the rate of transfer from air to leaf surface (the cuticle) or to the interior of the leaf via stomata. Moisture makes a considerable difference, in that transfer to a wet surface is generally faster than to a dry surface. Based on experimental data, dry deposition velocities for SO_2 used in modelling large scale deposition are typically assumed to be in the range 2–5 mm s^{-1}. Similar considerations apply to other pollutants which are subject to a significant rate of adsorption at the ground or onto vegetation (NO, NO_2, HNO_3, *etc.*). NO_2 has a lower value of 1 mm s^{-1} while HNO_3 deposits very rapidly and values up to 40 mm s^{-1} are assumed.[42]

5.3.2 Wet Deposition. The term 'wet deposition' is used to describe pollutants brought to ground either by rainfall or by snow. This mechanism can be further subdivided depending on the point at which the pollutant was absorbed into the water droplets. In-cloud absorption followed by precipitation is termed 'rain-out'; below cloud absorption, *i.e.* pollutants collected as the raindrops fall, is termed 'wash-out'. The rate of removal of a pollutant by wash-out will increase in proportion to the rainfall rate. Overall we can define a scavenging coefficient which is the fractional loss of the pollutant from the gas phase per second. For SO_2 the scavenging ratio is of the order of 10^{-6} s^{-1} in drizzle (≤ 1 mm h^{-1}) and an order of magnitude greater in heavy rain. Half of the gas below the clouds can be removed in several hours of heavy rain. The rates for the nitrogen oxides are lower due to their reduced solubility in water. In remote areas the majority of the wet deposition of sulfur appears to be due to rain-out. Wash-out becomes relatively more significant near the sources of pollution where the gas concentrations are high.

Another mechanism of deposition is when fog or cloud droplets are removed directly to the ground or to vegetation. This is termed 'occult deposition'. It becomes significant at elevated locations such as mountains or hill tops. There is the potential for more severe damage to foliage than with acid rain since, as was previously mentioned, polluted fog or cloud droplets can contain much higher concentrations of acidic pollutants than raindrops.

Dry deposition and wet deposition are both important in the total deposition of sulfur and oxidized nitrogen compounds. Dry deposition is

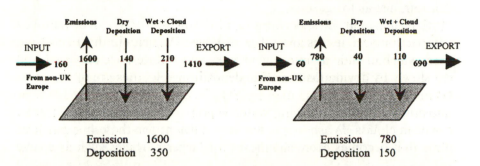

SULPHUR (kT S)

OXIDISED NITROGEN (kT N)

Figure 14 *Average annual emission and deposition of sulfur and oxidized nitrogen for the UK 1992–1994. The figures for the inputs are estimates of the European contribution to the total deposition figures*[42]

most significant where ground level concentrations are high, in other words close to the sources. Figure 14 shows the annual budgets for the UK between 1992 and 1994.[42] Deposition of reduced nitrogen (NH_3) is also important as discussed below. For sulfur, dry and wet deposition are respectively 40 and 60% of the total, for nitrogen the split is 27% dry and 73% wet. Modelling work suggests that about 45% of the wet S deposition and 25% of the dry S deposition in the UK comes from other European countries. The UK is however, a net exporter of air pollutants, while other countries of Europe, for example Norway and Sweden, are net importers. The UK is the largest single contributor to sulfur deposition in southern Norway although emissions from many other countries are transported there.

5.3.3 Deposition of Particles. The mechanism of deposition of particles depends on the particle size. Large particles with diameter greater than 10 μm fall slowly by gravitational settlement. The larger the particles, the more rapidly they fall. The sedimentation velocities for particles of density 2 g cm^{-3} are as follows:

Diameter (μm)	Velocity (m s^{-1})	Diameter (μm)	Velocity (m s^{-1})
5	1.5×10^{-3}	50	1.4×10^{-1}
10	6.1×10^{-3}	100	4.6×10^{-1}
20	2.4×10^{-2}	150	8.0×10^{-1}

Particles larger than 150 μm diameter, falling at over 1 m s^{-1}, remain airborne for such a short time that they do not need to concern us as air pollutants. Particles less than 5 μm have sedimentation velocities which are so low that their movement is determined by the natural turbulence of the air, just as for gases.

Intermediate particles, between 1 and 10 μm diameter, can be removed by impaction onto leaves and other obstacles. Particles in the 0.1 to 1 μm range, which include most of the nitrates and sulfates, are only removed very slowly by dry deposition. The deposition velocities are of the order of 1 mm s^{-1}, much lower than for SO_2. The most likely route for their removal is rain-out following water vapour condensation and droplet growth in clouds. Wash-out is not very efficient for these fine particles although it becomes more significant for larger particles such as coarse dust.

5.4   Acid Rain

5.4.1   Rainwater Composition.   Even in the absence of air pollutants, rain-water is slightly acidic (pH 5.6) due to atmospheric carbon dioxide. 'Acid rain' therefore refers to rain with a pH below about 5. Annual average acidities over much of central and eastern Europe correspond to pH values between 4.2 and 4.8 whilst rain in the Mediterranean region and the western extremes of Europe has a pH of 5.0 or above. Similarly the pH in the north-eastern USA is in the region 4.3–4.7 but increases to 5.0 or above in the mid-west and western states. Table 14 shows the annually averaged composition at three UK sites.[40] Barcombe Mills is in

Table 14   *Precipitation-weighted mean concentrations (μeq l^{-1}) and mean annual rainfall for three UK sites, 1992–94*[42]

Site		Barcombe Mills (SE England)	Thorganby (NE England)	Eskdalemuir (S Scotland)
Rainfall mm		733	494	1523
pH		4.72	4.16	4.68
Cations	H^+	19	69	21
	NH_4^+	26	68	19
	Na^+	125	51	65
	Mg^{2+}	34	17	18
	Ca^{2+}	24	40	10
Cations	Total	228	245	133
Anions[a]	SO_4^{2-}	52 (37)	86 (79)	36 (29)
	NO_3^-	25	42	18
	Cl^-	146 (<0)	108 (48)	74 (<0)
Anions	Total	223	236	128

[a] Note: Figures in brackets are non-seasalt sulphate and chloride based on a standard ionic composition of seawater and the assumption that all Na^+ is sea-salt derived. (<0) indicates that subtraction of the seasalt component leaves a negative number.

the south-east of England, Thorganby in the north-east and Eskdalemuir in southern Scotland. Thorganby is the UK monitoring site with the highest annual acidity of precipitation (pH 4.16). It is in a region close to several major coal-fired power stations and is unusual in having a very high non-sea-salt chloride arising from near-source HCl wet deposition. Excluding this site, sodium and chlorine in rain are predominantly provided by sea-salt. The acid has been partially neutralized by ammonia and other ions such as Ca^{2+} which may have originated as calcium carbonate, $CaCO_3$. The sulfate contribution to the acidity is larger than that of nitrate but the last decade has seen a progressive increase in the relative importance of nitrate due to the greater reductions in SO_2 emissions than NO_x emissions. This trend is expected to continue.

5.4.2 The Effects. The phrase 'acid rain' has come to be used very loosely to mean almost anything to do with acidification, whether or not rain is actually involved. Three particular effects have received most attention. Historically the first of these was the increased acidification of lakes and streams in Scandinavia leading to loss of fish and other aquatic organisms. This was attributed to the acidity of rain polluted by sulfur and nitrogen oxides and the UK was blamed as being the chief culprit. Similarly the north-eastern USA was blamed for acidification in Canada. Acidification of fresh waters in the UK itself was demonstrated later.[60]

The second type of environmental damage is damage to forests.[61,62] The effects became very noticeable in Germany in the early 1980s with the worst effects being noted in the south-west (the Black Forest) and on the eastern border with the Czech Republic, a country which shares the same problem. Since then other countries have reported similar phenomena. Although the reasons for the damage have been the subject of much debate, it seems likely to be a combination of factors. Predisposing factors include drought and high altitude, and in some cases disease plays a role. Some forests in the former Eastern European countries suffer from the effects of very high SO_2 levels but this is not the case elsewhere. Possible mechanisms include:

- the effect of ozone initiating an attack on cell walls, with subsequent further deterioration being due to acid rain or acid mists and fogs

[60] UK Acid Waters Review Group, 'Acidity in United Kingdom Fresh Waters', HMSO, London, 1986 and Second Report, 1988.
[61] UK Terrestrial Effects Review Group, 'The Effects of Acid Deposition on the Terrestrial Environment in the United Kingdom', HMSO, London, 1988.
[62] UK Terrestrial Effects Review Group, Second Report, 'Air Pollution and Tree Health in the United Kingdom', Department of the Environment, London, 1993.

leaching nutrients and resulting in the breakdown of chlorophyll. Reduced root growth and nutrient uptake follow.

- acidification of the ground with consequent effects on the soil chemistry including elevation of mobile aluminium levels which can damage the roots.
- excess deposition of nitrogen (as nitrate and ammonium) which can have a variety of effects. In the ground NH_4^+ can release H^+ during the process of being oxidized to NO_3^- by bacteria. The H^+ can then be leached out. From the point of view of acidification phenomena, ammonia should therefore not be regarded as an ally even though, prior to its transformation to nitrate, it reduces the rain-water acidity.

Given the complexity of the biological and chemical processes mentioned it is clear that control of the effects of 'acid rain' on biological systems must focus on all the relevant pollutants—SO_2, NO_x, NH_3, and hydrocarbons (as precursors of ozone).

The third problem associated with acid rain is the attack on stonework and the decay of famous cathedrals and other buildings constructed of limestone.[63] Both wet and dry deposition of sulfur dioxide are involved. Under moist conditions SO_2 or sulfuric acid will convert calcium carbonate to gypsum, $CaSO_4.2H_2O$. Since the sulfate is more soluble than the carbonate, the reacted stone can be removed by dissolution. The solid gypsum also occupies a larger volume than the original carbonate and this leads to spalling of material from the surface. In polluted urban areas a black 'crust' of soot and gypsum builds up on stonework not washed by rain. A combination of these factors leads to a rate of loss of stone which depends on the deposition rate of SO_2 to the surface. Deposition to a moist surface is more rapid than to a dry surface so the fraction of time the surface is wetted as well as the SO_2 concentration is important. It is generally assumed that in urban areas the dry deposition of SO_2 gas is the major factor rather than the acidity of rain itself.

5.4.3 Patterns of Deposition and Critical Loads Assessment. In order to highlight areas where acidity is causing damage, and therefore to enable more effective control measures, a *critical loads* approach has been introduced throughout Europe. A critical load for a particular receptor–pollutant combination is defined as the highest deposition load that the receptor can withstand without long term damage occurring. A gridded critical loads map is prepared based on the geology and other factors affecting the response of fresh waters or terrestrial ecosystems to

[63] UK Building Effects Review Group, 'The Effects of Acid Deposition on Buildings and Building Materials', HMSO, London, 1989.

pollutant input. Plates 4a and 4b show where in the UK the critical loads for acidity have been exceeded for soils and fresh waters respectively.[64] For soils, there is a regional pattern, with the highest exceedence occurring in the most sensitive regions, notably Wales, the Pennines, and the Scottish Highlands which are also regions of high rainfall. The exceedence for waters is more localized in particular parts of the Pennines, western Scotland, and northern and southern Wales. Future patterns of exceedence of critical loads can be predicted on the basis of expected reductions in pollutant emissions. The needs for more stringent control measures can then be assessed.

Following emissions reductions of SO_2 within the EU the total area of Europe with exceedences of critical loads for sulfur has been reduced by 50% between 1980 and 1994. Exceedences for acidity do occur, however, partly because similar reductions in NO_x and ammonia have not occurred.[45] The situation is similar in the US where a decreasing trend in sulfate ions has been seen in recent years but along with slight increases in nitrate ions.

Questions

1. Discuss the potential uncertainties which affect the predictions of global climate models.
2. Discuss the origins of turbulence in the atmospheric boundary layer, and its potential impact on pollution dispersion processes.
3. Choose two of the world's megacities (*i.e.* those with a population over 10 million) and compare their air quality problems. Include in your discussion any relevant differences in meteorological conditions and emissions profiles.
4. Contrast the factors influencing air pollution in major cities of developing countries with the situation in major cities in Europe and the USA.
5. Why have particulate emissions from vehicles become an increasing cause for concern in the last few years?
6. Describe, with reference to chemical emissions and meteorological processes, why the exceedences for ozone are much higher in the cities of Los Angeles and Athens than for other European and US cities.
7. Explain the reasons why defining abatement strategies for ozone in Europe is potentially difficult, and explain why the strategies might be different for different countries.

[64] http://www.environment-agency.gov.uk/s-enviro.html 'UK State of the Environment Report', January 1998.

8. Recent years in the UK have seen a slight increase in rural ozone but a slight decrease in average urban values. Explain these trends in terms of emissions and chemical processes.

9. Trace the path of NO_x emissions from their emission from a power station to the point of impact in a mountain lake. In your answer refer to all possible chemical and physical processes which may be undergone.

10. What factors determine the relative importance of wet and dry deposition for different pollutants?

11. Describe how critical loads are used to design abatement strategies for acidity. Highlight the problems associated with the critical loads approach.

12. VOCs arise from a very wide range of sources. Discuss practical steps that may be taken to reduce these emissions.

CHAPTER 3

Freshwaters

JOHN G. FARMER AND MARGARET C. GRAHAM

1 INTRODUCTION

As the Earth's population continues to increase rapidly, the growing human need for freshwater (*e.g.* for drinking, cooking, washing, carrying wastes, cooling machines, irrigating crops, receiving sewage and agricultural runoff, recreation, and industrial purposes) is leading to a global water resources crisis. Rees[1] has commented that 'there is a growing consensus that if current trends continue, water scarcity and deteriorating water quality will become the critical factors limiting future economic development, the expansion of food production, and the provision of basic health and hygiene services to millions of disadvantaged people in developing countries'.

The inventory of all water at the Earth's surface (Table 1)[2] shows that the oceans, ice caps and glaciers contain 98.93% of the total, with groundwater (1.05%) accounting for most of the rest, and lakes and rivers amounting to only 0.009% and 0.0001%, respectively.[2] The average annual water withdrawal for use by humans is currently about 8% ($0.032 \times 10^6 \text{ km}^3$) of renewable freshwater resources on a global basis.[3] Of this, 69% is used by agriculture, 23% by industry/power, and 8% for domestic purposes. Figures vary widely between and within continents. On a per capita basis, North America withdraws seven times as much freshwater as Africa. European countries withdraw an average

Inevitably there are minor overlaps between some of the topics covered in this chapter (particularly in the area of carbonate equilibria and complexation) with coverage in Chapter 4, 'The Oceans'. These have been tolerated to ensure no loss of continuity for the reader.

[1] J. Rees, *Science and Public Affairs*, 1997, **Winter**, 20.
[2] E. K. Berner and R. A. Berner, 'Global Environment: Water, Air, and Geochemical Cycles', Prentice-Hall, Upper Saddle River, NJ, 1996.
[3] T. G. Spiro and W. M. Stigliani, 'Chemistry of the Environment', Prentice-Hall, Upper Saddle River, NJ, 1996.

Table 1 *Inventory of water at the Earth's surface*

Reservoir	Volume 10^6 km^3 (10^{18} kg)	Per cent of total
Oceans	1400	95.96
Mixed layer	50	
Thermocline	460	
Abyssal	890	
Ice caps and glaciers	43.4	2.97
Groundwater	15.3	1.05
Lakes	0.125	0.009
Rivers	0.0017	0.0001
Soil moisture	0.065	0.0045
Atmosphere total[a]	0.0155	0.001
Terrestrial	0.0045	
Oceanic	0.0110	
Biosphere	0.002	0.0001
Approximate total	1459	

Global Environment: Water Air Geochemical Cycles by Berner/Berner, © 1996.
Adapted by permission of Prentice-Hall, Upper Saddle River, NJ.[2]
[a] As liquid volume equivalent of water vapour.

of 15% of their renewable freshwater resources each year (33% agriculture, 54% industry/power, 13% domestic), with Poland as high as 37% and Sweden as low as 2%. It appears that the practical limit to annual water withdrawals will be reached when 30–45% of available freshwater is needed to meet consumption requirements.[1] On current trends of population growth and the fact that per capita consumption is increasing twice as fast as population growth, the global supply crisis is projected to occur between 2025 and 2050, although much earlier for some individual countries.[1]

The quality of freshwater is also of concern. The recent report on 'Europe's Environment: The Dobris Assessment', summarized by Burke,[4] reveals that about 65% of Europe's population depends on groundwater, a resource described as overexploited in almost 60% of European industrial and urban centres and threatened by pollutants. The European Union standard for total pesticides ($0.5 \, \mu g \, l^{-1}$) is exceeded in soil water in 60–75% of agricultural land, while river and lake eutrophication caused by excessive phosphorus and nitrogen from agricultural, domestic, and industrial effluents is a problem across most of Europe. The Eastern European countries fare badly when judged

[4] M. Burke, *Environ. Sci. Technol.*, 1996, **30**, 162A.

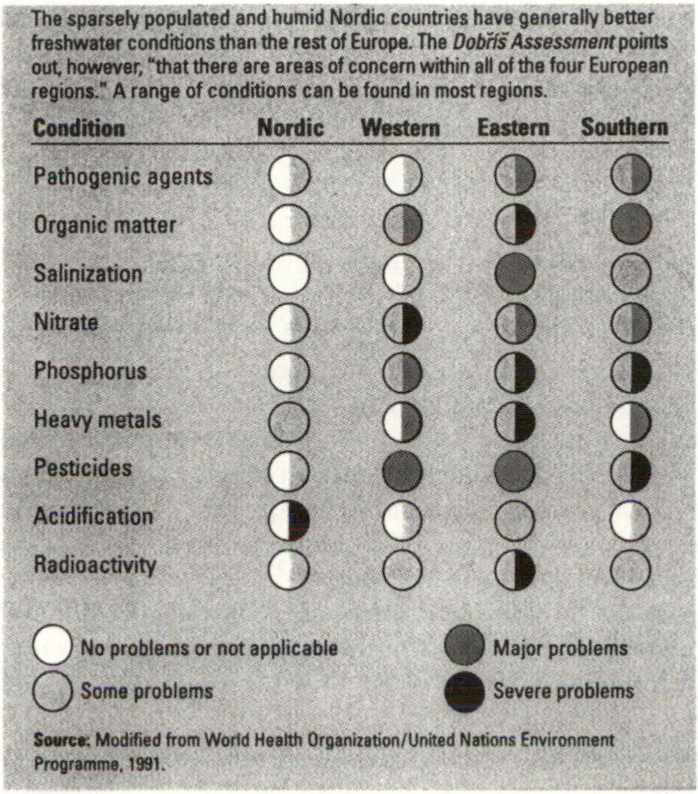

The sparsely populated and humid Nordic countries have generally better freshwater conditions than the rest of Europe. The *Dobris Assessment* points out, however, "that there are areas of concern within all of the four European regions." A range of conditions can be found in most regions.

Condition	Nordic	Western	Eastern	Southern
Pathogenic agents				
Organic matter				
Salinization				
Nitrate				
Phosphorus				
Heavy metals				
Pesticides				
Acidification				
Radioactivity				

○ No problems or not applicable   ● Major problems
○ Some problems   ● Severe problems

Source: Modified from World Health Organization/United Nations Environment Programme, 1991.

Figure 1   *Freshwater quality: a regional view*[4]

against the criteria of pathogenic agents, organic matter, salinization, nitrate, phosphorus, heavy metals, pesticides, acidification, and radio-activity, and even the Nordic countries of Norway, Sweden, and Finland, despite having generally better freshwater conditions than the rest of Europe, suffer from acidification (Figure 1).[4]

The effects of contaminants or pollutants on freshwater depend upon their chemical, physical, and biological properties, their concentrations and duration of exposure.[5] Aquatic life may be affected indirectly (*e.g.* through depletion of oxygen caused by biodegradation of organic matter) or directly, through exposure to toxic or carcinogenic chemicals, some of which may accumulate in organisms. Such toxicity may, however, be modified by the presence of other substances and the characteristics of the particular water body. For example, metal toxicity is affected by pH, which influences speciation, and dissolved organic

[5] Royal Commission on Environmental Pollution, 'Sixteenth Report: Freshwater Quality', HMSO, London, 1992.

carbon has been shown to reduce bioavailability by forming metal complexes and by subsequent adsorption to particulate matter in freshwater.[6] The toxicity of many heavy metals to fish is also inversely related to water hardness, with Ca^{2+} competing with free metal ions for binding sites in biological systems.

In biologically productive lakes which also develop a thermocline in summer, the bottom water may become depleted in oxygen, leading to a change from oxidizing to reducing conditions. As a result, the potential exists for remobilization of nutrients and metals from bottom sediments to the water column, one more example of how alterations in master variables in the hydrological cycle, in this case the redox status, can affect the fate and influence of pollutants.[7]

A further demonstration of the importance of fundamental properties of both pollutants and water bodies is provided by the behaviour of chemicals upon reaching a groundwater aquifer. Soluble chemicals, such as nitrate, move in the same direction as groundwater flow. A poorly soluble liquid which is less dense than water, such as petrol, spreads out over the surface of the water table and flows in the direction of the groundwater. Poorly soluble liquids which are denser than water, such as various chlorinated solvents, sink below the water table and may flow separately along low permeability layers encountered at depth in the aquifer and not necessarily in the same direction as that of the overlying groundwater.[5]

The rest of this chapter therefore consists of two major sections, first on fundamental aquatic chemistry of relevance to the understanding of pollutant behaviour in the aquatic environment and second on associated case studies and examples drawn from around the world, followed by a brief summary of water treatment methods.

2 FUNDAMENTALS OF AQUATIC CHEMISTRY

2.1 Introduction

2.1.1 Concentration and Activity. All natural waters contain dissolved solutes. In order to understand their chemical behaviour in aqueous systems it is first necessary to consider the relationship between solute concentration and solute activity.

The activity of a solute is a measure of its observed chemical behaviour in (aqueous) solution. Interactions between the solute and other species in solution lead to deviations between solute activity {i} and concentra-

[6] R. Renner, *Environ. Sci. Technol.*, 1997, **31**, 466A.
[7] 'Biogeodynamics of Pollutants in Soils and Sediments', ed. W. Salomons and W. M. Stigliani, Springer, Berlin, 1995.

tion [i]. An activity coefficient, γ_i, is therefore defined as a correction factor which interrelates solute activity and concentration.

$$\{i\} = \gamma_i\,[i] \tag{1}$$

Concentration and activity of a solute are only the same for very dilute solutions, *i.e.* γ_i approaches unity as the concentration of all solutes approaches zero. For non-dilute solutions, activity coefficients must be used in chemical expressions involving solute concentrations.† Although most freshwaters are sufficiently dilute to be potable (containing less than about 1000 mg l^{-1} total dissolved solids),[8] it cannot be assumed that activity coefficients are close to unity.

2.1.2 Ionic Strength. A major factor influencing the activity of solutes in aqueous solution is ionic strength. Ionic strength, I, is a measure of the charge density in solution and is defined as:

$$I = 0.5 \sum_i ([i]z_i^2) \tag{2}$$

where [i] is the concentration of solute ion, i, and z_i is the integral charge associated with solute ion, i.

Ionic strength, expressed in terms of mol l^{-1}, is most accurately determined by carrying out a total water analysis, *i.e.* quantification of the concentrations of all ionic species in solution and calculation of I using equation (2). It may, however, be estimated from measurements of either the total dissolved solids (TDS) in solution or, preferably, the specific conductance (SpC) of the solution, *e.g.* $I = 2.5 \times 10^{-5}$ TDS (mg l^{-1}) or $I = 1.7 \times 10^{-5}$ SpC (μS cm^{-1}).[9]

Various relationships between ionic strength and activity coefficients have been developed from theoretical considerations and subsequent empirical observations. Theoretical expressions are based on Debye-Hückel theory (a full treatment of ion activity in aqueous systems is given in Reference 10), but application is restricted to solutions of ionic strength < 0.1 M.

---

† Concentration is expressed on the molar scale in terms of mol l^{-1} of solvent or on the molal scale in terms of mol kg^{-1} of solvent. The molal scale gives concentrations that are independent of temperature and pressure. In this chapter the molar scale will be used on the basis that molarity and molality are almost identical for the low ionic strengths commonly associated with freshwaters.

[8] J. I. Drever, 'The Geochemistry of Natural Waters', 3rd Edn., Prentice-Hall, Englewood Cliffs, NJ, 1997.

[9] D. Langmuir and J. Mahoney, 'Practical Applications of Groundwater Geochemistry, Proceedings of the 1st Canadian/American Conference on Hydrogeology', ed. B. Hitchon and E. I. Wallick, National Water Well Association, Worthington, OH, 1985, p. 69.

$$\log \gamma_i = -0.5 z_i^2 \sqrt{I} \qquad \text{Debye–Hückel equation} \qquad (3)$$

$$\log \gamma_i = -0.5 z_i^2 \left(\frac{\sqrt{I}}{1 + 0.33 a \sqrt{I}} \right) \qquad \text{Extended Debye–Hückel equation} \qquad (4)$$

$$\log \gamma_i = -0.5 z_i^2 \left(\frac{\sqrt{I}}{1 + \sqrt{I}} \right) \qquad \text{Güntleberg approximation} \qquad (5)$$

A very dilute solution is defined as having an ionic strength $< 10^{-5}$ M and activity coefficients calculated by any of the above equations would have a value of unity. As most freshwaters have ionic strengths $> 10^{-5}$ M, activity coefficients should be calculated using equations (3)–(5). Where the ionic strength of the solution is $< 10^{-2.3}$ M,[10] activity coefficients can be calculated using the Debye–Hückel (DH) expression. In more concentrated solutions, values calculated using equation (3) differ from experimental data and this stems from the assumption that the solute ions are point charges, *i.e.* of infinitely small radius. The extended DH expression, equation (4), includes a parameter, a, which is related to ion size. For solutions of ionic strength $< 10^{-1}$ M either equation (4) or the Güntleberg approximation, equation (5), which incorporates an average value of 3 for a, should therefore be used.[10] The Güntleberg approximation is particularly useful in calculations where a number of ions are present in solution. It should be noted that single ion activity coefficients are not measurable and are only used to simplify calculations. Only products or ratios of ion activity coefficients are directly measurable. Using the above theory, but replacing z_i^2 with $z_+ z_-$, a mean ion activity coefficient can be calculated for an electrolyte (Example 1).

EXAMPLE 1 *Calculate the mean ion activity coefficient of (i) a 0.001 M and (ii) a 0.01 M aqueous solution of magnesium sulfate.*

Ionic strength, $I = 0.5 \sum [\text{i}] z_i^2$

(i) $I = 0.5[0.001 \times (+2)^2 + 0.001 \times (-2)^2]$ $= 0.004$ M
(ii) $I = 0.5[0.01 \times (+2)^2 + 0.01 \times (-2)^2]$ $= 0.04$ M

Log mean activity coefficients

(i) $\log \gamma_\pm = -0.5 |z_+ z_-| \sqrt{I} = -0.5 \times 2 \times 2 \times \sqrt{0.004}$ $= -0.13$

(ii) $\log \gamma_\pm = -0.5 |z_+ z_-| \dfrac{\sqrt{I}}{1 + \sqrt{I}} = -0.5 \times 2 \times 2 \times \dfrac{\sqrt{0.04}}{1 + \sqrt{0.04}}$ $= -0.33$

Mean activity coefficients

(i) $\gamma_\pm = 0.75$
(ii) $\gamma_\pm = 0.46$

[10] W. Stumm and J. J. Morgan, 'Aquatic Chemistry', 3rd Edn., John Wiley & Sons, New York, 1996.

These equations tend to underestimate activity coefficients for solutes in higher ionic strength solutions (>0.1 M) and various empirical correction factors have subsequently been added to the Güntleberg approximation. The Davies equation incorporates an empirical factor $-0.2I$ or $-0.3I$ within equation (5) to calculate activity coefficients for intermediate ionic strength aqueous solutions (0.1 to 0.7 M). This was developed further in the Specific Ion Interaction Theory (SIT),[11,12] which adds an ion- and electrolyte-specific correction factor to extend the range of application to 3.5 M. At low ionic strength these models are less accurate than the extended DH model because the DH expression used in Davies and SIT equations has a fixed average ion size parameter, $a = 3$ and $a = 4.6$, respectively. For high ionic strength solutions (*e.g.* brines), the Pitzer equations[13] add terms for binary and ternary interactions between solute species to a DH expression in calculations of activity coefficients. Although naturally occurring brines and some high ionic strength contaminated waters may require the more complicated expressions developed in the Davies, SIT, or Pitzer models, the use of equations (3)–(5) is justified for the ionic strengths of many freshwaters.

2.1.3 Equilibria and Equilibrium Constants. Many of the important reactions involving solutes in freshwaters can be described by equilibria. This approach means that an equilibrium constant, K, relating the activities of the solutes, can be defined for each stoichiometric equilibrium expression.

$$aA + bB \rightleftharpoons cC + dD \qquad (6)$$

$$K = \frac{\{C\}^c\{D\}^d}{\{A\}^a\{B\}^b} \qquad (7)$$

The equilibrium constant can also be defined in terms of the concentrations of the solutes.

$$K = \frac{(\gamma_C[C])^c(\gamma_D[D])^d}{(\gamma_A[A])^a(\gamma_B[B])^b} \qquad (8)$$

[11] I. Grenthe and H. Wanner, 'Guidelines for the Extrapolation to Zero Ionic Strength', Report NEA-TDB-2.1, F-91191, OECD, Nuclear Energy Agency Data Bank, Giv-sur Yvette, France, 1989.
[12] D. K. Nordstrom and J. L. Munoz, 'Geochemical Thermodynamics', Benjamin/Cummings, Menlo Park, CA, 1985.
[13] K. S. Pitzer, 'Thermodynamic Modelling of Geological Materials: Minerals, Fluids and Melts', ed. I. S. E. Carmichael and H. P. Eugster, Reviews in Mineralogy, Mineralogical Society of America, 1987, Vol. 17, p. 97.

For infinitely dilute solutions, { } = [ ] and the equilibrium constant can be written as:

$$K = \frac{[C]^c[D]^d}{[A]^a[B]^b} \qquad (9)$$

For solutions of fixed ionic strength, or, for example, where major ions in solution are present at concentrations several orders of magnitude greater than the solute, it can be assumed that the solute activity coefficients are also constants and can be incorporated into the equilibrium constant. The equilibrium constant for a fixed ionic strength aqueous solution is termed a constant ionic strength equilibrium constant, ^cK.

$$^cK = \frac{[C]^c[D]^d}{[A]^a[B]^b} \quad \text{where } ^cK = \frac{\gamma_A^a \gamma_B^b}{\gamma_C^c \gamma_D^d} K \qquad (10)$$

This again enables the use of concentrations rather than activities in equilibrium calculations.

It is sometimes advantageous to use a mixture of activities and concentrations and a mixed equilibrium constant, K', is defined as:

$$K' = \frac{\{C\}^c[D]^d}{[A]^a[B]^b} \quad \text{where } K' = \frac{\gamma_A^a \gamma_B^b}{\gamma_D^d} K \qquad (11)$$

For example, a mixed acidity constant is frequently used where pH has been measured according to the IUPAC convention as the activity of hydrogen ions but the concentrations of the conjugate acid–base pair are used.

The relationship between K and ^cK or K and K' as defined above can be used to calculate the effect of ionic strength of solution on the true equilibrium constant, K. ^cK and K' can be calculated using equations (3)–(5) together with experimentally determined values of K (Example 2).

EXAMPLE 2   *Use the Güntleberg approximation to calculate the mixed acidity constant, K', for a 1 litre solution containing 1×10^{-4} moles of an organic acid (K = 1.77×10^{-4}) and (i) 0.01 M, (ii) 0.1 M with respect to sodium chloride. (Hint: take logs of both sides of $K' = \dfrac{\gamma_{acid}}{\gamma_{anion}} K$ and substitute Güntleberg approximation expressions for γ_{acid} and γ_{anion}; use $z_{acid} = 0$ and $z_{anion} = -1$).*

$$pK' = pK + 0.5((z_{acid})^2 - (z_{anion})^2)\frac{\sqrt{I}}{(1 + \sqrt{I})}$$

(i) $= 3.75 + 0.5(0 - 1)(0.1/1.1)$ $= 3.70$

(ii) $= 3.75 + 0.5(0 - 1)(0.32/1.32)$ $= 3.63$

Clearly, the effect of increasing the ionic strength of solution is to decrease the value of pK', *i.e.* increasing K'. For $I = 0.1$ M, K' is 2.34×10^{-4} compared with 1.77×10^{-4} for an infinitely dilute solution.

2.2 Dissolution/Precipitation Reactions

Section 2.1 has provided relationships between solute activities and concentrations in solution in order that solute behaviour can be quantified. This section discusses dissolution and precipitation reactions that impart or remove solutes in natural waters and therefore modify the chemical composition of natural waters.

2.2.1 Physical and Chemical Weathering Processes. Initial steps involving physical weathering by thermal expansion and contraction or abrasion lead to the disintegration of rock. Disintegration increases the surface area of the rock which can, in the presence of water, undergo chemical weathering. Water acts not only as a reactant but also as a transporting agent of dissolved and particulate components and so weathering processes are extremely important in the hydrogeochemical cycling of elements. Mineral dissolution reactions often involve hydrogen ions from mineral or organic acids, *e.g.* acid hydrolysis of Na-feldspar in equation (14). Alternatively, the transfer of electrons (sometimes simultaneously with proton transfer) may promote the dissolution of minerals, *e.g.* reductive dissolution of biotite.

Chemical weathering of minerals results not only in the introduction of solutes to the aqueous phase but often in the formation of new solid phases. Dissolution is described as congruent, where aqueous phase solutes are the only products, or incongruent, where new solid phase(s) in addition to aqueous phase solutes are the products. These reactions can be represented by equilibria, where the equilibrium constant is related only to the activities of the aqueous solutes, *i.e.* an assumption is made that the activities of solid phases and that of water have the value of unity. Examples of weathering of primary minerals, those formed at the same time as the parent rock, are shown in equations (12)–(15):

Congruent dissolution of quartz:

$$SiO_{2(s)} + 2\,H_2O \rightleftharpoons Si(OH)_{4\ (aq)}^{0} \tag{12}$$
$$\text{quartz} \qquad\qquad \text{silicic acid}$$

$$K = \{Si(OH)_4^0\} \tag{13}$$

At pH < 9, the equilibrium between quartz and silicic acid can be represented as in equation (12) and the value of K at 298K is $1.05 \times 10^{-4} \, \mathrm{mol \, l^{-1}}$,[14] indicating that the solubility of quartz is low. At higher pH values, the dissociation of silicic acid results in the increased solubility of quartz (see Section 2.4.1).

Incongruent dissolution of Na-feldspar:

$$2 \, NaAlSi_3O_{8(s)} + 2 \, H^+ + 9 \, H_2O \rightleftharpoons 2 \, Na^+_{(aq)} + 4 \, Si(OH)^0_{4 \, (aq)} + Al_2Si_2O_5(OH)_{4(s)}$$
$$\text{Na-feldspar} \qquad\qquad\qquad\qquad\qquad \text{silicic acid} \qquad\qquad \text{kaolinite} \quad (14)$$

$$K = \{Si(OH)^0_4\}^4 \{Na^+\}^2 / \{H^+\}^2 \qquad (15)$$

Alteration of primary minerals such as the feldspars gives rise to secondary minerals such as kaolinite ($Al_2Si_2O_5(OH)_4$), smectites (*e.g.* $Na_{0.33}Al_{2.33}Si_{3.67}O_{10}(OH)_2$), and gibbsite ($Al(OH)_3$). As the composition of the new mineral phases and the potential for further alteration of secondary minerals are a function of the prevailing geochemical conditions, many equilibrium expressions must be used to fully describe chemical weathering processes in natural systems. For example, reactions leading to the formation of kaolinite from the primary mineral, Na-feldspar, as well as alteration of the secondary mineral, *e.g.* to gibbsite, must be considered. The composition of the aqueous phase is of major importance in determining the nature of the solid products formed during chemical weathering. In particular, the solution activities of silicic acid, metal ions, and hydrogen ions are key parameters influencing the formation and alteration of new solid phases (Example 3).[8]

2.2.2 Solubility. Dissolution of minerals during chemical weathering releases species into solution but aqueous phase concentrations are limited by the solubility of solid phases such as amorphous silica, gibbsite, and metal salts.

For example, with respect to the dissolution of magnesium sulfate and magnesium chloride:

$$MgSO_{4(s)} \rightleftharpoons Mg^{2+}_{(aq)} + SO^{2-}_{4(aq)} \qquad (16)$$

$$MgCl_{2(s)} \rightleftharpoons Mg^{2+}_{(aq)} + 2Cl^-_{(aq)} \qquad (17)$$

The equilibrium solubilities are defined by the respective solubility products, K_{SP}.

[14] J. W. Ball and D. K. Nordstrom, 'Users Manual for WATEQ4F, with Revised Thermodynamic Data Base and Test Cases for Calculating Speciation of Major, Trace and Redox Elements in Natural Waters', US Geological Survey Open File Report, Menlo Park, CA, 1991, p. 91.

$$K_{SP} = \{Mg^{2+}\}\{SO_4^{2-}\} \tag{18}$$

$$K_{SP} = \{Mg^{2+}\}\{Cl^-\}^2 \tag{19}$$

A solution is considered to be undersaturated, saturated or over-saturated with respect to a solid phase, for example magnesium sulfate, if $K_{SP} > \{Mg^{2+}\}\{SO_4^{2-}\}_{observed}$, $K_{SP} = \{Mg^{2+}\}\{SO_4^{2-}\}_{observed}$ or $K_{SP} < \{Mg^{2+}\}\{SO_4^{2-}\}_{observed}$, respectively.

EXAMPLE 3 *Investigate the stability relationship between gibbsite and kaolinite in natural waters at pH < 7.* $K_{kaolinite\text{-}gibbsite} = 10^{-4.4}$ *at 25 °C.*

Considering the formation of gibbsite from Na-feldspars (pH < 7):

$$NaAlSi_3O_8 + H^+ + 7H_2O \rightleftharpoons Na^+_{(aq)} + Al^{3+}_{(aq)} + 3OH^-_{(aq)} + 3Si(OH)_{4(aq)}$$

This equilibrium expression represents congruent dissolution of Na-feldspar but the solution quickly becomes saturated with respect to the solid phase $Al(OH)_3$ (gibbsite, $K_{SP} = 10^{-33.9}$).

e.g. $\quad Al^{3+}_{(aq)} + 3OH^-_{(aq)} \rightleftharpoons Al(OH)_{3(s)} \qquad -K_{SP} = 10^{33.9} = \dfrac{1}{\{Al^{3+}\}\{OH^-\}^3}$

The concentrations of these species may also be influential in biological systems where, for example, the presence of silicic acid may be responsible for the decreased toxicity of aluminium (see Section 3.1.2). The concentration of silicic acid in solution is a key parameter influencing the stability of gibbsite. In the presence of even low concentrations of silicic acid, *e.g.* 10^{-4} M, gibbsite is converted to kaolinite.

$$0.5Al_2Si_2O_5(OH)_{4(s)} + 2.5H_2O \rightleftharpoons Al(OH)_{3(s)} + Si(OH)_{4\ (aq)}$$

$$K = \{Si(OH)_4\} = 10^{-4.4}$$

In natural systems, the role of water as a transporting agent is important. Geochemical conditions promoting removal of silicic acid from the solid–water interface (*e.g.* water flow) favour the formation of gibbsite over kaolinite. (Adapted from Drever.[8])

2.2.3 Influence of Organic Matter. Natural organic matter is present in most natural systems. With respect to weathering processes, its importance in freshwaters, and indeed in soils and sediments in contact with freshwaters, can be attributed to the presence of organic acids, which include low molecular mass compounds, *e.g.* oxalic acid, and extremely complex, high molecular mass coloured compounds described as humic substances—a heterogeneous mixture of polyfunctional macro-molecules ranging in size from a few thousand to a few hundred thousand Da. In addition to inorganic acids, *e.g.* H_2CO_3, these provide hydrogen ions for the acid hydrolysis of minerals. As well as promoting dissolution, natural organic matter can influence the formation of new mineral phases.[15]

2.3 Complexation Reactions in Freshwaters

In this section, as a starting point, it is assumed that all species in solution are in hydrated form, *e.g.* $Al(H_2O)_6^{3+}$ where the six water molecules form the first co-ordination sphere of Al. The hydrated form is often represented as, for example, $Al_{(aq)}^{3+}$.

2.3.1 Outer and Inner Sphere Complexes. Outer sphere complexation involves interactions between metal ions and other solute species in which the co-ordinated water of the metal ion and/or the other solute species are retained. For example, the initial step in the formation of ion pairs, where ions of opposite charge approach within a critical distance and are then held together by coulombic attractive forces, is described as outer sphere complex formation.

$$Mg_{(aq)}^{2+} + SO_{4\ (aq)}^{2-} \rightleftharpoons (Mg^{2+}(H_2O)(H_2O)SO_4^{2-})_{(aq)} \qquad (20)$$

The formation of ion pairs is influenced by the nature of the oppositely charged ions, the ionic strength of solution, and ion charge. Ion pairs are generally formed between hard (low polarizability) metal cations, *e.g.* Mg^{2+}, Fe^{3+}, and hard anions, *e.g.* CO_3^{2-}, SO_4^{2-}, and ion pair formation is generally most significant in high ionic strength aqueous phases. In freshwaters, which are frequently of low ionic strength, ion charge is a key parameter. Greater coulombic interactions occur between oppositely charged ions with higher charge.

Inner sphere complexation involves interactions between metal ions and other species in solution which possess lone pairs of electrons. Inner sphere complexation involves the displacement of one or more coordinating water molecules and the transfer of at least one lone pair of electrons. Those species which possess electron lone pairs are termed ligands and complexation reactions may involve inorganic and/or organic ligands.

2.3.2 Hydrolysis. In aqueous systems, hydrolysis reactions are an important example of inner sphere complexation for many metal ions. The interaction between hydrated metal ions and water can be written as

---

[15] F. J. Stevenson, 'Humus Chemistry: Genesis, Composition, Reactions', John Wiley & Sons, New York, 1994.

a series of equilibria for which the stepwise equilibrium constants are denoted $*K_n$†.

$$Al^{3+}_{(aq)} + H_2O_{(l)} \rightleftharpoons Al(OH)^{2+}_{(aq)} + H^+_{(aq)} \qquad *K_1 = 10^{-4.95} \qquad (21)$$

$$Al(OH)^{2+}_{(aq)} + H_2O_{(l)} \rightleftharpoons Al(OH)^+_{2\,(aq)} + H^+_{(aq)} \qquad *K_2 = 10^{-5.6} \qquad (22)$$

$$Al(OH)^+_{2\,(aq)} + H_2O_{(l)} \rightleftharpoons Al(OH)^0_{3\,(aq)} + H^+_{(aq)} \qquad *K_3 = 10^{-6.7} \qquad (23)$$

$$Al(OH)^0_{3\,(aq)} + H_2O_{(l)} \rightleftharpoons Al(OH)^-_{4\,(aq)} + H^+_{(aq)} \qquad *K_4 = 10^{-5.6} \qquad (24)$$

Hydrolysis and, more generally, complex formation equilibria may be described by cumulative stability constants, β. Using equations (21)–(24):

$$Al^{3+}_{(aq)} + H_2O_{(l)} \rightleftharpoons Al(OH)^{2+}_{(aq)} + H^+_{(aq)} \qquad *\beta_1 = *K_1 = 10^{-4.95} \qquad (25)$$

$$Al^{3+}_{(aq)} + 2H_2O_{(l)} \rightleftharpoons Al(OH)^+_{2\,(aq)} + 2H^+_{(aq)} \quad *\beta_2 = *K_1*K_2 = 10^{-10.55} \qquad (26)$$

$$Al^{3+}_{(aq)} + 3H_2O_{(l)} \rightleftharpoons Al(OH)^0_{3(aq)} + 3H^+_{(aq)} \quad *\beta_3 = *K_1*K_2*K_3 = 10^{-17.25} \qquad (27)$$

$$Al^{3+}_{(aq)} + 4H_2O_{(l)} \rightleftharpoons Al(OH)^-_{4\,(aq)} + 4H^+_{(aq)} \quad *\beta_4 = *K_1*K_2*K_3*K_4 = 10^{-22.85} \quad (28)$$

Clearly, these hydrolysis reactions are dependent on hydrogen ion activity. The relationship between speciation and pH, which is influenced by the characteristics of the metal ion, will be discussed further in Section 2.4.1.

2.3.3 Inorganic Complexes.

The main inorganic ligands in oxygenated freshwaters, in addition to OH^-, are HCO_3^-, CO_3^{2-}, Cl^-, SO_4^{2-}, and F^-, and, under anoxic conditions, HS^- and S^{2-} (see Section 2.4.4). The stability of complexes formed between metal ions and inorganic ligands depends on the nature of the metal ion as well as the properties of

---

†The hydrolysis of hydrated metal ions can also be written as the interaction between the hydrated metal ion and the hydroxyl ligand. For example:

$$Al^{3+}_{(aq)} + OH^-_{(aq)} \rightleftharpoons Al(OH)^{2+}_{(aq)} \qquad\qquad K_1$$

K_1 and $*K_1$ are related as follows:

$$Al^{3+}_{(aq)} + OH^-_{(aq)} \rightleftharpoons Al(OH)^{2+}_{(aq)} \qquad\qquad K_1$$
$$H_2O_{(l)} \rightleftharpoons H^+_{(aq)} + OH^-_{(aq)} \qquad\qquad K_w$$

so

$$Al^{3+}_{(aq)} + H_2O_{(l)} \rightleftharpoons Al(OH)^{2+}_{(aq)} + H^+_{(aq)} \qquad\qquad *K_1 = K_1 K_w$$

The values for $*K_n$ are those obtained by Wesolowski and Palmer;[16] all values are given in unitless form but fundamentally are defined in terms of the products of the activities of ions on the molar or molal scales.

[16] D. J. Wesolowski and D. A. Palmer, *Geochim. Cosmochim. Acta*, 1994, **58**, 2947.

Table 2 *Major species in freshwaters*

Metal ion	Major species	$[M_{(aq)}^{x+}]/[M_{TOTAL\ (aq)}]$
Mg^{II}	Mg^{2+}	0.94
Ca^{II}	Ca^{2+}	0.94
Al^{III}	$Al(OH)_3^0$, $Al(OH)_2^+$, $Al(OH)_4^-$	1×10^{-9}
Mn^{IV}	MnO_2^0	–
Fe^{III}	$Fe(OH)_3^0$, $Fe(OH)_2^+$, $Fe(OH)_4^-$	2×10^{-11}
Ni^{II}	Ni^{2+}, $NiCO_3^0$	0.4
Cu^{II}	$CuCO_3^0$, $Cu(OH)_2^0$	0.01
Zn^{II}	Zn^{2+}, $ZnCO_3^0$	0.4
Pb^{II}	$PbCO_3^0$	0.05

Adapted from Sigg and Xue,[17] and reproduced with permission from W. Stumm and J. J. Morgan, 'Aquatic Chemistry', 3rd Edn., © John Wiley & Sons, Inc., 1996.

the ligand. Table 2 shows some of the major complexed metal species involving inorganic ligands in oxygenated freshwaters at pH 8.

2.3.4 Surface Complex Formation. Metal ions form both outer and inner sphere complexes with solid surfaces, *e.g.* hydrous oxides of iron, manganese, silicon and aluminium. In addition, metal ions, attracted to charged surfaces, may be held in a diffuse layer which, depending upon ionic strength, extends several nanometres from the surface into solution. Diffuse layer metal retention and outer sphere complex formation involve electrostatic attractive forces which are characteristically weaker than co-ordinative interactions leading to inner sphere surface complex formation. A number of factors influence metal interactions with surfaces, including the chemical composition of the surface, the surface charge, and the nature and speciation of the metal ion. The importance of the pH of the aqueous phase in these interactions will be discussed further in Section 2.4.1.

2.3.5 Organic Complexes. Dissolved organic matter consists of a highly heterogeneous mixture of compounds, including low molecular mass acids and sugars as well as the high molecular mass coloured compounds termed humic substances. Humic substances are secondary synthetic products derived from decaying organic debris. Although they are structurally poorly defined, it is accepted that large numbers of mainly oxygen-containing functional groups are attached to a flexible, predominantly carbon backbone. Individual organic molecules, in particular those of humic substances, can provide more than one functional group for complex formation with a hydrated metal ion. Ligands which

[17] L. Sigg and H. B. Xue, 'Chemistry of Aquatic Systems: Local and Global Perspectives', ed. G. Bidoglio and W. Stumm, Kluwer Academic, Dordrecht, 1994.

provide two and three functional groups are termed bidentate and tridentate respectively. The formation of complexes where more than one functional group from the same organic molecule is involved are more stable than those where functional groups are from discrete organic molecules. The concentration of dissolved organic matter in freshwaters is generally low (2–$6\ mg\ C\ l^{-1}$) and humic substances, comprising molecules which possess large numbers of functional groups in numerous different chemical environments, are implicated as the component of natural organic matter most important in metal binding.[15]

2.4 Species Distribution in Freshwaters

2.4.1 pH as a Master Variable. pH is one of the key parameters which influence species distribution in aqueous systems. Many equilibrium expressions contain a hydrogen ion activity term, *e.g.* in acid–base, complexation and surface charge formation equilibria. It is therefore useful to consider the relationship between the activity of the species of interest (*e.g.* contaminant metal ions, organic pollutants, naturally occurring inorganic and organic solutes, and weakly acidic functional groups on mineral surfaces) and the activity of hydrogen ions.

For example, an acid–conjugate base pair can be represented as HA and A^-, respectively:

$$HA \rightleftharpoons A^- + H^+ \qquad K = \{A^-\}\{H^+\}/\{HA\} \qquad (29)$$

The activity of acid initially in solution, C, termed the analytical activity, is equal to the sum of the equilibrium activities of the acid and its conjugate base.

$$C = \{HA\} + \{A^-\} \qquad (30)$$

Equations (29) and (30) can be combined to give an expression for the activity of the conjugate base in terms of the equilibrium constant, the analytical activity, which is also a constant, and the hydrogen ion activity.

$$K = \frac{\{A^-\}\{H^+\}}{(C - \{A^-\})} \qquad (31)$$

$$\{A^-\} = \frac{KC}{(K + \{H^+\})} \qquad (32)$$

Similarly an expression for the activity of the acid can be written:

$$\{HA\} = \frac{\{H^+\}C}{(K + \{H^+\})} \tag{33}$$

Alternatively, the activities of acid and conjugate base can be represented as a fraction of the analytical activity:

$$\alpha_0 = \frac{\{HA\}}{C} = \frac{\{H^+\}}{(K + \{H^+\})} \tag{34}$$

$$\alpha_1 = \frac{\{A^-\}}{C} = \frac{K}{(K + \{H^+\})} \tag{35}$$

$$\alpha_0 + \alpha_1 = 1 \tag{36}$$

α_0 and α_1 are termed dissociation fractions for the acid and conjugate base, respectively. The expressions for the dissociation fractions are independent of the analytical activity.

A plot of dissociation fraction against pH for an acid–conjugate base pair is shown in Figure 2.

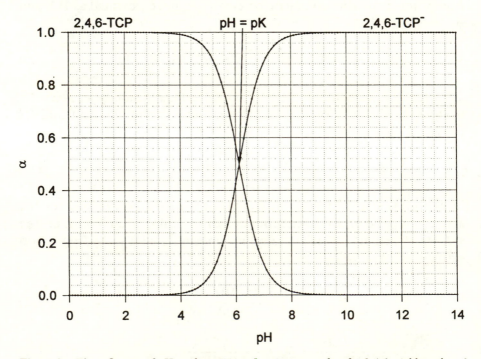

Figure 2   *The influence of pH on dissociation fractions α_0 and α_1 for 2,4,6-trichlorophenol*

More frequently log { } against pH is plotted using equations (37) and (38):

$$\log \{HA\} = \log \{H^+\} + \log C - \log (K + \{H^+\}) \tag{37}$$

$$\log \{A^-\} = \log K + \log C - \log (K + \{H^+\}) \tag{38}$$

The equilibrium pH can be obtained directly from the log { } *versus* pH plot by adding lines showing log $\{H^+\}$ and log $\{OH^-\}$ *versus* pH and by using an expression for charge balance:

$$\{H^+\} = \{A^-\} + \{OH^-\} \tag{39}$$

For a solution containing a weak acid, the charge balance relationship, equation (39), is satisfied at the equilibrium pH (Example 4). The same log { } *versus* pH plot can also be used to determine the equilibrium pH of a solution containing the salt of a weak acid, *e.g.* NaA, by using the appropriate charge balance expression. (Hint: $\{Na^+\} = C$, so $\{HA\} + \{H^+\} = \{OH^-\}$).

EXAMPLE 4 *An organic pollutant, 2,4,6-trichlorophenol (2,4,6-TCP), has been released into a natural water (assume* $I = 0$ *M). Illustrate graphically the relationship between (i) log {2,4,6-TCP} and pH; (ii) log {2,4,6-TCP$^-$} and pH. Use {2,4,6-TCP$_{TOTAL}$} = 4×10^{-4} M and K = $10^{-6.13}$.*

Graphical representation is best approached by dividing the pH range into two regions, pH < pK and pH > pK. The slope of the lines in each of these two regions can be determined by differentiating equations (32) and (33) with respect to pH.

Note that $\dfrac{d \log (\text{constant})}{d \, pH} = 0$; K is small relative to $\{H^+\}$ at pH < pK; K is large relative to $\{H^+\}$ at pH > pK.

At pH < pK:

$$\frac{d \log \{2,4,6\text{-TCP}\}}{d \, pH} = \frac{d \log \{H^+\}}{d \, pH} + \frac{d \log C}{d \, pH} - \frac{d \log \{H^+\}}{d \, pH}$$

$$\frac{d \log \{2,4,6\text{-TCP}^-\}}{d \, pH} = \frac{d \log K}{d \, pH} + \frac{d \log C}{d \, pH} - \frac{d \log \{H^+\}}{d \, pH}$$

$$\frac{d \log \{2,4,6\text{-TCP}\}}{d \, pH} = \frac{d \log \{H^+\}}{d \, pH} + 0 - \frac{d \log \{H^+\}}{d \, pH} = 0$$

$$\frac{d \log \{2,4,6\text{-TCP}^-\}}{d \, pH} = 0 + 0 - \frac{d \log \{H^+\}}{d \, pH} = +1$$

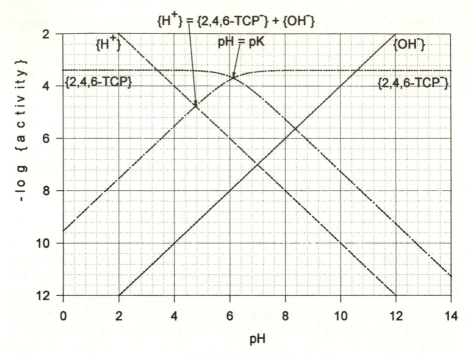

Graph to Example 4

At pH > pK:

$$\frac{\mathrm{d}\log\{2,4,6\text{-TCP}\}}{\mathrm{d}\,\mathrm{pH}} = \frac{\mathrm{d}\log\{\mathrm{H^+}\}}{\mathrm{d}\,\mathrm{pH}} + \frac{\mathrm{d}\log C}{\mathrm{d}\,\mathrm{pH}} - \frac{\mathrm{d}\log K}{\mathrm{d}\,\mathrm{pH}}$$

$$\frac{\mathrm{d}\log\{2,4,6\text{-TCP}^-\}}{\mathrm{d}\,\mathrm{pH}} = \frac{\mathrm{d}\log K}{\mathrm{d}\,\mathrm{pH}} + \frac{\mathrm{d}\log C}{\mathrm{d}\,\mathrm{pH}} - \frac{\mathrm{d}\log K}{\mathrm{d}\,\mathrm{pH}}$$

$$\frac{\mathrm{d}\log\{2,4,6\text{-TCP}\}}{\mathrm{d}\,\mathrm{pH}} = \frac{\mathrm{d}\log\{\mathrm{H^+}\}}{\mathrm{d}\,\mathrm{pH}} + 0 - 0 = -1$$

$$\frac{\mathrm{d}\log\{2,4,6\text{-TCP}^-\}}{\mathrm{d}\,\mathrm{pH}} = 0 + 0 - 0 = 0$$

$\log\{2,4,6\text{-TCP}\} = \log(4 \times 10^{-4})$ for pH < pK and $\log\{2,4,6\text{-TCP}^-\} = \log(4 \times 10^{-4})$ for pH > pK. Using the full expressions for $\log\{2,4,6\text{-TCP}\}$ and $\log\{2,4,6\text{-TCP}^-\}$, calculate several points on lines of slope $+1$ and -1 to complete the graph.

Carbonate Equilibria. Another example of the importance of acid–base behaviour, not only in aerated freshwaters but also in seawater (see Chapter 4), is the dissociation of diprotic acids such as carbonic acid, H_2CO_3. The hydration of dissolved $CO_{2(aq)}$ in natural waters gives rise to carbonic acid. The dissociation of carbonic acid not only influences the pH of the water but provides ligands which can complex trace metals. Example 5 illustrates the relationship between log { } and pH for a closed aqueous system which contains dissolved CO_2 (the presence of mineral phases has been ignored). By assuming that a system is closed to the atmosphere, it is possible to treat carbonic acid as a non-volatile acid and to consider only the hydration reaction which converts $CO_{2(aq)}$ to H_2CO_3. This means that the analytical activity of carbonic acid, C, is a constant for a given system. The dissociation of carbonic acid is described by equations (40) and (41) and the equilibrium constants, K_1 and K_2, are defined by equations (42) and (43).

$$H_2CO_3{*} \rightleftharpoons HCO_3^- + H^+ \tag{40}$$

$H_2CO_3{*}$ is used to represent the sum of $CO_{2(aq)}$ and H_2CO_3, *i.e.* to take account of the presence of $CO_{2(aq)}$ which is in equilibrium with H_2CO_3.

$$HCO_3^- \rightleftharpoons CO_3^{2-} + H^+ \tag{41}$$

$$K_1 = \frac{\{HCO_3^-\}\{H^+\}}{\{H_2CO_3{*}\}} \tag{42}$$

$$K_2 = \frac{\{CO_3^{2-}\}\{H^+\}}{\{HCO_3^-\}} \tag{43}$$

As before, the analytical activity of a weak acid can be written as the sum of undissociated and dissociated acid species:

$$C = \{H_2CO_3{*}\} + \{HCO_3^-\} + \{CO_3^{2-}\} \tag{44}$$

Combining equations (42)–(44), an expression for each of these ($H_2CO_3{*}$, HCO_3^-, and CO_3^{2-}) is obtained:

$$\{H_2CO_3{*}\} = C\left(1 + \frac{K_1}{\{H^+\}} + \frac{K_1 K_2}{\{H^+\}^2}\right)^{-1} \tag{45}$$

$$\{HCO_3^-\} = C\left(\frac{\{H^+\}}{K_1} + 1 + \frac{K_2}{\{H^+\}}\right)^{-1} \tag{46}$$

$$\{CO_3^{2-}\} = C\left(\frac{\{H^+\}^2}{K_1 K_2} + \frac{\{H^+\}}{K_2} + 1\right)^{-1} \tag{47}$$

The dissociation fractions, α_0, α_1, and α_2, are obtained by dividing each of equations (45)–(47) by the analytical activity, C (*cf.*

equation 44):

$$\alpha_0 = \left(1 + \frac{K_1}{\{H^+\}} + \frac{K_1 K_2}{\{H^+\}^2}\right)^{-1} \tag{48}$$

$$\alpha_1 = \left(\frac{\{H^+\}}{K_1} + 1 + \frac{K_2}{\{H^+\}}\right)^{-1} \tag{49}$$

$$\alpha_2 = \left(\frac{\{H^+\}^2}{K_1 K_2} + \frac{\{H^+\}}{K_2} + 1\right)^{-1} \tag{50}$$

where

$$\alpha_0 + \alpha_1 + \alpha_2 = 1 \tag{51}$$

Graphical representation of log { } versus pH can again be used to obtain the equilibrium pH (Example 5).

EXAMPLE 5 *Illustrate graphically the relationship between (i) log $\{H_2CO_3*\}$, (ii) log $\{H_2CO_3\}$, (iii) log $\{HCO_3^-\}$, and (iv) log $\{CO_3^{2-}\}$ and pH for a closed aqueous system. Use $\{carbonic\ acid_T\}$, $C = 2 \times 10^{-5}$ M, $K_1 = 5.1 \times 10^{-7}$ and $K_2 = 5.1 \times 10^{-11}$.*

The presence of dissolved CO_2 is taken into account by the following equilibrium:

$$CO_{2(aq)} + H_2O_{(l)} \rightleftharpoons H_2CO_{3(aq)} \quad K_{hydration} = \frac{\{H_2CO_3\}}{\{CO_2\}} = 1.54 \times 10^{-3} \text{ at } 298 \text{ K}$$

By defining $\{HCO_3*\}$ as $\{H_2CO_3*\} = \{CO_{2(aq)}\} + \{H_2CO_3\}$ and using the approximation $\{HCO_3*\} \sim \{CO_{2(aq)}\}$ because the hydration equilibrium lies far to the left, $K_{hydration} = \{H_2CO_3\}/\{H_2CO_3*\}$ and $\{H_2CO_3\} = \{H_2CO_3*\} \times 1.54 \times 10^{-3}$. In this way the true activity of carbonic acid can be determined.

By assuming a closed system, the total analytical activity of carbonic acid, C, is constant [this is not true for an open system (Example 6)]. Thus equations (42)–(44) can be combined to give equations (45)–(47). The graphical representation can be constructed by dividing the pH range into three regions, pH < pK_1, pK_1 < pH < pK_2, and pH > pK_2, and the procedure outlined in Example 4, *i.e.* obtaining logarithmic expressions for $\{H_2CO_3*\}$, $\{HCO_3^-\}$ and $\{CO_3^{2-}\}$ and differentiating each with respect to pH, can then be utilized to construct the plot of log { } against pH.

$$\{H_2CO_3*\} = C\left(1 + \frac{K_1}{\{H^+\}} + \frac{K_1 K_2}{\{H^+\}^2}\right)^{-1} = \frac{\{H^+\}^2 C}{\{H^+\}^2 + \{H^+\}K_1 + K_1 K_2}$$

At pH < pK_1:

$$\log\{H_2CO_3*\} = 2\log\{H^+\} + \log C - 2\log\{H^+\} = \log C$$

$$\frac{d\log\{H_2CO_3*\}}{d\,pH} = 0$$

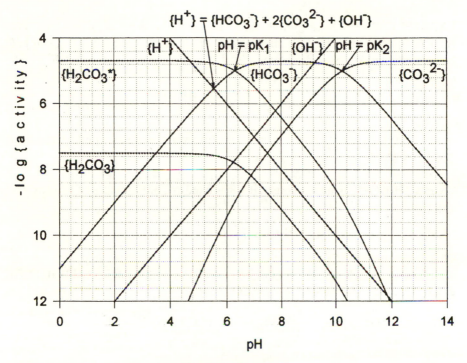

Graph to Example 5

At $pK_1 < pH < pK_2$:

$$\log \{H_2CO_3{}^*\} = 2 \log \{H^+\} + \log C - \log \{H^+\} - \log K_1$$

$$= \log \{H^+\} + \log C - \log K_1$$

$$\frac{d \log \{H_2CO_3{}^*\}}{d \, pH} = -1$$

At $pH > pK_2$:

$$\log \{H_2CO_3{}^*\} = 2 \log \{H^+\} + \log C - \log K_1 K_2$$

$$\frac{d \log \{H_2CO_3{}^*\}}{d \, pH} = -2$$

The charge balance expression (52) can be used to determine the equilibrium pH.

Example 6 extends the consideration of the carbonate system to include equilibration of an aqueous system with an atmosphere (*e.g.* air) containing $CO_{2(g)}$, *i.e.* an open system.

EXAMPLE 6 *Illustrate graphically the relationship between (i) log $\{H_2CO_3{}^*\}$, (ii) log $\{H_2CO_3\}$, (iii) log $\{HCO_3^-\}$, and (iv) log $\{CO_3^{2-}\}$ and pH for an open aqueous system. Use $p_{CO_2} = 3.5 \times 10^{-4}$ atm, $K_H = 4.8 \times 10^{-2}$ (at 288 K), $K_1 = 5.1 \times 10^{-7}$ and $K_2 = 5.1 \times 10^{-11}$.*

$$CO_{2(g)} + H_2O \rightleftharpoons H_2CO_3{}^* \qquad K_H = \frac{\{H_2CO_3{}^*\}}{p_{CO_2}} = 4.8 \times 10^{-2}$$

where K_H is the Henry's Law constant for atmosphere–aqueous phase equilibria involving gases.

Clearly $\{H_2CO_3{}^*\}$ is now a constant for a fixed partial pressure of CO_2 and is independent of pH, *i.e.*

$$\{H_2CO_3{}^*\} = K_H \times p_{CO_2}$$

This means that $\{H_2CO_3\}$ is also constant over the entire pH range.
Combining this additional equilibrium expression with equations (42) and (43) gives expressions for $\{HCO_3^-\}$ and $\{CO_3^{2-}\}$:

$$\{HCO_3^-\} = \frac{K_1 K_H p_{CO_2}}{\{H^+\}} \quad \text{and} \quad \{CO_3^{2-}\} = \frac{K_1 K_2 K_H p_{CO_2}}{\{H^+\}^2}$$

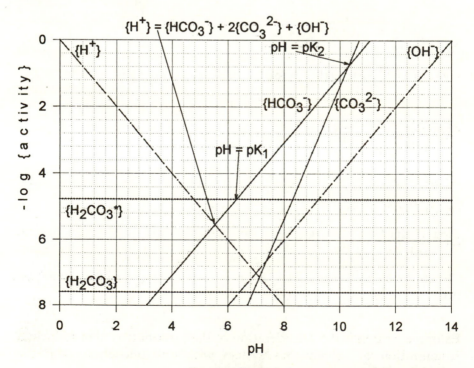

Graph to Example 6

Additionally, the expression for C becomes:

$C = K_H p_{CO_2}/\alpha_0$, which increases with increasing pH according to the function

$$\left(1 + \frac{K_1}{\{H^+\}} + \frac{K_1 K_2}{\{H^+\}^2}\right)$$

Plotting log { } *versus* pH for $\{HCO_3^-\}$ and $\{CO_3^{2-}\}$ gives lines of slope $+1$ and $+2$ respectively. The charge balance expression, which is the same as that for the closed system, can again be used to determine the equilibrium pH.

Alkalinity. An important definition arising from considerations of carbonate equilibria is that of alkalinity which is derived with reference to the charge balance expression for carbonic acid:

$$\{H^+\} = \{HCO_3^-\} + 2\{CO_3^{2-}\} + \{OH^-\} \tag{52}$$

Alkalinity is described as the acid neutralizing capacity of an aqueous system or equivalently as the amount of base possessed by the system:

$$\text{Alkalinity} = \{HCO_3^-\} + 2\{CO_3^{2-}\} + \{OH^-\} - \{H^+\} \tag{53}$$

It is important to understand that, by definition, alkalinity is independent of addition or removal of CO_2 (or H_2CO_3) from the system (*cf.* equation (52)—H_2CO_3 does not appear in the charge balance expression). Alkalinity is an important parameter in assessing the effects of environmental change on aqueous systems (see Section 3.4.1). Species other than carbonate can contribute to alkalinity and an alternative definition is:

$$\text{Alkalinity} = \{HCO_3^-\} + 2\{CO_3^{2-}\} + \{OH^-\} - \{H^+\}$$
$$= \{Na^+\} + \{K^+\} + 2\{Ca^{2+}\} + 2\{Mg^{2+}\} - \{Cl^-\}$$
$$- 2\{SO_4^{2-}\} - \{NO_3^-\} \tag{54}$$

pH Dependence of Complex Formation. Other equilibria such as metal complexation reactions can be considered as acid–base reactions and plots of log { } against pH also provide information about the dominant species present in solution under different geochemical conditions. Hydrolysis of metal cations occurs progressively with increasing pH, *e.g.* $M^{3+}_{(aq)}$, $M(OH)^{2+}_{(aq)}$, $M(OH)^+_{2(aq)}$, $M(OH)^0_{3(aq)}$, and $M(OH)^-_{4(aq)}$ are the hydrolysis products for many M^{III} cations (see Section 2.3.2). $M^{3+}_{(aq)}$ is generally found at low pH and only at extremely high pH is $M(OH)^-_4$ formed. The dependence of hydrolysis reactions on pH is illustrated further in Example 7 and in Section 3.1.2.

EXAMPLE 7 *What are the dominant species of Fe^{III} present in an aqueous solution in the absence of dissolved CO_2 at pH 1, 6, and 11? Use $\{Fe_T^{III}\} = 10^{-9}$ M and log $*\beta_1 = -3.05$, log $*\beta_2 = -6.31$, log $*\beta_3 = -13.8$ and log $*\beta_4 = -22.7$. Assume that no $Fe(OH)_3$ is precipitated,* i.e. *homogeneous solution.*

$$
\begin{array}{ll}
Fe^{3+} + H_2O \rightleftharpoons Fe(OH)^{2+} + H^+ & *\beta_1 \\
Fe^{3+} + 2H_2O \rightleftharpoons Fe(OH)_2^+ + 2H^+ & *\beta_2 \\
Fe^{3+} + 3H_2O \rightleftharpoons Fe(OH)_3^0 + 3H^+ & *\beta_3 \\
Fe^{3+} + 4H_2O \rightleftharpoons Fe(OH)_4^- + 4H^+ & *\beta_4
\end{array}
$$

$$\{Fe_T^{III}\} = \{Fe^{3+}\} + \{Fe(OH)^{2+}\} + \{Fe(OH)_2^+\} + \{Fe(OH)_3^0\} +$$
$$\{Fe(OH)_4^-\} = 10^{-9} \text{ M}$$

This equation can be solved for $\{Fe^{3+}\}$ at various pH values by using the equilibrium constants for each of the four equilibria above.

$$\{Fe_T^{III}\} = \{Fe^{3+}\}\left(1 + \frac{*\beta_1}{\{H^+\}} + \frac{*\beta_2}{\{H^+\}^2} + \frac{*\beta_3}{\{H^+\}^3} + \frac{*\beta_4}{\{H^+\}^4}\right)$$

e.g. at pH 2, $\{Fe^{3+}\} = 9.14 \times 10^{-10}$ M

The concentrations of all other species can be calculated using $\{Fe^{3+}\}$ at each selected pH value. Plotting $-\log \{ \}$ against pH illustrates that the dominant species of Fe^{III} are Fe^{3+}, $Fe(OH)_2^+$ and $Fe(OH)_4^-$ at pH 1, 6, and 11, respectively.

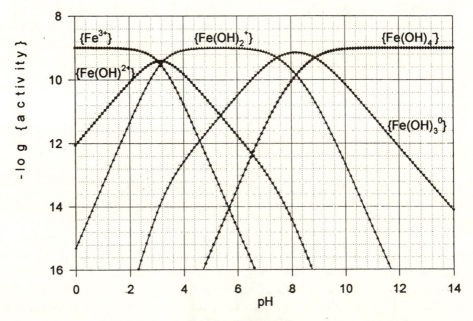

Graph to Example 7

It should be noted that, for higher values of $\{Fe_T^{III}\}$, equilibria for polymeric iron species should be included and that $Fe(OH)_3$ may precipitate. (The formation of polymeric iron species may be an intermediate step leading to the formation of the solid phase.[10])

Example 8 illustrates the influence of pH on species distribution for a $Cu^{II}–CO_2–H_2O$ system.

EXAMPLE 8 *What are the dominant species of Cu^{II} at pH 4, 9, and 12 in an aqueous solution where $C_{carbonate} = 2 \times 10^{-3}$ M and $\{Cu_T^{II}\} = 5 \times 10^{-8}$ M? Use $\log {}^*\beta_1 = -8.0$, $\log {}^*\beta_2 = -16.2$, $\log {}^*\beta_3 = -26.8$, $\log {}^*\beta_4 = -39.9$, $\log {}^*\beta_{1\,carbonate} = 6.77$, $\log {}^*\beta_{2\,carbonate} = 10.01$. Also required are the acidity constants for carbonic acid:*

$$K_1 = 4.67 \times 10^{-7} \text{ and } K_2 = 4.65 \times 10^{-11}.$$

It is first necessary to calculate $\{CO_3^{2-}\}$ for different pH values. These values can then be used together with the stability constants to solve for $\{Cu^{2+}\}$ at the different pH values. Then the activity of each species can be calculated and plots of $-\log \{\ \}$ against pH, $\{\ \}$ against pH or % distribution $\left(\dfrac{\{\ \}}{\{Cu_T^{II}\}} \times 100\right)$ illustrate that Cu^{2+}, $CuCO_3^0$, and $Cu(OH)_3^-$ are the dominant species at pH 4, 9, and 12, respectively.

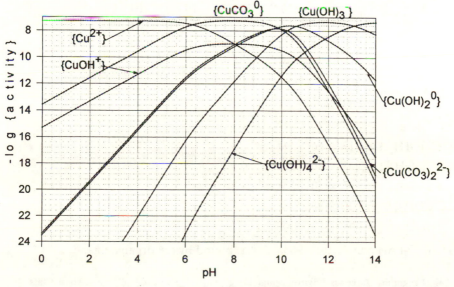

Graph to Example 8

At low pH values, ligands such as F^- and SO_4^{2-} may compete successfully with the hydroxyl ion for a metal cation (see Section 3.1.2). Most significantly, the formation of such complexes increases the solubility of the metal ion by several orders of magnitude over that predicted by the solubility product of the solubility-limiting solid phase, *e.g.* hydroxides. The presence of even trace concentrations of organic ligands has an even greater effect over a wide pH range on metal cation speciation (Example 9) and solubility, increasing the aqueous phase concentration by up to seven orders of magnitude.[18] Complexation clearly has an important role to play, not only in reducing metal toxicity (see Section 3.1.2), but in markedly increasing metal ion mobility.

EXAMPLE 9 *What are the dominant species of Cu^{II} at pH 8, in the absence of dissolved carbonate species, where $\{Cu_T^{II}\} = 5 \times 10^{-8}$ M and an organic acid $(K = 10^{-4.13}$, $\beta_{organic} = 10^{7.8}$ at pH 5) is present at 2×10^{-7} M? Use $\log *\beta_1 = -8.0$, $\log *\beta_2 = -16.2$, $\log *\beta_3 = -26.8$, $\log *\beta_4 = -39.9$.*

$$\{Cu_T^{II}\} = \{Cu^{2+}\} + \{Cu(OH)^+\} + \{Cu(OH)_2^0\} + \{Cu(OH)_3^-\} + \{Cu(OH)_4^{2-}\}$$
$$+ \{Cu\,Org^+\}$$

$$= \{Cu^{2+}\}\left(1 + \frac{*\beta_1}{\{H^+\}} + \frac{*\beta_2}{\{H^+\}^2} + \frac{*\beta_3}{\{H^+\}^3} + \frac{*\beta_4}{\{H^+\}^4}\right.$$

$$\left. + \beta_{organic}\{Organic_T\}\frac{K}{(K + \{H^+\})}\right)$$

$$5 \times 10^{-8} = \{Cu^{2+}\}\left(1 + \frac{10^{-8}}{10^{-5}} + \frac{10^{-16.2}}{10^{-10}} + \frac{10^{-26.8}}{10^{-15}} + \frac{10^{-39.9}}{10^{-20}}\right.$$

$$\left. + 10^{7.8} \times (2 \times 10^{-7}) \times 0.88\right)$$

$$= \{Cu^{2+}\}(1 + 10^{-3} + 10^{-6.2} + 10^{-11.8} + 10^{-19.9} + 11.12)$$

$$= \{Cu^{2+}\}12.12$$

$$\{Cu^{2+}\} = 4.13 \times 10^{-9}\,M$$
$$\{Cu(OH)^+\} = 4.13 \times 10^{-12}\,M$$
$$\{Cu(OH)_2^0\} = 2.61 \times 10^{-15}\,M$$
$$\{Cu(OH)_3^-\} = 6.55 \times 10^{-21}\,M$$
$$\{Cu(OH)_4^{2-}\} = 5.20 \times 10^{-29}\,M$$
$$\{Cu\,Org^+\} = 4.59 \times 10^{-8}\,M$$

Hence, at pH 5 more than 90% Cu^{II} is bound by the organic ligand.

[18] D. Langmuir, 'Aqueous Environmental Geochemistry', Prentice-Hall, Englewood Cliffs, NJ, 1997.

Influence of Ionic Strength. The ionic strength of solution will influence the activities of species and so equilibrium constants must be adjusted to take account of ionic strength as discussed in Section 2.1.2. This applies not only to the acid–base equilibrium constants but also to the dissociation constant for water and the stability constants for complex formation. The values of equilibrium constants also apply only at a specified temperature, *e.g.* 298 K.

2.4.2 pε as a Master Variable. Chemical speciation is also influenced by the redox conditions prevailing in natural waters. Although redox reactions are often slow, and therefore species are present at activities far from equilibrium, they are commonly represented by thermodynamic equilibrium expressions which can provide the boundary conditions towards which a system is proceeding.

pε, a parameter describing redox intensity, gives the hypothetical electron activity at equilibrium. It measures the relative tendency of a solution to accept or donate electrons with a high pε being indicative of a tendency for oxidation, while a low pε is indicative of a tendency for reduction. $p\varepsilon = -\log\{e^-\}$ is analogous to the pH scale $(pH = -\log\{H^+\})$ since a low value for pε is obtained where the hypothetical $\{e^-\}$ is large (pH is low where $\{H^+\}$ is large) and conversely a high value of pε is obtained where $\{e^-\}$ is small (pH is high where $\{H^+\}$ is small). pε is related to the electrode potential, E_H, by the expression $p\varepsilon = 16.9\ E_H$ (V). A full treatment of this relationship can be found in Reference 10. Although the electron activity is a hypothetical phenomenon, pε is a useful parameter to describe the redox intensity of natural systems and hence the distribution of species under prevailing redox conditions (Example 10).

EXAMPLE 10 *Calculate the pε values for the following solutions (298 K, I = 0):*

(i) A solution at pH 2 containing $\{Fe^{3+}\} = 10^{-4.5}$ M and $\{Fe^{2+}\} = 10^{-2.7}$ M where log K = 13.
(ii) A neutral solution containing $\{Mn^{2+}\} = 10^{-6}$ M in equilibrium with the solid phase, $Mn(IV)O_2$ and log K = 40.84.

(i) At pH 2, the hydrated metal ion has not undergone hydrolysis to any significant extent

$$Fe^{3+} + e^- \rightleftharpoons Fe^{2+} \qquad\qquad K = \frac{\{Fe^{2+}\}}{\{Fe^{3+}\}\{e^-\}}$$

$$p\varepsilon = \log K + \log \frac{\{Fe^{3+}\}}{\{Fe^{2+}\}}$$

$$= 13 + \log(10^{-4.5}/10^{-2.7})$$

$$= 13 - 1.8 = 11.2$$

(ii) $MnO_2 + 4H^+ + 2e^- \rightleftharpoons Mn^{2+} + 2H_2O$ $K = \dfrac{\{Mn^{2+}\}}{\{H^+\}^4\{e^-\}^2}$

$$p\varepsilon = 0.5\log K + 0.5\log\left(\dfrac{\{H^+\}^4}{\{Mn^{2+}\}}\right)$$

$$= 20.42 + 0.5\log\left(\dfrac{10^{-28}}{10^{-6}}\right)$$

$$= 20.42 - 11 = 9.42$$

For a fixed pH value (*e.g.* pH = 2), the species distribution (*e.g.* where $\{Fe_T\} = 1 \times 10^{-3}$ M) at different $p\varepsilon$ values can be illustrated by plotting $\log\{\ \}$ against $p\varepsilon$. This is analogous to the treatment of pH as a master variable.

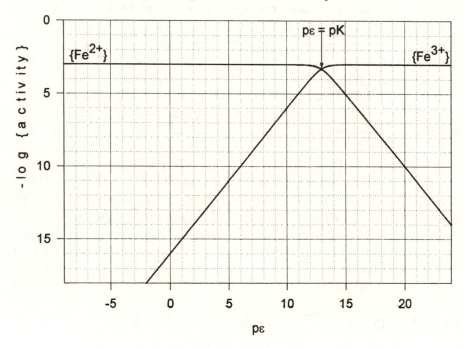

Graph to Example 10

Redox intensity or electron activity in natural waters is usually determined by the balance between processes which introduce oxygen (*e.g.* dissolution of atmospheric oxygen, photosynthesis) and those which remove oxygen (*e.g.* microbial decomposition of organic matter). Often these processes are controlled by the availability of inorganic nutrients such as phosphate and nitrate, *e.g.* as utilized in the formation of organic matter during photosynthesis (see Section 3.4):

$$106CO_2 + 16NO_3^- + HPO_4^{2-} + 122H_2O + 18H^+ \rightleftharpoons$$

$$C_{106}H_{263}O_{110}N_{16}P_1 + 138O_2 \qquad (55)$$

Table 3 *Redox processes—terminal electron acceptors*

Terminal electron acceptor	pε
O_2	13.75
NO_3^- (reduction to N_2)	12.65
Mn^{IV}	8.9
NO_3^- (reduction to NH_4^+)	6.15
Fe^{III}	-0.8
reducible organic matter	-3.01
SO_4^{2-}	-3.75
CO_2	-4.13

Adapted with permission from W. Stumm and J. J. Morgan, 'Aquatic Chemistry', 3rd Edn., © John Wiley & Sons, Inc., 1996.

The decay of organic matter produced in this manner leads to the subsequent consumption of oxygen, *e.g.* by respiration:

$$C_{106}H_{263}O_{110}N_{16}P_1 + 138O_2 \rightleftharpoons$$
$$106CO_2 + 16NO_3^- + HPO_4^{2-} + 122H_2O + 18H^+ \qquad (56)$$

The decay of organic matter requires the presence of a terminal electron acceptor and in equation (56) molecular oxygen is reduced to water. Other terminal electron acceptors present in natural waters include NO_3^-, Mn^{IV}, Fe^{III}, SO_4^{2-}, and CO_2. Once all molecular oxygen has been consumed, organic matter is decomposed via reactions involving other terminal electron acceptors in a series determined by the pε intensity as shown in Table 3.

The sequence of redox reactions involving organic matter, all of which are microbially mediated, can be thought of as progressing through decreasing levels of pε and so, for example, NO_3^- in denitrification reactions would be utilized as the electron acceptor before Mn^{IV} and SO_4^{2-} would be utilized before CO_2 (see Section 3.3.2).

Redox processes are important for elements which can exist in more than one oxidation state in natural waters, *e.g.* Fe^{II} and Fe^{III}, Mn^{II} and Mn^{IV}. These are termed redox-sensitive elements. The redox conditions in natural waters often affect the mobility of these elements since the inherent solubility of different oxidation states of an element may vary considerably. For example, Mn^{II} is soluble whereas Mn^{IV} is highly insoluble. In oxic systems, Mn^{IV} is precipitated in the form of oxides and oxyhydroxides. In anoxic systems, Mn^{II} predominates and is able to diffuse along concentration gradients both upwards and downwards in a water column. This behaviour gives rise to the classic concentration profiles observed for Mn (and Fe) at oxic–anoxic interfaces as illustrated in Figure 3.

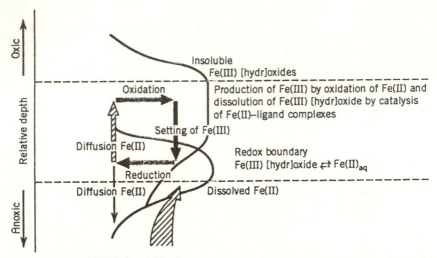

Figure 3   *Oxic–anoxic boundary in the water or sediment column*[10]
(Reproduced with permission from W. Stumm and J. J. Morgan, 'Aquatic
Chemistry', 3rd Edn., © John Wiley & Sons, Inc., 1996)

Other examples of redox-sensitive elements include natural and man-made heavy elements such as the actinides uranium, plutonium and neptunium, all of which can exist in multiple oxidation states in natural waters. Redox conditions in natural waters are also indirectly important for solute species associated with redox-sensitive elements. For example, dissolution of iron (hydr)oxides under reducing conditions may lead to the solubilization and hence mobilization of associated solid phase species, *e.g.* arsenate, phosphate (see Sections 3.2.1, 3.3.2 and 3.4.1).

2.4.3   pε–pH Relationships.   Many redox equilibria also involve the transfer of protons and so chemical speciation depends on both pε and pH. Stability relationships involving redox processes can be investigated via pε–pH diagrams. The region of interest for natural waters displayed in these diagrams is defined by the stability limits of water, *i.e.* where the total partial pressure of oxygen and hydrogen is no greater than 1 atm.

Upper limit:

$$0.5O_{2(g)} + 2e^- + H_2O_{(l)} \rightleftharpoons 2OH^-_{(aq)} \qquad \log K = 13.55 \tag{57}$$

Where $p_{O_2} = 1$, $\log p_{O_2} = 0$

$$\log K = 2 \log \{OH^-\} - 2 \log \{e^-\} - 0.5 \log p_{O_2} = 2 \log \frac{K_w}{\{H^+\}} - 2 \log\{e^-\} - 0$$

$$= 2 \log K_w - 2 \log \{H^+\} - 2 \log \{e^-\}$$

$$p\varepsilon = 20.78 - pH \tag{58}$$

Lower limit:

$$2H_2O_{(l)} + 2e^- \rightleftharpoons H_{2\,(g)} + 2OH^-_{(aq)} \qquad \log K = -28.0 \qquad (59)$$

Where $p_{H_2} = 1$, $\log p_{H_2} = 0$

$$\log K = 2 \log \{OH^-\} + \log p_{H_2} - 2 \log \{e^-\} = 2 \log \frac{K_w}{\{H^+\}} + 0 - 2 \log \{e^-\}$$

$$= 2 \log K_w - 2 \log \{H^+\} - 2 \log \{e^-\}$$

$$p\varepsilon = - pH \qquad (60)$$

For any system, stability relationships between solution phase species and solid phases can be used to construct $p\varepsilon$–pH diagrams representing species distribution within the upper and lower stability limits of water. The relationships between $p\varepsilon$ and pH for an Fe–O_2–H_2O system (*i.e.* absence of dissolved CO_2) at 298 K and 1 atm total pressure can be developed from the redox equilibria in Tables 4 and 5. Consider the main solid phases to be Fe, $Fe(OH)_3$, and $Fe(OH)_2$ (Example 11). The redox processes involved are illustrated in Table 4.

Using values of the equilibrium constants, K, and values of $\{Fe_T^{II}\}$ and $\{Fe_T^{III}\}$, $p\varepsilon$ can be calculated for different pH values. The relationships

Table 4 *Redox equilibria for Fe–O_2–H_2O system*

High and intermediate $p\varepsilon$

Low pH	Intermediate pH	
$Fe^{3+} + e^- \rightleftharpoons Fe^{2+}$	$Fe(OH)^{2+} + e^- + H^+ \rightleftharpoons Fe^{2+} + H_2O$	
	$Fe(OH)_2^+ + e^- + 2H^+ \rightleftharpoons Fe^{2+} + 2H_2O$	
	$Fe(OH)_{3(s)} + e^- + 3H^+ \rightleftharpoons Fe^{2+} + 3H_2O$	
	$Fe(OH)_{3(s)} + e^- + H^+ \rightleftharpoons Fe(OH)_{2(s)} + H_2O$	

Low $p\varepsilon$

Low pH	Intermediate pH	High pH
$Fe^{2+} + 2e^- \rightleftharpoons Fe_{(s)}$	$Fe(OH)_2 + 2e^- + 2H^+ \rightleftharpoons Fe_{(s)} + 2H_2O$	$Fe(OH)_4^- + e^- + 2H^+ \rightleftharpoons Fe(OH)_{2(s)} + 2H_2O$

Table 5 *Equilibria for the Fe–O_2–H_2O system involving only proton transfer*

High and intermediate $p\varepsilon$

Low pH	Intermediate pH	High pH
$Fe^{3+} + H_2O \rightleftharpoons Fe(OH)^{2+} + H^+$	$Fe(OH)^{2+} + H_2O \rightleftharpoons Fe(OH)_2^+ + H^+$	$Fe(OH)_{3(s)} + H_2O \rightleftharpoons Fe(OH)_4^- + H^+$
	$Fe(OH)_2^+ + H_2O \rightleftharpoons Fe(OH)_{3(s)} + H^+$	
	$Fe^{2+} + 2H_2O \rightleftharpoons Fe(OH)_{2(s)} + 2H^+$	

between $p\varepsilon$ and pH for redox equilibria involving proton and electron transfer give lines, *e.g.* $p\varepsilon = a\,pH + b$ (where a and b are constants), with slope = a. Redox equilibria involving only electron transfer give rise to horizontal lines, *i.e.* $p\varepsilon = b$, a = 0. Other processes involve only proton transfer (Table 5). Vertical lines arise from equilibria involving only proton transfer and pH = c (where c is a constant).

EXAMPLE 11   *Construct a $p\varepsilon$–pH diagram for the Fe–O$_2$–H$_2$O system where $\{Fe_T\}$ is 1×10^{-6} M. (Hint: at the boundary between Fe^{3+} and Fe^{2+}, for example, $\{Fe^{3+}\} = \{Fe^{2+}\}$).*

$Fe^{3+} + e^- \rightleftharpoons Fe^{2+}$ $\log K = +13.0$

$p\varepsilon = \log K + \log \dfrac{\{Fe^{3+}\}}{\{Fe^{2+}\}} = 13.0$

$Fe(OH)^{2+} + e^- + H^+ \rightleftharpoons Fe^{2+} + H_2O$ $\log K = +15.2$

$p\varepsilon = \log K - pH - \log \dfrac{\{Fe^{2+}\}}{\{Fe(OH)^{2+}\}} = 15.2 - pH$

$Fe(OH)_2^+ + e^- + 2H^+ \rightleftharpoons Fe^{2+} + 2H_2O$ $\log K = +18.7$

$p\varepsilon = \log K - 2pH - \log \dfrac{\{Fe^{2+}\}}{\{Fe(OH)_2^+\}} = 18.7 - 2pH$

$Fe(OH)_{3(s)} + e^- + 3H^+ \rightleftharpoons Fe^{2+} + 3H_2O$ $\log K = +17.9$
$p\varepsilon = \log K - \log \{Fe^{2+}\} - 3pH = 23.9 - 3pH$

$Fe(OH)_{3(s)} + e^- + H^+ \rightleftharpoons Fe(OH)_{2(s)} + H_2O$ $\log K = +5.5$
$p\varepsilon = \log K - pH = 5.5 - pH$

$Fe(OH)_4^- + e^- + 2H^+ \rightleftharpoons Fe(OH)_{2(s)} + 2H_2O$ $\log K = +22.2$
$p\varepsilon = \log K - 2pH - \log \{Fe(OH)_4^-\} = 16.2 - 2pH$

$Fe^{2+} + 2e^- \rightleftharpoons Fe_{(s)}$ $\log K = -13.8$
$p\varepsilon = 0.5 \log K + 0.5 \log \{Fe^{2+}\} = -9.9$

$Fe(OH)_{2(s)} + 2e^- + 2H^+ \rightleftharpoons Fe_{(s)} + 2H_2O$ $\log K = -1.4$
$p\varepsilon = 0.5 \log K - pH = -0.7 - pH$

$Fe^{3+} + H_2O \rightleftharpoons Fe(OH)^{2+} + H^+$ $\log K = -2.2$

$pH = -\log K + \log \dfrac{\{Fe(OH)^{2+}\}}{\{Fe^{3+}\}} = +2.2$

$Fe(OH)^{2+} + H_2O \rightleftharpoons Fe(OH)_2^+ + H^+$ $\log K = -3.5$

$pH = -\log K + \log \dfrac{\{Fe(OH)_2^+\}}{\{Fe(OH)^{2+}\}} = +3.5$

$Fe(OH)_2^+ + H_2O \rightleftharpoons Fe(OH)_{3(s)} + H^+$ $\log K = +0.8$
$pH = -\log K - \log \{Fe(OH)_2^+\} = +5.2$

$Fe(OH)_{3(s)} + H_2O \rightleftharpoons Fe(OH)_4^- + H^+$ $\log K = -16.7$
$pH = -\log K + \log \{Fe(OH)_4^-\} = +10.7$

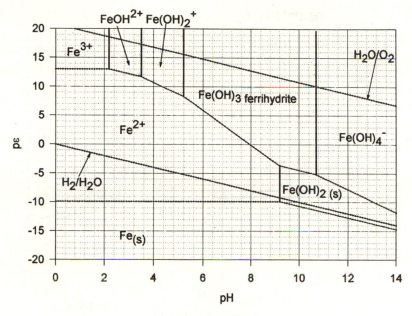

Graph to Example 11

$$Fe^{2+} + 2H_2O \rightleftharpoons Fe(OH)_{2(s)} + 2H^+ \qquad\qquad \log K = -12.4$$
$$pH = -0.5\log K - 0.5\log\{Fe^{2+}\} = +9.2$$

Plotting each of these lines for the applicable ranges in pε and pH enables the equilibrium distribution of iron species to be determined under any set of environmental conditions. All redox reactions are, however, bounded by the stability of water (equations 57–60).

The pε–pH diagram in Example 11 can be extended to incorporate (hydr)oxide solid phases, *e.g.* hematite and magnetite, and/or carbonate solid phases for waters containing dissolved CO_2. Example 12 includes $FeCO_{3(s)}$ in addition to the solid phases of Fe used in Example 11.

EXAMPLE 12 *Construct a pε–pH diagram for the Fe–O_2–CO_2–H_2O closed system where $\{Fe_T\}$ is 1×10^{-6} M and $C = 1 \times 10^{-3}$ M.*

In addition to the expressions presented in Example 11, the following equilibria need to be considered:

$$FeCO_{3(s)} + H^+ \rightleftharpoons Fe^{2+} + HCO_3^- \qquad\qquad \log K = -0.6$$
$$pH = \log K - \log\{Fe^{2+}\} - \log\{HCO_3^-\} = -0.6 + 6 + 3 = +8.4$$

$$FeCO_{3(s)} + H^+ + 2e^- \rightleftharpoons Fe_{(s)} + HCO_3^- \qquad\qquad \log K = -14.4$$
$$p\varepsilon = 0.5\log K - 0.5\,pH - 0.5\log\{HCO_3^-\} = -7.2 - 0.5\,pH + 1.5$$
$$= -5.7 - 0.5\,pH$$

$$FeCO_{3(s)} + 2H_2O \rightleftharpoons Fe(OH)_{2(s)} + H^+ + HCO_3^- \qquad\qquad \log K = -13.0$$
$$pH = -\log K + \log\{HCO_3^-\} = 13.0 - 3 = +10.0$$

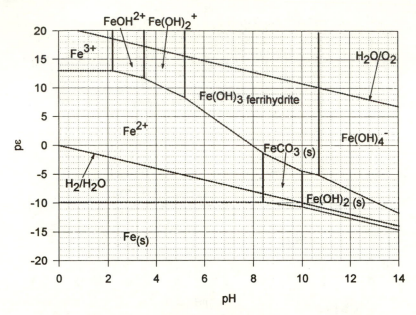

Graph to Example 12

$$Fe(OH)_{3(s)} + HCO_3^- + 2H^+ + e^- \rightleftharpoons FeCO_{3(s)} + 3H_2O \qquad \log K = +18.5$$
$$p\varepsilon = \log K + \log \{HCO_3^-\} - 2\,pH = 18.5 - 3 - 2\,pH = 15.5 - 2pH$$

$$Fe(OH)_4^- + HCO_3^- + 3H^+ + e^- \rightleftharpoons FeCO_{3(s)} + 4H_2O \qquad \log K = +35.2$$
$$p\varepsilon = \log K + \log \{HCO_3^-\} + \log \{Fe(OH)_4^-\} - 3pH$$
$$\quad = 35.2 - 3 - 6.0 - 3pH = 26.2 - 3pH$$

It is important also to consider other important redox equilibria which may influence iron speciation, such as the reduction of sulfate to sulfide which determines the stability field of pyrite (FeS_2). A $p\varepsilon$–pH diagram (Example 13) can be constructed for the overall system, Fe–O_2–H_2O–'S'–CO_2, by combining redox equilibria for S species (most importantly SO_4^{2-}, H_2S, and HS^-) with those already shown in Examples 11 and 12.

EXAMPLE 13   *Construct a $p\varepsilon$–pH diagram for the Fe–O_2–'S'–CO_2–H_2O open system where $\{Fe_T\}$ is 1×10^{-6} M, $\{S_{(aq)T}\} = 1 \times 10^{-2}$ M, $p_{CO_2} = 1 \times 10^{-3}$ atm and $K_H = 3.39 \times 10^{-2}$ at 298 K.*

In addition to the expressions in Example 11 the following equilibria are needed:

$$FeCO_{3(s)} + 2H^+ \rightleftharpoons Fe^{2+} + H_2CO_3^* \qquad \log K = +5.8$$
$$pH = 0.5\log K - 0.5\log\{Fe^{2+}\} - 0.5\log K_H p_{CO_2} = 2.9 + 3.0 + 2.24 = 8.1$$

$$Fe(OH)_{3(s)} + H_2CO_3^* + H^+ + e^- \rightleftharpoons FeCO_{3(s)} + 3H_2O \qquad \log K = +12.1$$
$$p\varepsilon = \log K + \log K_H p_{CO_2} - pH = 12.1 - 4.47 - pH = 7.6 - pH$$

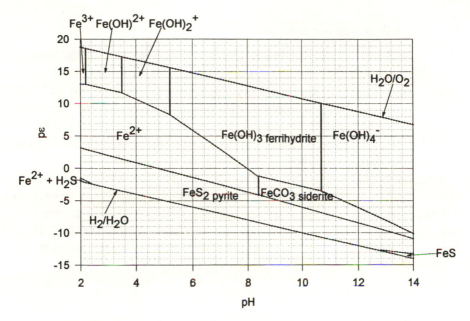

Graph to Example 13

$$Fe(OH)_4^- + H_2CO_3^* + 2H^+ + e^- \rightleftharpoons FeCO_{3(s)} + 4H_2O \qquad \log K = +28.4$$
$$p\varepsilon = \log K + \log K_H p_{CO_2} + \log\{Fe(OH)_4^-\} - 2\,pH$$
$$= 28.4 - 4.47 - 6 - 2\,pH = 17.9 - 2\,pH$$

In the stability field of sulfate:

$$2SO_4^{2-} + Fe^{2+} + 16H^+ + 14e^- \rightleftharpoons FeS_{2(s)} + 8H_2O \qquad \log K = +86.8$$
$$p\varepsilon = 6.2 + \tfrac{1}{7}\log\{SO_4^{2-}\} + \tfrac{1}{14}\log\{Fe^{2+}\} - \tfrac{8}{7}pH = 5.5 - \tfrac{8}{7}pH$$

$$FeCO_{3(s)} + 2SO_4^{2-} + 14e^- + 18H^+ \rightleftharpoons FeS_{2(s)} + H_2CO_3 + 8H_2O$$
$$\log K = +85.4$$
$$p\varepsilon = 6.1 - \tfrac{9}{7}pH - \tfrac{1}{14}\log K_H p_{CO_2} + \tfrac{1}{7}\log\{SO_4^{2-}\} = 6.1 - \tfrac{9}{7}pH$$

At lower pε values (in the H_2S and HS^- stability fields):

$$FeS_{2(s)} + 4H^+ + 2e^- \rightleftharpoons Fe^{2+} + 2H_2S_{(aq)} \qquad \log K = -5.0$$
$$p\varepsilon = -2.5 - 0.5\log\{Fe^{2+}\} - \log\{H_2S\} - 2pH = 2.5 - 2pH$$

$$FeS_{2(s)} + H^+ + 2e^- \rightleftharpoons FeS_{(s)} + HS^- \qquad \log K = -14.6$$
$$p\varepsilon = 0.5\log K - 0.5\log\{HS^-\} - 0.5pH = -6.3 - 0.5pH$$

It should be remembered that these diagrams are based on equilibrium conditions and that redox reactions in natural waters may be far from equilibrium. The diagrams can, however, be used to predict species distribution in a system of a given composition when it reaches equilibrium. Equilibrium constants should again be adjusted for ionic strength effects.

2.5 Modelling Aquatic Systems

This section has discussed the fundamental importance of the master variables pε and pH in determining species distribution under the conditions prevailing in natural systems (pε −10 to +17 and pH 4 to 10). These equilibria also provide the basis for many of the currently available geochemical computer models (*e.g.* WATEQ4F,[14] PHREEQC,[19] and MINTEQA2[20]) which are being used to predict the geochemical behaviour of metal ions, organic pollutants *etc.*

Computer codes have been developed to predict the distribution of metal ions between different aqueous species (including complexes), to predict species distribution in an aqueous solution taking account of adsorption/ion exchange processes (not discussed here), and to formulate 'best-case' and 'worst-case' scenarios for contaminant transport.

3 CASE STUDIES

3.1 Acidification

3.1.1 Diatom Records. The onset and flourishing of the Industrial Revolution, with its dependence for energy upon the combustion of fossils fuels, especially coal, released huge quantities of sulfur dioxide and nitrogen oxides to the atmosphere, which ultimately were returned to the land surface in the form of acid precipitation (H_2SO_4, HNO_3). The effect of this upon freshwater lakes, especially in poorly buffered catchment areas (*e.g.* low-carbonate soils, granitic bedrock) was a significant decrease in pH and a concomitant decline in biological productivity, often leading to clear acid waters devoid of fish. That the acidification of freshwater lakes is a comparatively recent (post-1850) phenomenon linked to acid deposition has been demonstrated by the diatom record in the sediments at the bottom of acidified lakes.[21] Diatoms are algae with silicified cell walls. Depending upon the acidity of the water, some species flourish more than others. Upon deposition, these siliceous remains are preserved in the sediment column. The dramatic change in pH after 1850 inferred from the diatom record in dated sediments from the Round Loch of Glenhead in the Galloway region of south-west Scotland rules out long term acidification processes as the cause.[22]

[19] D. L. Parkhurst, 'Users Guide to PHREEQC – A Computer Programme for Speciation, Reaction-path, Advective Transport, and Inverse Geochemical Calculations', US Geological Survey Water Resources Inv. Report, 1995, 95.
[20] J. D. Allison, D. S. Brown, and K. J. Novo-Gradac, 'MINTEQA2, A Geochemical Assessment Data Base and Test Cases for Environmental Systems: Version 3.0 User's Manual', Report EPA/600/3–91/-21, US EPA, Athens, GA, 1991.
[21] R. W. Battarbee, R. J. Flower, A. C. Stevenson, and B. Rippey, *Nature*, 1985, **314**, 350.
[22] V. J. Jones, A. C. Stevenson, and R. W. Battarbee, *Nature*, 1986, **322**, 157.

3.1.2 Aluminium. The effect of such acidification upon the mobilization, behaviour, and subsequent biological impact of aluminium in freshwaters has been spectacular. Constituting 8.13% of the Earth's crust, aluminium is the third most abundant element and is largely associated with crystalline aluminosilicate minerals of low solubility. In the absence of strong acid inputs, processes of soil development normally lead to only a small fraction of aluminium becoming available to participate in biogeochemical reactions. Thus, in the Northern Temperate Region, mobilization of aluminium from upper to lower mineral horizons by organic acids (or H_2CO_3) leached from foliage or forest floors often leads to the formation of $Al(OH)_3$ in a process known as podzolization. Under elevated partial pressures of CO_2, however, dissociation of H_2CO_3 produces H^+, which may solubilize aluminium, while HCO_3^- serves as a counterion for transport of cationic $Al_{(aq)}^{III}$ through soil. Upon discharge to a surface water, CO_2 degasses, resulting in aluminium hydrolysis and precipitation as $Al(OH)_3$. Thus, in most natural waters, concentrations of dissolved aluminium are generally low due to the relatively low solubility of natural aluminium minerals under circumneutral pH values. In the case of strong acid (H_2SO_4, HNO_3) inputs, however, especially to sensitive regions with small pools of basic cations (Ca^{2+}, Mg^{2+}, Na^+, K^+) and an inability to retain inputs of strong acid anions (SO_4^{2-}, NO_3^-, Cl^-), acidic cations (H^+, Al^{III}) are transported along with acid anions from soil to surface water.[23]

The speciation of dissolved aluminium in surface waters is highly pH-dependent (see Section 2.3.2), with, in a simple system (considering only monomeric aluminium species), in equilibrium with solid phase $Al(OH)_3$, the aquated ion $Al_{(aq)}^{3+}$ predominating at pH < 3, progressive hydrolysis to $Al(OH)^{2+}$ and $Al(OH)_2^+$ in the pH range 4.5 to 6.5, followed by $Al(OH)_3^0$ from pH 6.5 to 7, and then $Al(OH)_4^-$ predominating as the pH increases beyond 7. From pH 4.5 to 7.5, the solubility of aluminium is low and in this range it is often precipitated as $Al(OH)_3$. Below pH 4.5 and above pH 7.5 the concentration of aluminium in solution increases rapidly. In a part experimental fractionation, part speciation modelling study of aluminium in acidified Adirondack surface waters in the north-eastern USA, Driscoll and Schecher[23] found that, in the absence of organic ligands, $Al_{(aq)}^{3+}$ was significant at pH < 5, and that the inorganic monomeric complexes of aluminium predominated at pH > 5. In particular, Al–F complexes were the dominant form at pH 5–6 whereas at pH > 6, Al–OH species became the major form. In the presence of organic ligands, however, they found that, over a broad pH range of 4.3–7.0, alumino-organic complexes were a major component of

[23] C. T. Driscoll and W. D. Schecher, *Environ. Geochem. Health*, 1990, **12**, 28.

the monomeric aluminium. It is the chemically labile inorganic mono-meric aluminium that has been identified as the toxic agent for fish which, in acidified lakes and streams at aluminium concentrations of 100–200 μg l^{-1}, are unable to maintain their osmoregulatory balance and are susceptible to respiratory problems from coagulation of mucus and $Al(OH)_3$ on their gills at their physiological pH of 7.2.[24]

With respect to human exposure to aluminium in drinking water, it must be remembered that aluminium is often deliberately added, in the form of $Al_2(SO_4)_3$, to water supplies at treatment works to remove coloration by organic compounds in reservoirs in upland catchments. This it achieves by hydrolysing to a gelatinous, high surface area, precipitate of $Al(OH)_3$, which helps to remove the coloured organic colloids. It has been suggested, partially because of observed water aluminium-related dialysis dementia in some chronic renal patients on artificial kidney dialysis machines in the 1960s and 1970s, that exposure to low concentrations of aluminium in drinking water, for which there is a current EC Maximum Admissible Concentration of 200 μg l^{-1}, might be implicated in Alzheimer's Disease.[25] There is no proof of this at present, although it is known that the incidence of senile dementia on the remote Pacific island of Guam, where bauxite is mined and the local water contains elevated levels of aluminium, is a hundred-fold greater than in the USA. It has been mooted that, for humans, a high intake of silicic acid (H_4SiO_4) could reduce the bioavailability of aluminium by two orders of magnitude for, at slightly alkaline intestinal pH, hydroxyaluminosilicates are stable and unavailable for absorption.[26] This would mimic the situation in nature, where, in the absence of a strong acid input, the bioavailability of aluminium is kept low by the formation of hydroxyaluminosilicate species.

3.1.3 Acid Mine Drainage and Ochreous Deposits. One of the major pollution problems affecting freshwaters is that of acid waters discharged from coal and metal mines.[27] In an active underground coal mine, pumping lowers the water table. The deeper strata thus become exposed to air with the result that pyrite (FeS_2) present is subject to oxidation, generating acid conditions.

$$2FeS_{2(s)} + 2H_2O + 7O_2 \rightleftharpoons 2FeSO_{4(s)} + 2H_2SO_{4(l)} \qquad (61)$$

[24] N. J. Bunce, 'Environmental Chemistry', Wuerz, Winnipeg, 2nd Edn., 1994.
[25] 'Aluminium in Food and the Environment', ed. R. C. Massey and D. Taylor, Royal Society of Chemistry, Cambridge, 1989.
[26] J. D. Birchall, *Chem. Br.*, 1990, **26**, 141.
[27] R. J. Pentreath, 'Issues in Environmental Science and Technology', ed. R. E. Hester and R. M. Harrison, Royal Society of Chemistry, Cambridge, 1994, Vol. 1, p. 121.

When mining and pumping cease, the water table returns to its natural level. While the flooding of the mine stops the direct oxidation of pyrite, it does bring the sulfuric acid and iron sulfates into solution. Some of the ferrous ion (Fe^{2+}) may be oxidized to the ferric ion (Fe^{3+}), a slow process at low pH, but one which can be catalysed by bacteria, and the ferric ion may react further with pyrite.[28]

$$4Fe^{2+} + O_{2(g)} + 4H^+ \rightleftharpoons 4Fe^{3+} + 2H_2O \qquad (62)$$

$$FeS_{2(s)} + 14Fe^{3+} + 8H_2O \rightleftharpoons 15Fe^{2+} + 2SO_4^{2-} + 16H^+ \qquad (63)$$

On reaching the surface, the acidic mine drainage water is mixed with air and oxygenated water, leading to rapid oxidation from the ferrous to the ferric form and the precipitation of the characteristic unsightly orange ochreous deposits of iron (oxy)hydroxides, *e.g.* ferrihydrite ($Fe(OH)_3$), goethite (FeOOH), observed along many stream and river beds.

$$4Fe^{2+} + O_{2(g)} + 4H^+ \rightleftharpoons 4Fe^{3+} + 2H_2O \qquad (64)$$

$$Fe^{3+} + 3H_2O \rightleftharpoons Fe(OH)_{3(s)} + 3H^+ \qquad (65)$$
ferrihydrite

$$Fe^{3+} + 2H_2O \rightleftharpoons FeOOH_{(s)} + 3H^+ \qquad (66)$$
goethite

The overall process may be represented as:

$$2FeS_{2(s)} + \tfrac{15}{2}O_{2(g)} + 7H_2O_{(l)} \rightleftharpoons 2Fe(OH)_{3(s)} + 4H_2SO_{4(l)} \qquad (67)$$

Micro-organisms (*e.g. Thiobacillus thiooxidans, Thiobacillus ferrooxidans, Metallogenium*) are involved as catalysts in many of the oxidizing reactions, which would be extremely slow under the prevailing conditions of low pH (*e.g.* <4.5).[29] The combination of acid waters and coatings can have devastating effects upon aquatic biota, with depletion of free-swimming and bottom-dwelling organisms, the loss of spawning gravel for fish, and direct fish mortalities. Typical approaches to treatment of acid mine drainage have included methods to remove the iron floc by oxidation and to adjust the pH through the use of limestone filter beds.

3.1.4 *Acid Mine Drainage and Release of Heavy Metals.*
The production of sulfuric acid from sulfide oxidation in mines can also lead to the leaching of metals other than iron, with the result that the emerging acidic waters may be laden with heavy metals. Thus in the former metalliferous mining area of south-west England, where there are now many abandoned mine workings, the closure and flooding of the famous Wheal Jane tin mine in 1992 led to a highly acidic cocktail of dissolved

[28] J. E. Andrews, P. Brimblecombe, T. D. Jickells, and P. S. Liss, 'An Introduction to Environmental Chemistry', Blackwell Science, Oxford, 1996.
[29] C. F. Mason, 'Biology of Freshwater Pollution', Longman, Harlow, 3rd Edn., 1996.

metals (Cu, Pb, Cd, Sn, As) entering the Carnon and Fal rivers at 7–15 million litres per day and spreading throughout the surrounding estuaries and coastal waters.[30]

In the USA, the EPA has identified over 31 000 hazardous waste sites, with the largest complex of 'Superfund' sites to be remediated in western Montana, in the Clark Fork River Basin where there have been more than 125 years of copper and silver mining and smelting activities. Moore and Luoma[31] have characterized three types of contamination resulting from large-scale metal extraction: primary, consisting of wastes produced during mining, milling, and smelting and deposited near their source of origin; secondary, resulting from transport of contaminants away from these sites by rivers or through the atmosphere to soils, groundwaters, rivers, *etc.*; and tertiary, where contaminants may be remobilized many kilometres away from their point of origin.

Some of the largest waste deposits occur in tailing ponds containing acid mine water. In the Clark Fork River Complex, it is estimated that ponds contain approximately 9000 t arsenic, 200 t cadmium, 90 000 t copper, 20 000 t lead, 200 t silver, and 50 000 t zinc. These metals enter streams and rivers as solutes and particulates and contaminate sediments in the river and reservoirs far downstream from the primary sources (Figure 4).[31] Downstream concentrations follow an exponential decline when viewed over several hundred kilometres. The sediment of Milltown Reservoir, more than 200 km from the mines and smelters at Butte and Anaconda, is highly contaminated with various metals and arsenic. It is believed that oxidation–reduction processes release arsenic from the reservoir sediments and cause contamination of an aquifer from which water drawn through wells, now closed, contains arsenic at concentrations higher than the EPA drinking water standards.[31]

When open-pit mining ended in the Clark Fork River Complex in 1982, pumping was discontinued, with the result that water began filling underground shafts and tunnels and the 390 m deep Berkeley pit. Contaminated acid water containing individual metal and sulfate concentrations thousands of times those in uncontaminated water could flow into an adjacent alluvial aquifer and eventually over the rim of the pit. Large-scale hard-rock mining in the western United States, especially Nevada, has greatly increased in recent years, with the result that deep 'pit lakes' are likely to form as open-pit metal mines intersecting groundwater are depleted and shut down.[32] Pit lakes in high sulfide rock will tend to have poor quality acidic water, although oxidized rock that contains significant carbonate will produce better quality, near-

[30] L. E. Hunt and A. G. Howard, *Mar. Pollut. Bull.*, 1994, **28**, 33.
[31] J. N. Moore and S. N. Luoma, *Environ. Sci. Technol.*, 1990, **24**, 1279.
[32] G. C. Miller, W. B. Lyons, and A. Davis, *Environ. Sci. Technol.*, 1996, **30**, 118A.

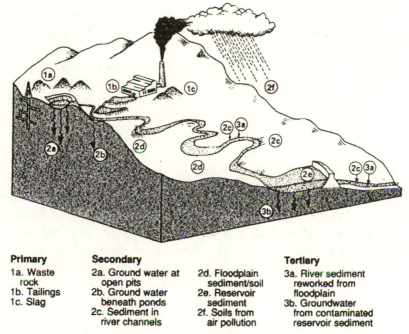

Primary	Secondary		Tertiary
1a. Waste rock	2a. Ground water at open pits	2d. Floodplain sediment/soil	3a. River sediment reworked from floodplain
1b. Tailings	2b. Ground water beneath ponds	2e. Reservoir sediment	3b. Groundwater from contaminated reservoir sediment
1c. Slag	2c. Sediment in river channels	2f. Soils from air pollution	

Figure 4 *Types of contamination resulting from large scale metal extraction*[31] (Reprinted with permission from *Environ. Sci. Technol.*, **24**, 1279. © 1990 American Chemical Society)

neutral pit lake water. Most of the larger existing pit lakes currently contain water that does not meet standards for drinking water, agricultural water quality, or aquatic life.[32] Factors such as the oxygen status of the lake, pH, the hydrogeologic flow system, composition of the wall-rock, evapoconcentration, biological activity, and hydrothermal inputs are all important to the modelling of future water quality and impact.

Remediation approaches in circumstances like those described here have been based on a variety of physical, chemical, and biological systems. These include the construction of ponds where sufficient organic matter is available to establish anaerobic conditions and immobilize at least some of the metals (*e.g.* Cu, Pb, Cd, Zn) as sulfides. Similar passive treatment of tailings-impacted groundwater has also employed precipitation of metal sulfides as the key clean-up step.[33] Active chemical treatment with lime, producing sludges, and biosorption by reeds, wetlands, *etc.*, are other methods which have been tried. Although there have been considerable technical improvements around the world in recent years in metal extraction and smelting, waste treatment, and containment, the regulatory requirements are becoming much more stringent. For example, the new National Pollutant Discharge Elimina-

[33] R. J. Allan, 'Proceedings of the 10th International Conference on Heavy Metals in the Environment', CEP, Edinburgh, 1995, Vol. 1, p. 293.

tion System permits of the US EPA stipulate $>90\%$ survival of selected test organisms for 48 hours in water discharges and effluents. While fathead minnows and water fleas, sensitive at the mg l^{-1} level, were formerly used, the new standard will be the much more sensitive arthropod, *Ceriodaphnia dubia*, which is affected at the μg l^{-1} level. It is understood that most mines and mine–mill discharges in Missouri's New Lead Belt may well fail this test.[34]

3.2   Metals in Water

3.2.1   Arsenic in Groundwater.   In what has been described as the largest arsenic poisoning epidemic in the world, hundreds of thousands of people in West Bengal, India, have been seriously affected by arsenic poisoning resulting from the consumption of water drawn from tube wells sunk some 20 to 150 metres below the ground into aquifers.[35] In West Bengal and other parts of the world (*e.g.* Bangladesh, Chile, Mexico, Argentina, Ghana, Mongolia, and Taiwan) where inorganic arsenic concentrations in drinking water are elevated (mg l^{-1} against WHO recommended limits of 10 μg l^{-1}), exposure has resulted in hyperpigmentation, hyperkeratosis, and cancer of the skin.[36,37]

Sulfide minerals are one of the most important natural sources of arsenic in groundwater. Oxidation of arsenopyrite (FeAsS), in analogous fashion to pyrite (FeS$_2$), may release high concentrations of arsenic into solution.[38]

$$4FeAsS_{(s)} + 13O_{2(g)} + 6H_2O \rightleftharpoons 4Fe^{2+} + 4AsO_4^{3-} + 4SO_4^{2-} + 12H^+ \quad (68)$$

In Bengal, it has been suggested that large-scale withdrawal of groundwater for irrigation produces seasonal fluctuation of the water table, which in turn results in intake of oxygen into the pore waters of sediments that are arsenic-rich in the form of arsenopyrite.[39] The exact speciation of soluble inorganic arsenic, *e.g.* as the undissociated forms or different oxyanions of the acids $H_3As^{III}O_3$ and $H_3As^VO_4$, will be dependent upon the prevailing pε and pH. Sorption of arsenic, especially

[34] N. L. Gale, B. G. Wixson, D. Forciniti, and D. Murphy, 'Proceedings of the 15th Annual European Meeting of the Society for Environmental Geochemistry and Health', British Geological Survey, Keyworth, 1997.

[35] T. R. Chowdhury, B. Kr. Mandal, G. Samanta, G. Kr. Basu, P.P. Chowdhury, C. R. Chanda, N. Kr. Karan, D. Lodh, R. Kr. Dhar, D. Das, K. C. Saha, and D. Chakraborti, 'Arsenic: Exposure and Health Effects', ed. C. O. Abernathy, R. L. Calderon, and W. R. Chappell, Chapman & Hall, London, 1997, p. 93.

[36] 'Arsenic: Exposure and Health', ed. W. R. Chappell, C. O. Abernathy, and C. R. Cothern, Science and Technology Letters, Northwood, 1994.

[37] 'Arsenic: Exposure and Health Effects', ed. C. O. Abernathy, R. L. Calderon, and W. R. Chappell, Chapman & Hall, London, 1997.

[38] P. L. Smedley, W. M. Edmunds, and K. B. Pelig-Ba, 'Environmental Geochemistry and Health', ed. J. D. Appleton, R. Fuge, and G. J. H. Ball, Geological Society, London, 1997, p. 163.

[39] P. Bagla and J. Kaiser, *Science*, 1996, **274**, 174.

pentavalent arsenate (*e.g.* AsO_4^{3-}, $HAsO_4^{2-}$, $H_2AsO_4^-$), onto ferric hydroxide ($Fe(OH)_3$) produced under oxidizing conditions may, however, restrict its mobility and availability, although fertilizer phosphate present in groundwater could perhaps displace the arsenate. Arsenite, especially as $H_3As^{III}O_3$ ($pK_1 = 9.2$), the predominant form under reducing conditions at pH < 9.2, is much less strongly sorbed. As an alternative to arsenopyrite oxidation, it is possible that elevated arsenic concentrations in Bengal groundwater could result from the release of adsorbed arsenate during the dissolution of ferric hydroxide under reducing conditions (see Section 3.3.2).

Although trivalent inorganic arsenic, with its propensity for binding to the SH group of enzymes, is acknowledged to be more toxic to humans than pentavalent inorganic arsenic, it must be recognised that As^V can be converted to As^{III} in the human body as part of the reduction/biomethylation pathway of excretion:

$$As^V \rightarrow As^{III} \rightarrow MMAA \rightarrow DMAA \tag{69}$$

where MMAA ($CH_3AsO(OH)_2$, monomethylarsonic acid) and DMAA (($CH_3)_2AsO(OH)$, dimethylarsinic acid) are less toxic metabolites, the latter predominating as the major form of arsenic excreted in urine.[36,37] It is possible that such methylation could also occur naturally by microbial action in freshwaters, although reported occurrences suggest that the effect is small.[40]

3.2.2 Lead in Drinking Water. The naturally soft, slightly acidic, plumbosolvent water of the Loch Katrine water supply for the Glasgow area was recognized many years ago to release lead from the lead pipes and tanks in the domestic plumbing of the Victorian and subsequent (even post-World War II) eras.[41]

$$O_{2(g)} + 4H^+ + 4e^- \rightleftharpoons 2H_2O \qquad E^0 = 1.229 \text{ V} \tag{70}$$

$$Pb^{2+} + 2e^- \rightleftharpoons Pb_{(s)} \qquad E^0 = -0.126 \text{ V} \tag{71}$$

$$O_{2(g)} + 4H^+ + 2Pb_{(s)} \rightleftharpoons 2Pb^{2+} + 2H_2O \qquad E^0 = 1.355 \text{ V} \tag{72}$$

Essentially, elemental lead becomes soluble in acidic conditions due to its oxidation by dioxygen. Furthermore, compounds such as carbonate and hydroxycarbonate compounds of lead, *i.e.* $PbCO_3$ and $Pb_3(CO_3)_2(OH)_2$, that may coat the pipes, will dissolve under acid conditions.

In view of the concern over detrimental effects of exposure to lead upon human health, in particular the possible impact upon intelligence and behaviour of young children, steps were taken in the mid-1970s to

[40] W. R. Cullen and K. J. Reimer, *Chem. Rev.*, 1989, **89**, 713.
[41] M. R. Moore, W. N. Richards, and J. G. Sherlock, *Environ. Res.*, 1985, **38**, 67.

reduce the lead content of tap water in Glasgow and other susceptible areas, which often exceeded the WHO maximum guideline at the time of $100\ \mu g\ l^{-1}$. The method chosen was adjustment of pH to 8–9 by lime dosing. The effects of liming the Glasgow water supply were quite dramatic. Whereas pre-1978, when the pH was 6.3, only 50% of random daytime samples were $<100\ \mu g\ l^{-1}$ in lead, the figure increased to 80% during 1978–1980, when the adjusted pH was 7.8. After 1980, when the pH was increased further to 9.0, 95% of samples were $<100\ \mu g\ l^{-1}$.[41] It appears that carbonate and hydroxycarbonate lead compounds present in the coatings on the pipes were stabilized. Significant reductions in blood lead levels of key exposed groups (*e.g.* pregnant women) were also observed.

Since 1989, as regulatory upper limits for lead in drinking water have fallen, *e.g.* to $50\ \mu g\ l^{-1}$ (EC) and now to $10\ \mu g\ l^{-1}$ (WHO), orthophosphate has been added to the water supply in Glasgow to precipitate insoluble lead compounds such as $Pb_3(PO_4)_2$ and $Pb_5(PO_4)_3OH$. This has resulted in a fall in the proportion of households with water lead $>10\ \mu g\ l^{-1}$ from 49% in 1981 to 17% in 1993.[42] Despite this improvement, an estimated 13% of infants were still exposed via bottle feeds to tap water lead concentrations in excess of $10\ \mu g\ l^{-1}$ and it seems very unlikely that further treatment of the water supply will be able to guarantee water lead concentrations $<10\ \mu g\ l^{-1}$.

3.2.3   Cadmium in Irrigation Water.   In the 1950s in Japan, many people, especially menopausal women suffering from malnutrition, low vitamin D intake and calcium deficiency, suffered a condition known as Itai-Itai (Ouch-Ouch) disease, with symptoms ranging from lumbago-type pains to multiple fractures of softened bones. The cause was irrigation water from a river (Jintsu) chronically contaminated with dissolved cadmium from a zinc mining and smelting operation. Contaminated rice from the irrigated paddy fields was eaten and the Ca^{2+} in bones replaced by Cd^{2+}, an ion of the same charge and size.

Although this is a particularly extreme example of acute effects resulting from high exposure to this non-essential element, concern has been expressed about chronic effects (*e.g.* kidney damage, hypertension) from possible enhanced exposure of humans through increased application of sewage sludge to agricultural land, in view of EC-enforced cessation of dumping at sea by 1999. Compared with other heavy metals, cadmium exhibits an especially high mean sludge concentration ($20\ mg\ kg^{-1}$) relative to mean soil concentration ($0.4\ mg\ kg^{-1}$).[43] In

[42] G. C. M. Watt, A. Britton, W. H. Gilmour, M. R. Moore, G. D. Murray, S. J. Robertson, and J. Womersley, *Brit. Med. J.*, 1996, **313**, 979.
[43] P. O'Neill, 'Environmental Chemistry', Chapman & Hall, 2nd Edn., London, 1993.

acidic soils the concentration of Cd^{2+} available for uptake by plants can be substantial,[44] as it adsorbs only weakly onto clays, whereas at pH > 7 it readily precipitates, *e.g.* as $CdCO_3$ for which the solubility product, $K_{SP} = 1.8 \times 10^{-14}$, is indicative of the advantages of liming. Similarly, in drinking water, the presence of dissolved carbonate at a concentration of 5×10^{-4} M can reduce the solubility of cadmium from 637 mg l^{-1} to 0.11 mg l^{-1}, in line with evidence that hard water, with a high calcium content, can protect against cadmium.[45]

3.2.4 Selenium in Irrigation Water. In 1983 high rates of embryonic deformity and death, attributed to selenium toxicosis, were found in wild aquatic birds at the Kesterson National Wildlife Refuge in the San Joaquin Valley of California.[46] Kesterson Reservoir was a regional evaporation pond facility where drainage waters, often containing high levels of salts and trace elements (including selenium), from irrigated farmland had been collected since 1978.

With evapotranspiration, greatly in excess of precipitation, bringing soluble salts to the surface of farmland in the arid climate of the west-central San Joaquin Valley, and crop productivity, after irrigation, threatened by shallow saline groundwater near the root zone, grids of sub-surface tile drains were constructed to divert the saline waters to a collective drain (San Luis), which flowed into Kesterson Reservoir. It was the geologic setting of the San Joaquin Valley as well as the climate, however, which led not only to the soil salinization but also to the presence of selenium (in the form of SeO_4^{2-}) in Kesterson inflow waters at concentrations in excess of the US EPA designation of 1000 $\mu g\, l^{-1}$ for selenium as a toxic waste, never mind its 10 $\mu g\, l^{-1}$ limit for drinking water and the <2.3 $\mu g\, l^{-1}$ limit subsequently suggested for the protection of aquatic life.

To the west of the San Joaquin River (Figure 5), selenium in the soils is believed to be of natural origin. During the Jurassic and Cretaceous periods, there was deposition of marine sediments comprising sandstones, shales, and conglomerates, including seleno-sulfides of iron ($FeS_2 + FeSe_2$). Subsequent uplifting of these sediments produced the Coast Ranges and subjected the sediments to weathering under oxidizing conditions. With resultant acid neutralized by the carbonate component of the sediments, the predominant form of selenium produced from oxidation of selenide, Se^{-II}, under alkaline conditions, is selenate, $Se^{VI}O_4^{2-}$, in preference to selenite, $Se^{IV}O_3^{2-}$, or biselenite, $HSe^{IV}O_3^-$.[47]

[44] J. E. Fergusson, 'The Heavy Elements: Chemistry, Environmental Impact and Health Effects', Pergamon Press, Oxford, 1990.
[45] J. E. Fergusson, 'Inorganic Chemistry and the Earth', Pergamon Press, Oxford, 1982.
[46] T. S. Presser, *Environ. Management*, 1994, **18**, 437.
[47] R. H. Neal, 'Heavy Metals in Soils', ed. B. J. Alloway, Blackie, Glasgow, 2nd Edn., 1995, p. 260.

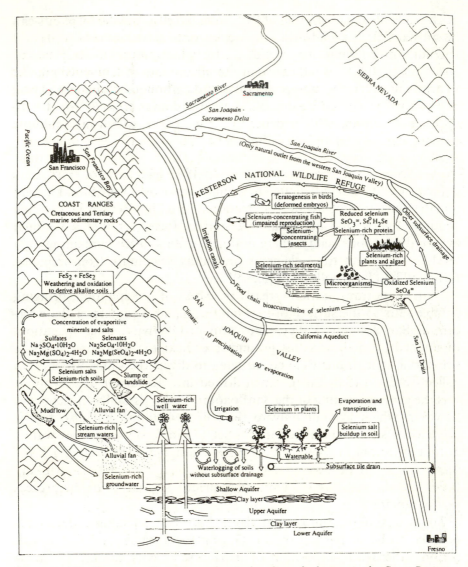

Figure 5   *The Kesterson effect: biogeochemical cycling of selenium in the Coast Ranges, San Joaquin Valley, and Kesterson National Wildlife Refuge of California*[46] (Reproduced with permission from *Environ. Management*, **18**, 437. © 1994 Springer Verlag)

Thus the soils on the Coast Ranges, which are alkaline, contain significant amounts of soluble mineral salts (*e.g.* $Na_2SO_4.10H_2O$, $Na_2Mg(SO_4)_2.4H_2O$), including selenates (*e.g.* $Na_2SeO_4.10H_2O$, $Na_2Mg(SeO_4)_2.4H_2O$).

The chemistry and mobility of selenite and selenate differ greatly in soils. Selenite is adsorbed by specific adsorption processes (*e.g.* on clays and hydrous metal oxides) and to a much greater extent than selenate, which is adsorbed, like sulfate, by comparatively weak non-specific processes. Thus, in aerated alkaline soils, such as those of semi-arid regions like the west-central San Joaquin Valley, mobile $Se^{VI}O_4^{2-}$ will be the dominant form and, through runoff, be capable of entering the groundwaters and sub-surface drainage waters, ultimately being removed to Kesterson Reservoir. There it is taken up by the biota, with devastating effects, and also converted to a range of selenium-containing species (*e.g.* $Se^{IV}O_3^{2-}$, Se^0, H_2Se^{-II}, Se-rich protein), with much ultimately being deposited in the sediments. There, over geological time, natural diagenetic processes would presumably lead again to the formation of reduced seleno-sulfide species.

In the meantime, however, considerable effort is being devoted to the clean-up of Kesterson Reservoir, which was drained in 1987, one year after the input of drainage water was stopped. Natural processes, stimulated by the addition of nutrients, of summertime microbial conversion and volatilization to the atmosphere of methylated forms such as dimethylselenide, $(CH_3)_2Se$, and downwards leaching of $Se^{VI}O_4^{2-}$ by percolating water in winter have helped to dissipate 68–88% of total selenium from the topsoil (0–15 cm) over a period of eight years.[48] Similarly, pilot bioreactors have been set up to convert drainage water SeO_4^{2-} into commercially useful selenium-containing products.

3.2.5 Aquatic Contamination by Gold Ore Extractants. The separate use of mercury and cyanide has led to the contamination of freshwater systems.

Mercury in the Amazon Basin. With the price of gold soaring in the late 1970s and early 1980s, there developed a modern-day gold rush in South America.[49] In the relatively inaccessible Amazon area in Brazil, there were soon up to a million people operating on an informal basis. The technique which is used to extract the placer gold which has accumulated in many stream and alluvial deposits is based on amalgamation with mercury. Typically, high pressure water hoses are used to dislodge alluvium, which is then taken into a large sluice by a motor suction pump. The sluices are lined with sacking into which mercury is added to amalgamate the gold particles, which lodge in the sacking. In other versions, no mercury is used in the sluices. Instead, material is collected from the sluice lining and crushed in an oil drum, where the mercury is then introduced, followed by manual panning.

The effects of all this activity can be seen in the increased turbidity, with clear rivers turning a muddy brown colour, and increased fish

[48] M. Flury, W. T. Frankenberger Jr., and W. A. Jury, *Sci. Total Environ.*, 1997, **198**, 259.
[49] D. Cleary and I. Thornton, 'Issues in Environmental Science and Technology', ed. R. E. Hester and R. M. Harrison, Royal Society of Chemistry, Cambridge, 1994, Vol. 1, p. 17.

mortality. Furthermore, releases of mercury to the aquatic environment have resulted in elevated concentrations of mercury in sediments and fish. Commonly eaten fish species from the Madeira river were found to contain mean mercury levels of 0.7 mg kg^{-1} and up to 3.8 mg kg^{-1} in the Tapajos Valley.[50] Given the bioaccumulation and biomagnification of mercury in many fish species and the possible microbial methylation of mercury deposited in sediments to the even more toxic lipid-soluble methyl mercury, CH_3Hg^+, there is a real prospect of enhanced dietary uptake of the most toxic form of mercury by the local populations. This would be in addition to exposure from inhalation of mercury vapour during the handling of mercury and from subsequent post-amalgamation treatment steps, which usually involve refinement of the extracted gold by heating to drive the more volatile mercury off. It is thought that there is an annual discharge of about 100 tonnes of mercury into the Amazon ecosystem and perhaps the same amount released as mercury vapour. Thus there could be an increased risk of the neurological damage and foetal deformities commonly associated with mercury poisoning, as most notably exemplified in the fishing village population of Minamata, Japan, in the 1950s when a chemical manufacturing plant discharged mercury salts and CH_3Hg^+ into the bay, to be taken up by fish and shellfish.

Cyanide.   Another method of extracting gold that can lead to contamination of freshwater is cyanide heap leaching, which is used to extract tiny amounts of gold from huge volumes of low-grade deposits of gold-bearing rock. In this case, ore is dug from open-pit mines, crushed, and spread on asphalt or plastic pads. The heaps are then sprayed with cyanide, which dissolves the gold.

$$4Au_{(s)} + 8NaCN + 2H_2O + O_{2(g)} \rightarrow 4NaOH + 4NaAu(CN)_2 \quad (73)$$

The gold-bearing cyanide solutions run off on impermeable asphalt or plastic to collection vats and are then treated, for example with zinc, to extract gold.

$$2NaAu(CN)_2 + Zn_{(s)} \rightarrow Na_2Zn(CN)_4 + 2Au_{(s)} \quad (74)$$

One major environmental problem with this technique is that cyanide is highly toxic. Although attempts are made to loop the cyanide through a closed system to try to ensure that none is lost, cyanide does commonly escape to surface waters and groundwater. While cyanide quickly

[50] L. D. Lacerda, O. Malm, J. R. D. Guimardes, W. Salomons, and R.-D. Wilken, 'Biogeodynamics of Pollutants in Soils and Sediments', ed. W. Salomons and W. M. Stigliani, Springer, Berlin, 1995, p. 213.

decomposes in oxygenated, acidic surface water, it can persist at toxic levels for much longer periods in groundwater and thus pose a longer term threat.

3.3 Historical Pollution Records and Perturbatory Processes in Lakes

3.3.1 Records—Lead in Lake Sediments. Under ideal circumstances, freshwater lake sediments can preserve the record of temporal variations in anthropogenic input of contaminants via atmospheric deposition, catchment runoff, effluent inflow, and dumping from industrial, transportation, mining, agricultural, and waste disposal sources.[51] Prerequisites are transfer of contaminants associated with settling inorganic particulates and/or biotic detritus from the water column to the sediments, no disturbance of sediments by physical mixing, slumping or bioturbation after deposition, no post-depositional degradation or mobility of the contaminants, and the establishment of a reliable time axis (*e.g.* via the use of the naturally occurring radionuclide ^{210}Pb, half-life 22.35 years, and the nuclear fallout radionuclide ^{137}Cs, half-life 30.2 years).

The derived record of lead deposition to sediments in the southern basin of Loch Lomond, Scotland, is shown in Figure 6.[52] The noticeable change in the ^{206}Pb/^{207}Pb ratio of the lead deposited from the late 1920s onwards is attributed to car exhaust emissions of lead depleted in ^{206}Pb as a result of the use of lead ores from Broken Hill, Australia, in the manufacture of alkyllead additives for the UK market. On the basis of the published literature, there appears to be reasonable confidence that, in the absence of post-depositional mixing, sediments do, in general, preserve a record of the deposition of lead to the sediments.[53,54] Many such records in North America and Western Europe now show concentrations of lead declining towards the sediment surface as a consequence of the withdrawal of leaded petrol and also a reduction in other emissions.[55]

3.3.2 Perturbatory Processes in Lake Sediments. The extent to which conditions for preservation of records are met depends upon the characteristics of the specific individual systems and contaminants under study. In reviewing the perturbation of historical pollution

[51] D. H. M. Alderton, 'Historical Monitoring', MARC Report No. 31, London, 1985, p. 1.
[52] J. G. Farmer, L. J. Eades, A. B. MacKenzie, A. Kirika, and A. E. Bailey-Watts, *Environ. Sci. Technol.*, 1996, **30**, 3080.
[53] J. R. Graney, A. N. Halliday, G. J. Keeler, J. O. Nriagu, J. A. Robbins, and S. A. Norton, *Geochim. Cosmochim. Acta*, 1995, **59**, 1715.
[54] H. C. Moor, T. Schaler, and M. Sturm, *Environ. Sci. Technol.*, 1996, **30**, 2928.
[55] E. Callender and P. C. Van Metre, *Environ. Sci. Technol.*, 1997, **31**, 424A.

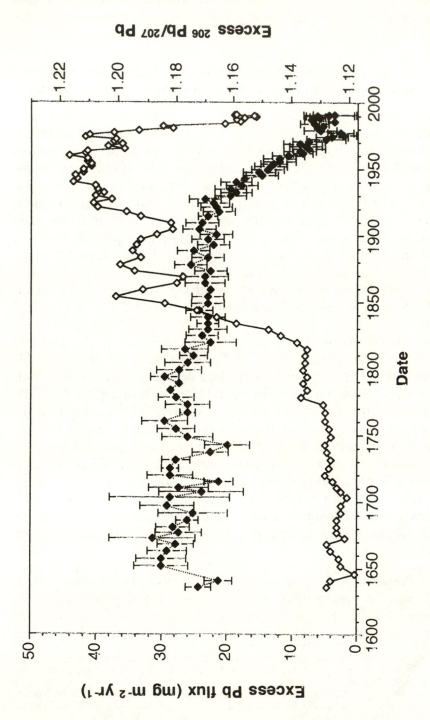

Figure 6 *Excess (i.e. anthropogenic) lead fluxes (◇) and associated $^{206}Pb/^{207}Pb$ ratios (◆), derived from analysis of thin sections of a sediment core from the southern basin of Loch Lomond, Scotland, versus calendar year[52] (Reproduced with permission from Environ. Sci. Technol., **30**, 3080. © 1996 American Chemical Society)*

records in aquatic sediments, Farmer[56] has suggested that in the case of heavy metal behaviour in freshwater lakes there should ideally be investigation of

- different systems of varying status (*e.g.* oligotrophic, eutrophic, acid)
- seasonal influences
- modes of element introduction and transport
- element associations in the water column
- depositional and redistributional processes and rates (*e.g.* via radionuclide and stable isotope studies) in the sediments
- concentration profiles and speciation in the overlying water, solid sediment, and pore water
- element/solid phase associations (*e.g.* via sequential chemical extraction procedures)
- authigenic mineral identification in the sediments

While opinions may differ as to whether particular sediment columns have been disturbed by mixing or the extent to which certain elements are vertically mobile, it is accepted that post-depositional redox-controlled cycling can affect the vertical profiles of manganese and iron.[57,58] The driving force is the microbiological decomposition of organic matter (represented as CH_2O) through bacterial utilization of O_2 and inorganic oxidizing agents in thermodynamically favoured sequence.[59]

$$\begin{array}{lcr}
 & & \Delta G \\
 & & (\text{kJ mol}^{-1}\,CH_2O) \\
CH_2O + O_{2(g)} \rightleftharpoons CO_{2(g)} + H_2O & -475 & (75) \\
5CH_2O + 4NO_3^- \rightleftharpoons 2N_{2(g)} + 4HCO_3^- + CO_{2(g)} + 3H_2O & -448 & (76) \\
CH_2O + 3CO_{2(g)} + H_2O + 2MnO_{2(s)} \rightleftharpoons 2Mn^{2+} + 4HCO_3^- & -349 & (77) \\
CH_2O + 7CO_{2(g)} + 4Fe(OH)_{3(s)} \rightleftharpoons 4Fe^{2+} + 8HCO_3^- + 3H_2O & -114 & (78) \\
2CH_2O + SO_4^{2-} \rightleftharpoons H_2S_{(g)} + 2HCO_3^- & -77 & (79) \\
2CH_2O \rightleftharpoons CH_{4(g)} + CO_{2(g)} & -58 & (80) \\
\end{array}$$

Upon dissolution of oxides and (oxy)hydroxides of manganese and iron under reducing conditions at depth, divalent cations of these elements can diffuse upwards through the pore waters to be oxidized and precipitated in near-surface oxic layers, leading to the characteristic

[56] J. G. Farmer, *Environ. Geochem. Health*, 1991, **13**, 76.
[57] W. Davison, *Earth Sci. Rev.*, 1993, **34**, 119.
[58] C. L. Bryant, J. G. Farmer, A. B. MacKenzie, A. E. Bailey-Watts, and A. Kirika, *Limnol. Oceanogr.*, 1997, **42**, 918.
[59] R. A. Berner, 'Early Diagenesis', Princeton University Press, Princeton, NJ, 1980.

near-surface enrichment of manganese and iron in the sediments of many well oxygenated lakes (see Section 2.4.3). Phosphorus and arsenic are also recycled, but in their case it is the negatively charged anions that are released into solution from an association with solid phase ferric (hydr)oxides, which dissolve under sufficiently reducing conditions. Upward diffusion of reduced species leads ultimately to surface enrichment via co-precipitation or adsorption of phosphate and arsenate on iron (hydr)oxides in surface layers. As a result, arsenic concentrations as high as $675\ mg\ kg^{-1}$ can be found near the surface of Loch Lomond sediments, compared with background values of $15–50\ mg\ kg^{-1}$ at depth.[60]

For lakes which have undergone significant acidification, it has been suggested that heavy metals could be released from surface sections by pH-dependent dissolution, resulting in sub-surface maxima in sedimentary heavy metal concentrations. In two Canadian acid lakes, however, Carignan and Tessier[61] found that downward diffusive fluxes of dissolved zinc from overlying waters into anoxic pore waters were responsible for the pronounced sub-surface sediment maxima in solid phase zinc, presumably as the insoluble sulfide.

Traditionally, heavy metals such as lead, zinc, copper, and cadmium have been considered diagenetically (*i.e.* as a consequence of chemically and biologically induced changes in prevailing sedimentary conditions, *e.g.* pε, anionic composition, *etc.*) immobile and fixed in the sediment after deposition, partly as a result of the formation of comparatively insoluble sulfides under reducing conditions at depth. This may well be an oversimplified view, as interactions and formation of complexes between metals and dissolved organic matter (*e.g.* humic substances) may maintain dissolved metal concentrations at levels greater than those predicted by simple solubility product calculations (see Sections 2.3.5 and 2.4.1).[62] A combination of geochemical modelling, incorporating interactions of metals with mineral phases and naturally occurring organic matter,[63] and the development of increasingly sophisticated analytical technology, such as DGT (diffusive gradients in thin films) for the high-resolution determination on a sub-millimetre scale of metals in sedimentary pore waters,[64] offers considerable promise for greater understanding of redox-driven cycling of trace elements in lakes.[65]

[60] J. G. Farmer and M. A. Lovell, *Geochim. Cosmochim. Acta*, 1986, **50**, 2059.
[61] R. Carignan and A. Tessier, *Science*, 1985, **228**, 1524.
[62] H. Elderfield and A. Hepworth, *Mar. Pollut. Bull.*, 1975, **6**, 85.
[63] E. Tipping, *Computer Geosci.*, 1994, **20**, 973.
[64] H. Zhang, W. Davison, S. Miller, and W. Tych, *Geochim. Cosmochim. Acta*, 1995, **59**, 4181.
[65] J. Hamilton-Taylor and W. Davison, 'Physics and Chemistry of Lakes', ed. A. Lerman, D. M. Imboden, and J. R. Gat, Springer, Berlin, 1995, p. 217.

Finally, it should be noted that sediment cores are becoming increasingly used to investigate historical perspectives of environmental contamination by anthropogenic organic compounds, such as organochlorine pesticides, polychlorinated biphenyls, polycyclic aromatic hydrocarbons, polychlorinated dibenzo-*p*-dioxins, and polychlorinated dibenzofurans.[66-68]

3.3.3 Onondaga Lake. Onondaga Lake, near Syracuse, New York State, has been called 'one of the world's most polluted lakes' and provides a textbook case of the impact of industrial processes on the environment.[69] Rich natural resources, especially the local brine springs, made the Syracuse region an ideal location for chemical manufacturing. Unfortunately, the exploitation of these, despite the implementation of significant technological advances and, for the time, environmentally responsible operation of factories, had a devastating impact upon the chemistry of the lake.

The Solvay Process, which was introduced for the manufacture of the industrially important chemical, 'soda ash', *i.e.* sodium carbonate (Na_2CO_3), along the west shore of Onondaga Lake in 1884, was cheaper and less polluting than the existing Leblanc Process. Essentially, it made use of two cheap and plentiful naturally occurring substances in the area—NaCl from the deep brine springs and $CaCO_3$ from limestone outcroppings—in a simple reaction which yielded two useful products.

$$CaCO_3 + 2NaCl \rightleftharpoons Na_2CO_3 + CaCl_2 \qquad (81)$$

In practice, this overall reaction required a series of individual reactions and chemical intermediates, one of the most important involving ammonia in a step to separate $NaHCO_3$ from NH_4Cl by fractional crystallisation at 273 K.

$$NH_{3(g)} + CO_{2(g)} + NaCl_{(aq)} + H_2O_{(l)} \rightleftharpoons NaHCO_{3(s)} + NH_4Cl_{(aq)} \qquad (82)$$

The $NaHCO_3$ was subsequently heated to yield Na_2CO_3 while NH_3, the most expensive compound used, was regenerated from the decomposition of NH_4Cl. So the very efficient Solvay Process, one of the first

[66] D. Swackhamer, 'Issues in Environmental Science and Technology', ed. R. E. Hester and R. M. Harrison, Royal Society of Chemistry, Cambridge, 1996, Vol. 6, p. 137.

[67] M. F. Simcik, S. J. Eisenreich, K. A. Golden, S.-P. Liu, E. Lipiatou, D. L. Swackhamer, and D. T. Long, *Environ. Sci. Technol.*, 1996, **30**, 3039.

[68] R. F. Pearson, D. L. Swackhamer, S. J. Eisenreich, and D. T. Long, *Environ. Sci. Technol.*, 1997, **31**, 2903.

[69] A. T. Schwartz, D. M. Bunce, R. G. Silberman, C. L. Stanitski, W. J. Stratton, and A. P. Zipp, 'Chemistry in Context', ed. C. Wheatley, W. C. Brown Publishers, Dubuque, IA, 1997, p. 219.

industrial processes to regenerate intermediates, maximized profits and, in theory, prevented environmental harm from unwelcome discharges.

Unfortunately, however, sales of $CaCl_2$ from the Onondaga Lake plant failed to match those of Na_2CO_3, with the result that excess $CaCl_2$ was allowed to be released into a tributary of the lake. In addition, substantial quantities of unmarketable salts (mainly $CaCl_2$) from the Solvay Process were dumped daily into the lake. Furthermore, waste slurries pumped into diked beds along the lake shoreline resulted in substantial leaching of Ca^{2+}, Na^+, and Cl^- ions into the lake by rainwater runoff. The Ca^{2+} ions reacted with CO_3^{2-} ions, resulting from

$$H_2O_{(l)} + CO_{2(g)} \rightleftharpoons H_2CO_{3(aq)} \rightleftharpoons 2H^+_{(aq)} + CO_{3(aq)}^{2-} \tag{83}$$

to precipitate $CaCO_3$ via

$$Ca^{2+}_{(aq)} + CO_{3(aq)}^{2-} \rightleftharpoons CaCO_{3(s)} \tag{84}$$

The effect of the deposition of $CaCO_3$, at up to 1.7×10^4 t per year, was to increase the sedimentation rate of the lake several-fold, with a large $CaCO_3$ delta where the major tributary, Ninemile Creek, flows into the lake, and create a layer of $CaCO_3$ about 1 m thick over the bottom sediments. As a result, Onondaga Lake became effectively a saturated solution of $CaCO_3$, with a pH of 7.6–8.2 and, most unusually for a lake in the north-eastern USA, immune to both acid precipitation and phosphate-induced eutrophication, the latter largely being avoided as a consequence of

$$3Ca^{2+}_{(aq)} + 2PO_{4(aq)}^{3-} \rightleftharpoons Ca_3(PO_4)_{2(s)} \tag{85}$$

Since the closure of the Solvay Plant in 1980, the water quality has improved as salt concentrations have decreased from 3500 mg l^{-1} to 1000 mg l^{-1}.

The other major industrial process which has polluted Onondaga Lake has been the production of sodium hydroxide (NaOH) and chlorine, two of the world's most important chemicals, from sodium chloride via the electrolysis of NaCl solution and subsequent amalgamation of the sodium metal produced with the mercury cathode [*i.e.* Na(Hg)], prior to spraying into water to yield NaOH. With a single electrolysis cell containing up to 4 t Hg and a single plant containing dozens of cells, the escape of mercury (*e.g.* through leakage) at a rate of 5–10 kg per day between 1946 and 1970 into the lake resulted in mercury-contaminated sediments (>40 mg kg^{-1}) and fish, much of the mercury in the latter being in the form of the highly toxic CH_3Hg^+. Despite the closure of the chlor-alkali plant in the late 1980s, mercury is still flowing into the lake from tributaries, although concentrations of soluble mercury compounds in the lake are declining.[69]

3.4 Nutrients in Water and Sediments

3.4.1 Phosphorus and Eutrophication. Eutrophication can be considered as the excessive primary production of algae and higher plants through enrichment of waters by inorganic plant nutrients, usually nitrogen and phosphorus. The latter, in the form of phosphate, is normally the limiting nutrient because the amount of biologically available phosphorus is small in relation to the quantity required for algal growth.[70] Sources of nutrients can be discrete (*e.g.* specific sewage outfall) or diffuse (*e.g.* farmland fertilizers). Eutrophic lakes, highly productive and often turbid owing to the presence of algae, can be contrasted with oligotrophic lakes, which exhibit low productivity and are clear in summer. There have been many examples of unsightly algal blooms affecting freshwater bodies throughout the world, from Lake Erie in North America to the Norfolk Broads in East Anglia, England.[29] Public concern has increased along with the reported incidences of toxicity of the bloom-forming organisms, in particular the cyanobacteria (blue-green algae), which have been implicated in fish fatalities, for example in Loch Leven, Scotland.[71]

The chemical form of phosphorus in the water column available for uptake by biota is important. The biologically available phosphorus is usually taken to be 'soluble reactive phosphorus (orthophosphate)', *i.e.* that which, upon acidification of a water sample, reacts with added molybdate to yield molybdophosphoric acid which is then reduced with $SnCl_2$ to the intensely coloured molybdenum blue complex and is determined spectrophotometrically ($\lambda_{max} = 882$ nm).[72] Reduction in inputs of phosphate, for example from point sources or by creating water meadows and buffer strips to contain diffuse runoff, has obviously been one of the major approaches to stemming eutrophication trends and encouraging the restoration of affected lakes. That this has not always been successful, however, can be attributed in many cases to the release/recycling of phosphorus previously deposited to and incorporated within the bottom sediments of the lake systems in question.[73]

[70] D. M. Harper, 'Eutrophication of Freshwaters: Principles, Problems and Restoration', Chapman & Hall, London, 1992.
[71] S. G. Bell and G. A. Codd, 'Issues in Environmental Science and Technology', ed. R. E. Hester and R. M. Harrison, Royal Society of Chemistry, Cambridge, 1996, Vol. 5, p. 109.
[72] J. Murphy and J. R. Riley, *Analyt. Chim. Acta*, 1962, **27**, 31.
[73] M. W. Marsden, *Freshwater Biol.*, 1989, **21**, 139.

The potential mobility and bioavailability of sedimentary phosphorus are to a large extent governed by the chemical associations and interactions of phosphorus with different sedimentary components.[74] Bearing in mind the removal or transport of phosphorus to the sediments, especially important phases are likely to be 'organic phosphorus' from deposited, dead, decaying biota, and 'sorbed orthophosphate' on inorganic particulates (*e.g.* iron (hydr)oxides). Many sequential extraction schemes have been developed to investigate phosphorus fractionation in lake sediments. Perhaps the most sophisticated is that of Psenner *et al.*[75] who identify, operationally define, and separate the following fractions: 'labile, loosely bound or adsorbed' (NH_4Cl-extractable); 'reductant-soluble, mainly from iron (hydr)oxide surfaces' (buffered dithionite-extractable); 'adsorbed to metal oxides (*e.g.* Al_2O_3)' (NaOH-extractable), subsequently distinguishable from 'organic' (also NaOH-extractable); 'apatite-bound' (HCl-extractable); and 'residual' (persulfate-digestible).

Many such studies of sedimentary phosphorus profiles, also incorporating pore water measurement of soluble reactive phosphate, have demonstrated that redox-controlled dissolution of iron (hydr)oxides under reducing conditions at depth releases orthophosphate to solution. This then diffuses upwards (and downwards) from the pore water maximum to be re-adsorbed or co-precipitated with oxidized Fe^{III} in near-surface oxic sections. The downwards decrease in solid phase 'organic' phosphorus indicates increasing release of phosphorus from deposited organic matter with depth, some of which will become associated with hydrous iron and other metal oxides, added to the pool of mobile phosphorus in pore water or contribute to 'soluble unreactive phosphorus'. The characteristic reactions involving inorganic phosphorus in the sediments of Toolik Lake, Alaska, are shown in Figure 7.[76] If, at depth, the concentrations of Fe^{2+} and phosphate are high enough, authigenic vivianite ($Fe_3(PO_4)_2.8H_2O$) may precipitate out.

With redox control largely responsible for phosphorus mobility in sediments, what might the consequences of oxygen depletion in the hypolimnion be? If conditions in the surface sediments are not sufficiently oxidizing to precipitate iron (hydr)oxides and thereby adsorb the phosphate (*i.e.* the redox boundary for iron may be in the overlying water column), the phosphate from previously deposited sediments would stream off into the water column and promote eutrophication. This process is called internal loading of phosphorus.

[74] J. G. Farmer, A. E. Bailey-Watts, A. Kirika, and C. Scott, *Aquat. Conserv.*, 1994, **4**, 45.
[75] R. Psenner, B. Bostrom, M. Dinka, K. Pettersson, R. Pucsko, and M. Sager, *Arch. Hydrobiol. Ergebn. Limnol.*, 1988, **30**, 98.
[76] J. C. Cornwell, *Arch. Hydrobiol.*, 1987, **109**, 161.

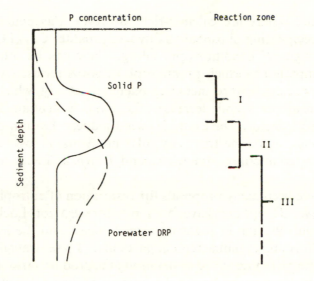

$$I \quad 22Fe^{2+} + 66H_2O + HPO_4^{2-} \rightleftharpoons$$

$$[Fe(OH)_3]_{22} - HPO_4^{2-}(s) + 66H^+ + 22e^-$$

$$II \quad [Fe(OH)_3]_{22} - HPO_4^{2-}(s) + 66H^+ + 22e^- \rightleftharpoons$$

$$22Fe^{2+} + 66H_2O + HPO_4^{2-}$$

$$III \quad 3Fe^{2+} + 2PO_4^{3-} + 8H_2O \rightleftharpoons Fe_3(PO_4)_2 \cdot 8H_2O(s)$$

Figure 7   *Characteristic reactions involving phosphorus in the sediments of Toolik Lake, Alaska. The primary processes controlling porewater phosphorus concentrations are adsorption to and desorption from iron oxyhydroxides and the precipitation of authigenic vivianite*[76]
(Reproduced with permission from *Arch. Hydrobiol.*, **109**, 161. © 1987 E. Schweizerbart'sche Verlagsbuchhandlung)

Redox control may not be the only process affecting release of phosphorus from sediments. During the enhanced photosynthesis of algal blooms, the pH of lake water increases as CO_2 is used up and HCO_3^- increases. Thus, in summer, both Lough Neagh[77] in Northern Ireland and Lake Glanningen[78] in Sweden have shown an increase in water phosphate concentration. It seems likely that OH^- is exchanging with sorbed phosphate in alkaline lakes, thus releasing phosphorus from association with iron and aluminium (hydr)oxides (see Section

[77] B. Rippey, 'Interactions between Sediments and Freshwater', ed. H. L. Golterman, Dr. W. Junk B. V., The Hague, 1977, p. 349.
[78] S-O. Ryding and C. Forsberg, 'Interactions between Sediments and Freshwater', ed. H. L. Golterman, Dr. W. Junk B. V., The Hague, 1977, p. 227.

2.4.3). If the waters are calcium-rich, however, this could have the effect of precipitating phosphate as hydroxyapatite $(Ca_{10}(PO_4)_6(OH)_2)$ under the high pH conditions prevailing. There can be other factors, such as temperature, which promote phosphorus release. An increase in temperature can lead to increased bacterial activity, which increases oxygen consumption and decreases the redox potential. Turbulence may also be a factor. For example, wind-induced bottom currents in shallow lakes could destroy any pH gradients across a buffered sediment–water interface and re-suspend P-rich sediment in water of high pH.[79]

There have been many proposals for restoration of eutrophic lakes.[80] For example, 34 options have been put forward for Loch Leven,[81] Scotland, with the aim of reducing algal biomass and the incidence of the bloom-forming cyanobacteria in particular. These strategies fall into two categories: (i) those aimed at stemming the production of algae in the first place, including reduction in the supplies of light and nutrients, and (ii) those aimed at reducing existing algal biomass, including physical methods such as increased flushing of the loch and harvesting of blooms, chemical treatment with algicides, and biological methods involving viruses, parasitic fungi and grazing protozoans, rotifers, and micro-crustaceans. Thus far, progress has largely been restricted to reducing external inputs of phosphorus from point sources to the loch, an approach which could perhaps be supplemented by the creation of water meadows and/or buffer strips to reduce inputs from diffuse runoff. Elsewhere, there was a surge in wetlands construction in the 1980s, using a mixture of plants to clean water contaminated with nitrates and phosphates.[82] One of the largest wetlands restoration projects is planned for the Florida Everglades, to the south and east of the rich farmlands around Lake Okeechobee, where runoff waters enriched in phosphorus from fertilizers have disrupted the flora and fauna. As water flows through cattails and sawgrass, phosphorus concentrations are expected to decline from 170 μg l^{-1} to 50 μg l^{-1}. It should be noted that the addition of ferric sulfate or chloride to mop up phosphate in eutrophic lakes,[29] may not work in the long term as there may be subsequent release of deposited phosphate from sediments under reducing conditions.

[79] L. Hakanson and M. Jansson, 'Principles of Lake Sedimentology', Springer, Berlin, 1983.
[80] A. J. D. Ferguson, M. J. Pearson, and C. S. Reynolds, 'Issues in Environmental Science and Technology', ed. R. E. Hester and R. M. Harrison, Royal Society of Chemistry, Cambridge, 1996, Vol. 5, p. 27.
[81] A. E. Bailey-Watts, I. D. M. Gunn, and A. Kirika, 'Loch Leven: Past and Current Water Quality and Options for Change', Report to the Forth River Purification Board, Institute of Freshwater Ecology, Edinburgh, 1993.
[82] P. Young, *Environ. Sci. Technol.*, 1996, **30**, 292A.

A novel use of phosphate to counter acidification has emerged in recent years.[83] Traditional remedial methods which have been adopted include the direct liming of lakes, *e.g.* in Sweden,[84] or of catchments, *e.g.* around Loch Fleet in south-west Scotland.[85] Although helpful, such approaches based on neutralization are costly and usually need to be repeated at regular intervals. Furthermore, the resulting Ca-rich waters may turn out to support biota quite different from those found in natural softwater lakes. An alternative approach has recently been tried by Davison and co-workers on Seathwaite Tarn, an upland reservoir in the English Lake District.[83] Phosphate fertilizer was added to stimulate primary productivity and thereby increase the assimilation of nitrate. This generates base according to the equation:

$$106CO_{2(g)} + 138H_2O + 16NO_3^- \rightleftharpoons (CH_2O)_{106}(NH_3)_{16} + 16OH^- + 138O_{2(g)}$$

$$(86)$$

As concentrations of nitrate are increasingly high in acid waters, the addition of modest amounts of phosphate may generate sufficient base to combat acidity without inducing excessive productivity. An increase in pH of 0.5 and a marked increase in biological productivity at all levels were observed over the three-year period of the experiment. In the longer term, additional quantities of base should be generated by the anoxic decomposition of organic material accumulating on the lake bed (through the dissimilative reduction of inorganic oxidants such as nitrate, sulfate or iron (hydr)oxides present in the sediments—equations (76) to (79)). If oxygen is the electron acceptor there is no net gain of base (equation (75)), there is no advantage in adding nitrate because 1 mole of nitrate is required to generate 1 mole of base, and the generated base (contributing to alkalinity) should not be confused with the temporary rise in pH associated with CO_2 consumption which affects neither alkalinity nor acidity.

3.4.2 Nitrate in Groundwater.

3.4.2 Nitrate in Groundwater. Principal sources of nitrate in water are runoff and drainage from land treated with agricultural fertilizers and also deposition from the atmosphere as a consequence of NO_x released from fossil fuel combustion. The nitrogen present in soil organic matter may also be released as nitrate through microbial action. Nitrate's role as a nutrient contributes significantly to blooms of algae, which upon death

[83] W. Davison, D. G. George, and N. J. A. Edwards, *Nature*, 1995, **377**, 504.
[84] P. Nyberg and E. Thornelof, *Water, Air, Soil Pollut.*, 1988, **41**, 3.
[85] 'Restoring Acid Waters: Loch Fleet, 1984–1990', ed. G. Howells and T. R. K. Dalziel, Elsevier, London, 1992.

are decomposed first by aerobic bacteria, thereby depriving fish and other organisms of oxygen.[29,81]

There has long been concern expressed over the presence of nitrate in drinking water at concentrations exceeding the EC guideline of 50 mg l^{-1} because of the risk of methaemoglobinaemia (blue baby syndrome). Here, NO_3^- is reduced in the baby's stomach to NO_2^- which, on absorption, reacts with oxyhaemoglobin to form methaemoglobin. There have also been fears over the reaction of NO_2^- with secondary amines from the breakdown of meat or protein to produce carcinogenic N-nitroso compounds but there is as yet no clear evidence of a link between stomach cancer and nitrate in water.[86] Nevertheless, there is growing concern over the contamination of groundwater by nitrate, for example in regions such as the Sierra Pelona Basin, California.[87] There the local groundwaters, the major source of drinking water from private water wells located near each private residence in the rural communities, often exceed the US EPA maximum contaminant level for drinking water of 10 mg l^{-1} (NO3-N). Isotopic investigations, based upon $^{15}N/^{14}N$, have confirmed the predominance of anthropogenic, organic human and/or animal waste and decay of irrigation-enhanced vegetation rather than natural nitrate sources.

3.5   Organic Matter and Organic Chemicals in Water

3.5.1   BOD and COD.   The solubility of oxygen in water in equilibrium with the atmosphere at 25 °C is 8.7 mg l^{-1}. Causes of oxygen depletion include decomposition of biomass (*e.g.* algal blooms) and the presence of oxidizable substances (*e.g.* sewage, agricultural runoff, factory effluents) in the water. The addition of oxidizable pollutants to streams produces a typical sag in the dissolved oxygen concentration. The degree of oxygen consumption by microbially mediated oxidation of organic matter in water is called the Biochemical (or Biological) Oxygen Demand (BOD) (equation (75)). Another index is the Chemical Oxygen Demand (COD), which is determined by using the powerful oxidizing agent, dichromate ($Cr_2O_7^{2-}$), to oxidize organic matter,

$$2Cr_2O_7^{2-} + 3CH_2O + 16H^+ \rightleftharpoons 4Cr^{3+} + 3CO_{2(g)} + 11H_2O \qquad (87)$$

followed by back-titration of excess added dichromate with Fe^{2+},

$$Cr_2O_7^{2-} + 6Fe^{2+} + 14H^+ \rightleftharpoons 2Cr^{3+} + 6Fe^{3+} + 7H_2O \qquad (88)$$

---

[86] T. M. Addiscott, 'Issues in Environmental Science and Technology', ed. R. E. Hester and R. M. Harrison, Royal Society of Chemistry, Cambridge, 1996, Vol. 5, p. 1.
[87] A. E. Williams, L. J. Lund, J. A. Johnson, and Z. J. Kabala, *Environ. Sci. Technol.*, 1998, **32**, 32.

As $Cr_2O_7^{2-}$ oxidizes substances not oxidized by O_2, the COD is usually greater than the BOD and to some extent overestimates the threat posed to oxygen content.

3.5.2 Synthetic Organic Chemicals. A large number of organic compounds are synthesized for agricultural use, mainly as pesticides, and for industrial use in solvents, cleaners, degreasers, petroleum products, plastics manufacture, *etc.*[88] Many organic micropollutants percolate into the soil and accumulate in aquifers or surface waters. They can contaminate drinking water sources via agricultural runoff to surface waters or percolation into groundwaters, industrial spillages to surface waters and groundwaters, runoff from roads and paved areas, industrial waste water effluents leaching from chemically treated surfaces, domestic sewage effluents, atmospheric fallout, and as leachate from industrial and domestic landfill sites.[88]

Pesticides. Many pesticides in aquifers have resisted degradation and are more likely to persist there because of reduced microbial activity, absence of light, and lower temperatures.[89] Numerous aquifers, for example in eastern England, have been found to exceed the Maximum Admissible Concentration (MAC) guidelines for total pesticides in drinking water ($0.5\ \mu g\ l^{-1}$), due primarily to the presence of herbicides of the carboxy acid and basic triazine groups. Gray[88] has listed the 12 pesticides most often found in UK drinking waters in two groups, frequently occurring (atrazine, chlortoluon, isoproturon, MCPA, mecoprop, simazine) and commonly occurring (2,4-D, dicamba, dichlorprop, dimethoate, linuron, 2,4,5-T). A major seasonal source of insecticide in freshwaters is sheep dipping. With organophosphorus compounds, which themselves replaced the more persistent organochlorine insecticides such as lindane, now falling out of favour because of health risks to users, the use of synthetic pyrethroids as alternatives has resulted in the death of aquatic organisms in UK rivers. Pyrethroids, which do not persist in the environment and are largely non-toxic to mammals, are toxic to invertebrates and are estimated to be at least 100 times more toxic in the aquatic environment than organophosphorus pesticides. It appears to be the pouring of waste dip into holes in the ground that has caused the problem.[90] To protect the aquatic environment, the Environment Agency in England assesses water quality against Environmental Quality Standards (EQSs). Defined as the concentration of a substance which must not be exceeded within the aquatic environment in order to

[88] N. F. Gray, 'Drinking Water Quality', John Wiley & Sons, Chichester, 1994.
[89] K. R. Eke, A. D. Barnden, and D. J. Tester, 'Issues in Environmental Science and Technology', ed. R. E. Hester and R. M. Harrison, Royal Society of Chemistry, Cambridge, 1996, Vol. 5, p. 43.
[90] F. Pearce, *New Scientist*, 11 January, 1997, 4.

protect it for its recognized uses, EQSs are specific to individual substances, including pesticides.

Polychlorinated Biphenyls (PCBs).   The Great Lakes, the largest body of freshwater in the world, with long hydraulic residence times, long food chains, and multiple sources of PCBs, have been the focal point of PCB research in aquatic systems. The distribution of PCBs between the dissolved and particulate phases is dependent on the concentration of suspended particulate matter, dissolved and particulate organic carbon concentrations, and the extent to which the system is at equilibrium. In the 1990s, water concentrations of up to $0.6 \, \mathrm{ng \, l^{-1}}$ have been observed in the most contaminated lakes (Michigan, Erie, and Ontario), which may be compared with the US EPA Great Lakes Water Quality Guidance criteria of $0.017 \, \mathrm{ng \, l^{-1}}$. Sediment concentrations peaked in the 1970s, the maximum period of PCB production in the USA. Following a ban upon their North American production during the 1970s, there was a significant decline in the PCB concentrations of Great Lakes' fish from the mid-1970s to the mid-1980s but the rate of decrease has since slowed or stopped for all lakes.[66] In common with other organochlorine compounds, it is the stability, persistence, volatility, and lipophilicity of PCBs which lead to considerable biomagnification along the food chain, often far from the place of release.

Endocrine Disruptors.   Concern has recently been expressed over the possible role of various synthetic organic chemicals (*e.g.* organochlorine pesticides, PCBs, phthalates, alkylphenolethoxylates, alkylphenols, *etc.*) as disruptors of endocrine systems of wildlife and perhaps even of humans.[91] Reproductive changes in male alligators from Lake Apopka, Florida, embryonic death, deformities, and abnormal nesting behaviour in fish-eating birds in the Great Lakes region and the occurrence of hermaphroditic fish near sewage outfalls on some British rivers have been attributed to postulated oestrogenic or anti-androgenic effects of some of these chemicals.[90–92] Most alkylphenolethoxylates (APEs), which are used as detergents, emulsifiers, wetting agents, and dispersing agents, enter the aquatic environment after disposal in wastewater. During biodegradation treatment of the latter, the APEs are transformed into more toxic, short chain ethoxylates, alkylphenol carboxylic acids, and alkylphenols.[93] The threshold concentration of nonylphenol in water for production of the female egg

[91] T. Colborn, J. P. Meyers, and D. Dumanoski, 'Our Stolen Future', Little, Brown, Boston, 1996.
[92] M. Lee, *Chem. Br.*, 1996, **32**, 5.
[93] R. Renner, *Environ. Sci. Technol.*, 1997, **31**, 316A.

yolk protein, vitellogenin, in male rainbow trout is $\sim 10 \ \mu g \ l^{-1}$.[94] It also appears that nonylphenol concentrations in many European rivers are up to 10 times higher than those found in US rivers, which are typically $< 1 \ \mu g \ l^{-1}$.[93] It has been suggested, however, that natural oestrogens (*e.g.* oestrone and 17β-oestradiol) may be responsible for the observed effects, perhaps after conversion of inactive excreted metabolites to active forms by bacterial enzyme action during sewage treatment.[95]

Industrial Solvents. Six are widely used in the UK—dichloromethane (DCM), trichloromethane, 1,1,1-trichloroethane (TCA), tetrachloromethane, trichloroethene (TCE) and tetrachloroethene or perchloroethylene (PCE).[88] Although there has been a steady decrease in use over the past 20 years, four (TCE, PCE, TCA, and DCM) are frequently found in drinking water. The major pollution threat is when solvents are discharged directly into or onto the ground, due to illegal disposal or accidental spillage. Since groundwater is not exposed directly to the atmosphere, the solvents are not able to escape by evaporation, so it is the aquifers that are most at risk from these chemicals. Pollution of groundwater by industrial solvents is a very widespread problem (*e.g.* Netherlands, Italy, USA, UK[96]) with aquifers underlying urbanized areas such as Milan, Birmingham, London or New Jersey containing high concentrations of all solvents. The stability of TCE in particular and the inability of these solvents to evaporate readily mean that such contamination will last for many decades.[88]

A typical example of groundwater contaminated by synthetic organic chemicals from jet fuel and degreasing solvents (*e.g.* halogenated methanes, ethanes, and ethenes, including TCE, PCE, and vinyl chloride) can be found at Otis Air Force Base on Cape Cod, Massachusetts, USA, where the Cape's sole-source aquifer has been affected. The groundwater plumes, some in excess of 5 km in length, moving at ~ 0.5 m per day, contaminate 30 million litres of the Cape's drinking water every day. New reactive wall (Ni-Fe) technology is being employed to try to reduce concentrations of 5–150 $\mu g \ l^{-1}$ TCE and PCE below the local drinking water limits of 0.5 $\mu g \ l^{-1}$.[97]

MTBE. Methyl *tert*-butyl ether (MTBE), an oxygen-containing compound used as a fuel additive since the 1970s, moves quickly through

[94] S. Jobling, D. Sheahan, J. A. Osborne, P. Matthiessen, and J. P. Sumpter, *Environ. Toxicol. Chem.*, 1996, **15**, 194.
[95] J. Kaiser, *Science*, 1996, **274**, 1837.
[96] M. O. Rivett, D. N. Lerner, and J. W. Lloyd, *Water Environ. Management*, 1990, **4**, 242.
[97] E. L. Appleton, *Environ. Sci. Technol.*, 1996, **30**, 536A.

soil, is highly soluble and does not biodegrade easily. It has been found in shallow groundwater in Denver, New England, and elsewhere in the USA, most notably in Santa Monica, California, where a 1996 spill of MTBE-containing gasoline resulted in the contamination of wells at concentrations as high as 610 μg l^{-1}, well above the state's advisory limit of 35 μg l^{-1}.[98] In a preliminary assessment of the occurrence and possible sources of MTBE in groundwater in the United States in 1993/94, as part of the US Geological Survey's National Water-Quality Assessment programme, Squillace and co-workers found that, out of 60 volatile organic chemicals determined, MTBE was the second most frequently detected chemical in samples of shallow ambient groundwater from urban areas.[99] MTBE was detected above 0.2 μg l^{-1} in 17% of shallow groundwater samples from urban areas, 1.3% of shallow groundwater from agricultural areas, and 1.0% of deeper groundwater samples from major aquifers. Only 3% of the shallow wells sampled in urban areas had MTBE concentrations >20 μg l^{-1}, the estimated lower limit of the US EPA draft drinking water health advisory. As MTBE generally was not found in shallow urban groundwater, along with benzene, toluene, ethylbenzene, or xylene, which are commonly associated with petrol spills, it was concluded that possible sources of MTBE in groundwater include leaking storage tanks and non-point sources such as recharge of precipitation and stormwater runoff. As MTBE is thought to be potentially carcinogenic to humans, its use poses an interesting dilemma for regulators, given that it helps to reduce carbon monoxide emissions from cars.

4   TREATMENT

4.1   Purification of Water Supplies

Other than those already discussed in Section 3.1.2, *i.e.* coagulation of colloids using Al$_2$(SO$_4$)$_3$, and in Section 3.2.2, *i.e.* pH adjustment using lime, measures for the purification of drinking water supplies are largely outwith the scope of this Chapter. The main unit processes are sum-marised in Table 6.[100]

[98] M. Cooney, *Environ. Sci. Technol.*, 1997, **31**, 269A.
[99] P. J. Squillace, J. S. Zogorski, W. G. Wilber, and C. V. Price, *Environ. Sci. Technol.*, 1996, **30**, 1721.
[100] H. Fish, 'Understanding Our Environment: An Introduction to Environmental Chemistry and Pollution', ed. R. M. Harrison, Royal Society of Chemistry, Cambridge, 2nd Edn., 1992, p. 53.

Table 6 *Summary of treatment processes used in purification of public water supply*

Process	Purpose
Raw water storage (short term)	Sedimentation. Die-off of faecal organisms. Balancing of intake water quality. Raw water reserve
Raw water storage (long term)	Oxidation of organic matter. Partial removal of NO_3^-, HCO_3^-, PO_4^{3-}, SiO_2, by algal uptake
Chemical precipitation using $Al_2(SO_4)_3$ or activated SiO_2, or Fe salts plus polyelectrolytes	Coagulation, flocculation, and settlement of turbidity and colour
Microstraining	Straining through very fine mesh rotary screens
Rapid filtration	Rapid up- or down-flow filtration through sand
Slow sand filtration	Filtration plus bio-oxidation by slow gravity downward flow
Chlorination and/or ozonation or UV	Disinfection
Softening by lime, lime-soda, or ion exchange	Removal of Ca and Mg hardness (no longer fashionable)
Activated carbon treatment by powder addition before filtration or passage through active carbon filters	Reduction in residual organic matter
Desalination by flash distillation or reverse osmosis	Production of freshwater from saltwater or superpurification of wastewater

Adapted with permission from 'Understanding Our Environment', 2nd Edn., p. 85. © 1992 The Royal Society of Chemistry.

4.2 Waste Treatment

Methods for the conventional primary, secondary, and tertiary treatment of wastewaters and sewage are listed in Table 7. Less conventional treatments, such as constructed wetlands where the localized diffusion of oxygen and release of organic nutrients by the roots of mixed plants enables fast degradative activity by both aerobic and anaerobic organisms as well as plant assimilation of pollutants, have been introduced in recent years.[101] In addition, reedbed systems employing a single plant species, most commonly *Phragmites australis* in Europe, are becoming increasingly popular, especially for industrial applications, as they are simple to operate and cheap to maintain.[101]

[101] R. Cobban, D. Gregson, and P. Phillips, *Chem. Br.*, 1998, **34**, 40.

Table 7 *Conventional methods of waste treatment*

Treatment stage	Method
Preliminary treatment	Coarse solid and grit screening
Primary treatment	Suspended solids sedimentation
Secondary treatment—relies on micro-organisms extracting and chemically transforming nutrients from primary effluents	Anaerobic and aerobic digestion Activated sludge processes
Tertiary treatment—chemically, physically, or biologically removing nutrients, including phosphorus and nitrogen. Because these methods necessitate removing minutely sized trace quantities of compounds from large volumes of wastewater they are usually more expensive	Membrane filtration—removing particles that are too small for ordinary filtration. Generally this involves the pressurized flow of liquid across a membrane. There are three types of membrane filtration systems, which differ mainly in the size of particles they remove. (i) Microfiltration retains and concentrates particles the size of paint pigments and bacteria (ii) Ultrafiltration is used to reduce the biochemical oxygen demand of wastewater by removing substances such as sugars, fats, oils, and greases (iii) Reverse osmosis removes the smallest particles and can be used to retain substances such as dissolved salts or metal ions Ion exchange resins—used to exchange ions from solution Extended aeration with microbial uptake—aerobic micro-organisms are used to digest the last remaining nutrients found in wastewater Precipitation with iron or aluminium—mixing wastewater with iron or aluminium sulfates causes a chemical reaction forcing pollutants to come out of solution so that they can be collected by sedimentation or filtration

Adapted with permission from *Chem. Br.*, **34**(2), 40. © 1998 The Royal Society of Chemistry.

Questions

1. Calculate the ionic strength of a solution which is 0.01 M with respect to potassium sulfate and 0.002 M with respect to magnesium chloride. Use this value of ionic strength and the Güntleberg approximation to determine the mixed acidity constant, K', for methanoic acid ($K = 10^{-3.75}$).

2. Construct a plot to show how $-\log\{\text{activity}\}$ varies with pH for all species in one litre of groundwater containing 3×10^{-4} moles of 4-chlorophenol ($K' = 10^{-9.18}$). Annotate the plot to show the equilibrium pH of the groundwater.

3. Calculate the equilibrium pH of rainwater in equilibrium with $CO_{2(g)}$ at a partial pressure of 0.00035 atm. (Hint: the charge balance equation can be approximated by $\{H^+\} = \{HCO_3^-\}$)

	$\log K$ (288 K)
$CO_{2(g)} + H_2O \rightleftharpoons H_2CO_3^*$	-1.34
$H_2CO_3^* \rightleftharpoons H^+ + HCO_3^-$	-6.35
$HCO_3^- \rightleftharpoons H^+ + CO_3^{2-}$	-10.33

4. Construct a $p\varepsilon$–pH diagram for arsenic species present in natural waters. Assume that sulfur species are absent and that the total dissolved concentration of arsenic is 10^{-6} M. From the diagram, predict the major arsenic species in:

(i) groundwaters ($p\varepsilon$ -2 to $+6$, pH 4.8 to 8)
(ii) alkaline surface waters ($p\varepsilon$ $+2.5$ to $+10$, pH 7.2 to 9)

	$\log K$ (298 K)
$H_3AsO_4 \rightleftharpoons H_2AsO_4^- + H^+$	-2.24
$H_2AsO_4^- \rightleftharpoons HAsO_4^{2-} + H^+$	-6.76
$HAsO_4^{2-} \rightleftharpoons AsO_4^{3-} + H^+$	-11.60
$H_3AsO_3 \rightleftharpoons H_2AsO_3^- + H^+$	-9.23
$H_2AsO_3^- \rightleftharpoons HAsO_3^{2-} + H^+$	-12.10
$H_3AsO_3 + H_2O \rightleftharpoons H_3AsO_4 + 2H^+ + 2e^-$	-19.46
$H_3AsO_3 + H_2O \rightleftharpoons H_2AsO_4^- + 3H^+ + 2e^-$	-21.70
$H_3AsO_3 + H_2O \rightleftharpoons HAsO_4^{2-} + 4H^+ + 2e^-$	-28.46
$H_2AsO_3^- + H_2O \rightleftharpoons HAsO_4^{2-} + 3H^+ + 2e^-$	-19.23
$H_2AsO_3^- + H_2O \rightleftharpoons AsO_4^{3-} + 4H^+ + 2e^-$	-30.83
$HAsO_3^{2-} + H_2O \rightleftharpoons AsO_4^{3-} + 3H^+ + 2e^-$	-18.73
$As_{(s)} + 3H_2O \rightleftharpoons H_3AsO_3 + 3H^+ + 3e^-$	-12.17
$As_{(s)} + 3H_2O \rightleftharpoons H_2AsO_3^- + 4H^+ + 3e^-$	-21.40
$As_{(s)} + 3H_2O \rightleftharpoons HAsO_3^{2-} + 5H^+ + 3e^-$	-33.50

5. Draw, describe, and compare the concentration profiles of the chemical species of manganese, iron, arsenic, and phosphorus likely to be found in the solid and solution phases of unmixed sediments at the bottom of a seasonally anoxic, eutrophic freshwater lake during (i) summer and (ii) winter.

6. With respect to human health and ecological impact, discuss the short and long term consequences of the products of the 20th Century synthetic organic chemical industry upon the quality of surface waters and groundwaters.

7. Explain the chemistry underlying (i) the major detrimental effects of metal mining upon rivers and groundwaters and (ii) associated preventive or remedial measures.
8. Critically discuss the effectiveness of measures to counter the effects of (i) acidification and (ii) eutrophication in freshwater lakes.

FURTHER READING

E. K. Berner and R. A. Berner, 'Global Environment: Water, Air and Geochemical Cycles', Prentice-Hall, Englewood Cliffs, NJ, 1996.

'Freshwater Quality: Defining the Indefinable?', ed. P. J. Boon and D. L. Howell, The Stationery Office, Edinburgh, 1997.

J. I. Drever, 'The Geochemistry of Natural Waters', Prentice-Hall, Englewood Cliffs, NJ, 1997.

U. Forstner, 'Contaminated Sediments', Springer, Berlin, 1989.

A. J. Horne, 'Limnology', McGraw-Hill, New York, 2nd Edn., 1994.

A. G. Howard, 'Aquatic Environmental Chemistry', Oxford University Press, Oxford, 1998.

D. Langmuir, 'Aqueous Environmental Geochemistry', Prentice-Hall, Englewood Cliffs, NJ, 1997.

'The Fresh Waters of Scotland', ed. P. S. Maitland, P. J. Boon, and D. S. McLusky, John Wiley & Sons, Chichester, 1994.

C. F. Mason, 'Biology of Freshwater Pollution', Longman, 3rd Edn., 1996.

'Groundwater Contaminants and their Migration', ed. J. Mather, D. Banks, S. Dumpleton and M. Fermor, Geological Society, London, Special Publication No.128, 1998.

F. M. Morel and J. G. Hering, 'Principles and Applications of Aquatic Chemistry', John Wiley & Sons, New York, 1993.

J. Pankow, 'Aquatic Chemistry Concepts', Lewis, Michigan, 1991.

W. Salomons and U. Forstner, 'Metals in the Hydrocycle', Springer, Berlin, 1984.

W. Stumm, 'Chemistry of the Solid–Water Interface', John Wiley & Sons, New York, 1992.

W. Stumm and J. J. Morgan, 'Aquatic Chemistry', John Wiley & Sons, 3rd Edn., New York, 1996.

CHAPTER 4

The Oceanic Environment

STEPHEN J. DE MORA

1 INTRODUCTION

The World Ocean is a complex mixture containing all the elements, albeit in minute amounts in some cases. Seawater contains dissolved gases and, apart from some exceptional environments, is consequently both well oxygenated and buffered at a pH of about 8. There are electrolytic salts, the ionic strength of seawater being approximately 0.7, and multitudinous organic compounds in solution. At the same time, there is a wide range of inorganic and organic particles in suspension. These comfortable distinctions become quite confused in seawater because some molecules present in true solution are sufficiently large to be retained by a filter. Moreover, surface adsorption allows particles to scavenge dissolved elements and accumulate coatings of organic material from solution. Some elements, particularly those with biochemical functions, may be rapidly removed from solution. Concurrently, reactions involving geological time scales are proceeding slowly. Yet despite this apparent complexity, many aspects of the composition of seawater and chemical oceanography can now be explained with recourse to the fundamental principles of chemistry. This chapter serves to bridge the gap between those with environmental expertise and those with a traditional chemical background.

1.1 The Ocean as a Biogeochemical Environment

A traditional approach utilized in geochemistry, and now also in environmental chemistry, is to consider the system under investigation a reservoir. For a given component, the reservoir has sources (inputs) and sinks (outputs). The system is said to be at equilibrium, or operating under steady-state conditions, when a mass balance between inputs and outputs is achieved. An imbalance could signify that an important source or sink has been ignored. Alternatively, the system may be perturbed, possibly anthropogenically mediated, and therefore be changing toward a new equilibrium state.

Processes within the reservoir that affect the temporal and spatial distribution of a given component are transportation and transformations. Both physics and biology within the system play a role. Clearly, transport effects are dominated by the hydrodynamic regime. Although transformations could involve chemical (dissolution, redox reactions, speciation changes) or geological (sedimentation) processes, biological activity can control nutrient and trace metal distributions. Furthermore, the biota influences concentrations of O_2 and CO_2 which in turn determine the pH and pε (*i.e.* the redox potential see p. 157), respectively. For these reasons, some fundamental aspects of descriptive physical and biological oceanography are included in this chapter.

In terms of biogeochemical cycling, the ocean constitutes a large reservoir. The surface area is 361.11×10^6 km^2, encompassing nearly 71% of the earth's surface. The average depth is 3.7 km, but depths in the submarine trenches can exceed 10 km. The ocean contains about 97% of the water in the global hydrological cycle. A schematic representation of the ocean reservoir is presented in Figure 1. The material within it can be operationally defined, usually based on filtration, as dissolved or particulate. The ocean is divided into two layers, with distinct surface and deep waters. The boundary regions are also distinguished, as the composition in these regions can be quite different from that of bulk seawater. Furthermore, interactions within these environments can alter the mass transfers across the boundary. The rationale for such features will be presented in subsequent sections.

Material supplied to the ocean originates from the atmosphere, rivers, glaciers, and hydrothermal waters. The relative importance of these pathways depends upon the component considered and geographic location. River runoff commonly constitutes the most important source. Transported material may be either dissolved or particulate, but discharges are into surface waters and confined to coastal regions. Hydrothermal waters are released from vents on the sea floor. Such hydrothermal waters are formed when seawater circulates into the fissured rock matrix and, under conditions of elevated temperature and pressure, compositional changes in the aqueous phase occur due to seawater–rock interactions. This is an important source of some elements, such as Li, Rb, and Mn. The atmosphere supplies particulate material globally to the surface of the ocean. In recent years, this has been the most prominent pathway to the world ocean for Pb, identified by its isotopic signature as originating from petrol additives. Wind-borne

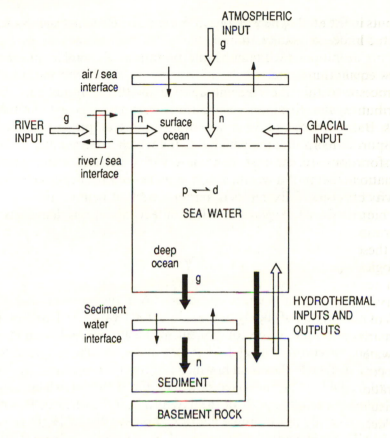

Figure 1 *A schematic representation of the ocean reservoir. The source and sink fluxes are designated as* g *and* n, *referring to gross and net fluxes, thereby indicating that interactions within the boundary regions can modify the mass transfer. Within seawater, the* $p \rightleftharpoons d$ *term signifies that substances can undergo particulate–dissolved interactions. However, it must be appreciated that several transportation and transformation processes might be operative* (From Chester[1])

transport is greatest in low latitudes and the Sahara Desert is known to act as an important source of dust. In contrast, glacial activity makes little impact on the world ocean. Glacier-derived material tends to comprise physically weathered rock residue, which is relatively insoluble. In addition, the input is largely confined to polar regions, with Antarctica responsible for approximately 90% of the material.

Sedimentation acts as the major removal mechanism. This is essentially a geological process in coastal environments but is biologically mediated in the open sea through the sinking of shells of microorganisms and faecal pellets. However, volatilization and subsequent

[1] R. Chester, 'Marine Geochemistry', Unwin Hyman, London, 1990, p. 698.

evasion to the atmosphere can be important for elements such as Se and Hg that undergo bioalkylation.

Some definitions facilitate the interpretation of chemical phenomena in the ocean. Conservative behaviour signifies that the concentration of a constituent (or absolute magnitude of a property) varies only due to mixing processes. Components or parameters that behave in this manner can be used as conservative indices of mixing. Examples are salinity and potential temperature, the definitions for which are presented in subsequent sections. In contrast, non-conservative behaviour indicates that the concentration of a constituent may vary as a result of biological or chemical processes. Examples of parameters that behave non-conservatively are dissolved oxygen and pH. Residence time, τ, is defined as:

$$\tau = \frac{A}{(\mathrm{d}A/\mathrm{d}T)}$$

where A is the total amount of constituent A in the reservoir and $\mathrm{d}A/\mathrm{d}T$ can be either in rate of supply or the rate of removal of A. This represents the average lifetime of the component in the system and is, in effect, a reciprocal rate constant (see Chapter 6). Finally, the photic zone refers to the upper surface of the ocean in which photosynthesis can occur. This is typically taken to be the layer down to the depth at which sunlight radiation has declined to 1% of the magnitude at the surface. This might typically be > 100 m for visible light or photosynthetically active radiation (PAR) but generally < 20 m for UV wavelengths, of recent interest due to the enhanced input invoked by stratospheric ozone depletion (see Chapter 2).

1.2   Properties of Water and Seawater

Water is a unique substance, with unusual attributes because of its structure. The molecule consists of a central oxygen atom with two attached hydrogen atoms forming a bond angle of about 105°. As oxygen is more electronegative than hydrogen, it attracts the shared electrons to a greater extent. In addition, the oxygen atom has a pair of lone orbitals. The overall effect produces a molecule with a strong dipole moment, having distinct negative (O) and positive (H) ends. While there are several important consequences, two will be considered here. Firstly, the positive H atoms of one molecule are attracted towards the negative O atom in adjacent molecules giving rise to hydrogen bonding. This has important implications with respect to a number of physical properties, especially those relating to thermal characteristics. Secondly, the large dipole moment ensures that water is a very polar solvent.

Considering firstly the physical properties, water has much higher freezing and melting points than would be expected for a molecule of molecular weight 18. Water has high latent heats of evaporation and fusion. This means that considerable energy is required to stimulate phase changes, the energy being utilized in hydrogen bond rupturing. Moreover, it has a high specific heat and is a good conductor of heat. Consequently, heat transfer in water by advection and conduction gives rise to uniform temperatures. The density of pure water exhibits anomalous behaviour. In ice, O atoms have 4 H atoms orientated about them tetrahedrally. These units are packed together with a hexagonal symmetry. At the freezing point, 0 °C, ice is less dense than water. Heating breaks hydrogen bonds, and molecules can achieve slightly closer packing, which causes the density to increase. The maximum density occurs at 4 °C because at higher temperatures thermal expansion compensates for this compression effect. As will be discussed later, seawater differs in this respect. Thus, fresh ice floats on water, which in part explains how rivers and lakes can freeze over but remain liquid at depth. With respect to other properties, water has a high surface tension that is manifest in stable droplet formation and has a relatively low molecular viscosity and therefore is quite a mobile fluid.

Water is an excellent solvent. It is extremely polar and can dissolve a wider range of solutes and in greater amounts than any other substance. Water has a very high dielectric constant, a measure of the solvent's ability to keep apart oppositely charged ions. The solvating characteristics of individual ions influence their behaviour in solution, *i.e.* in terms of hydration, hydrolysis, and precipitation. Although water exhibits amphoteric behaviour, electrolytic dissociation is quite small. Furthermore, dissociation gives equal ion concentrations to both H_3O^+ and OH^- and so pure water is neutral. The amphoteric behaviour enhances dissolution of introduced particulate matter through surface hydrolysis reactions.

While the concept will be considered in detail below, the term salinity ($S‰$) is introduced here as a measure of the salt content of seawater, a typical value for oceanic waters being 35 g kg^{-1}. In an oceanographic context, the most important consequence of the addition of salt to water is the effect on density. However, many of the characteristics outlined above are also altered. The addition of electrolytes can cause a small increase in the surface tension. This effect is not commonly observed in seawater due to the presence of surfactants, which decrease the surface tension and so facilitate foam formation. As illustrated in Figure 2, the presence of salt does depress the temperature of maximum density and the freezing point of the solution relative to pure water. Thus, seawater

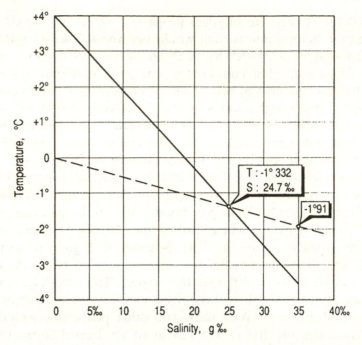

Figure 2   *The temperature of maximum density (—) and freezing point (– –) of seawater*
as a function of dissolved salt content
(From Tchernia[2])

with a typical salt content of 35 g kg^{-1} freezes at approximately $-1.9\,°C$
and the resulting ice is more dense than the solution. However, more
often than the formation of sea ice itself, the freezing process tends to
produce fresh ice overlying a more concentrated brine solution. Salts
can be precipitated at much lower temperatures, *i.e.* mirabilite
($Na_2SO_4.2H_2O$) at $-8.2\,°C$ and halite ($NaCl$) at $-23\,°C$. Some brine
inclusions and salt crystals can become incorporated into the ice.

From an oceanographic perspective, the fundamental properties of
seawater are temperature, salinity, and pressure (*i.e.* depth dependence).
Together, these parameters control the density of the water, which in
turn determines the buoyancy of the water and pressure gradients. Small
density differences integrated over oceanic scales cause considerable
pressure gradients and result in currents.

Surface water temperatures are extremely variable, obviously influenced by location and season. The minimum temperature found in polar
latitudes approaches $-2\,°C$. Equatorial oceanic waters can reach $30\,°C$.
Temperature variations with depth are far from consistent. A region in
which mixing is prevalent, as observed especially in the surface waters,

[2] P. Tchernia, 'Descriptive Regional Oceanography', Pergamon Press, Oxford, 1980, p. 253.

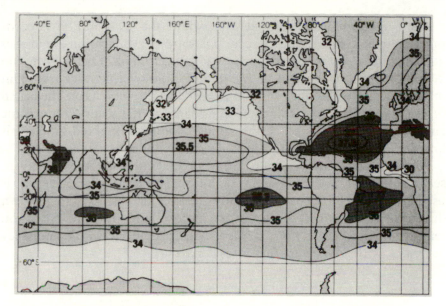

Figure 3 *The distribution of mean annual salinity in the surface waters of the ocean* (From The Open University[3])

produces a layer in which the temperature is relatively constant. The zone immediately beneath normally exhibits a sharp change in temperature, known as the thermocline. The thermocline in the ocean extends down to about 1000 m within equatorial and temperate latitudes. It acts as an important boundary in the ocean, separating the surface and deep layers and limiting mixing between these two reservoirs.

Below the thermocline, the temperature changes only little with depth. The temperature in seawater is non-conservative because adiabatic compression causes a slight increase in the *in situ* temperature measured at depth. For instance in the Mindanao Trench in the Pacific Ocean, the temperatures at 8500 m and 10 000 m are 2.23 °C and 2.48 °C, respectively. The term potential temperature is defined to be the temperature that the water parcel would have if raised adiabatically to the ocean surface. For the examples above, the potential temperatures are 1.22 °C and 1.16 °C, respectively. Potential temperature is a conservative index.

Salinity in the surface waters in the open ocean ranges between 33 and 37 (Figure 3), the main control being the balance between evaporation and precipitation. The highest salinities occur in regional seas where the evaporation rate is extremely high, namely the Mediterranean Sea (38–39) and the Red Sea (40–41). Within the World Ocean, the salinity is

[3] The Open University, 'Seawater: its Composition, Properties and Behaviour', Open University & Pergamon Press, Oxford, 1989, p. 165.

greatest in latitudes of about 20° where the evaporation exceeds precipitation. Lower salinities occur poleward as evaporation diminishes and near the equator where precipitation is very high. Local effects can be important, as evident in the vicinity of large riverine discharges that dilute the salinity. Salinity variations with depth are related to the origin of the deep waters and so will be considered in the section on oceanic circulation. A zone in which the salinity exhibits a marked gradient is known as a halocline.

Whereas the density of pure water is 1.000 g ml^{-1}, the density of seawater (S‰ = 35) is about 1.03 g ml^{-1}. The term 'sigma-tee', σ_t, is used to denote the density (actually the specific gravity and hence a dimensionless number) of water at atmospheric pressure based on temperature and salinity *in situ*. Density increases, and so the buoyancy decreases, with an increase in σ_t. It is defined as:

$$\sigma_t = (\text{specific gravity}_{S‰,T} - 1) \times 1000$$

In a plot of temperature against salinity (a *T–S* diagram), constant σ_t appear as curved lines which denote waters of constant pressure and are known as isopycnals. A zone in which the pressure changes greatly is known as a pycnocline. Within the water column, a pycnocline therefore separates waters with distinctive temperature and salinity characteristics, usually indicative of different origins. A *T–S* diagram can also be used to estimate the properties resulting from the mixing of two water masses. As noted above, the temperature is not a conservative property, and therefore σ_t is also non-conservative. To circumvent the associated difficulties of interpretation, an analogous term known as the potential density, σ_θ, is defined on the basis of potential temperature instead of *in situ* temperature. The σ_θ is therefore a conservative index.

1.3   Salinity Concepts

Salinity is a measure of the salt content of seawater. Developments in analytical chemistry have led to an historical evolution of the salinity concept. Intrinsically it would seem to be a relatively straightforward task to measure. This is true for imprecise determinations that can be quickly performed using hand-held refractometers. The salinity affects seawater density and thus the impetus for high precision in salinity measurements came from physical oceanographers.

The first techniques utilized for the determination of salinity, involving the gravimetric analysis of salt left after evaporating seawater to dryness, were fraught with difficulties. Variable amounts of water of crystallization might be retained. Some salts, such as $MgCl_2$, can decompose leaving residues of uncertain composition. Other constituents, especially organic material, might be volatilized or oxidized. Overall, such methods led to considerable inconsistencies and inaccuracies. The second set of procedures for salinity measurement made use of the observation from the *Challenger* expedition of 1872–76 that sea salt composition was apparently invariant. Hence, the total salt content could be calculated from any individual constituent, such as Cl^- that could be readily determined by titration with Ag^+. At the turn of the century, Knudsen defined salinity to be the weight in grams of dissolved inorganic matter contained in 1 kg of seawater, after bromides and iodides were replaced by an equivalent amount of chloride and carbonate was converted to oxide. Clearly from the adopted definition, the analytical technique was not specific to Cl^- and so the term chlorinity was introduced. Chlorinity (Cl‰) is the chloride concentration in seawater, expressed as $g\ kg^{-1}$, as measured by Ag^+ titration (*i.e.* ignoring other halide contributions by assuming Cl^- to be the only reactant). The relationship of interest was that between S‰ and Cl‰, as given as:

$$S‰ = 1.805Cl‰ + 0.030$$

As a calibrant solution for the $AgNO_3$ titrant, Standard Seawater was prepared that had certified values for both chlorinity and salinity. Unfortunately, the above salinity–chlorinity relationship was derived based on only nine seawater samples that were somewhat atypical. It has since been redefined using a much larger set of samples representative of oceanic waters to become:

$$S‰ = 1.80655Cl‰$$

The third category of salinity methodologies was based on conducto-metry, as the conductivity of a solution is proportional to the total salt content. Standard Seawater, now also certified with respect to conduc-tivity, provides the appropriate calibrant solution. The conductivity of a sample is measured relative to the standard and converted to salinity in practical salinity units (psu). Note that although psu has replaced the outmoded ‰, usually units are ignored altogether in modern usage. These techniques continue to be the most widely used methods because conductivity measurements can provide salinity values with a precision of ±0.001 psu. Highly precise determinations require temperature control of samples and standards to within $\pm0.001\ °C$. Application of a non-specific technique like conductometry relies upon the assumption that the sea salt matrix is invariant, both spatially and temporally. Thus, the technique cannot be reliably employed in marine boundary environ-

ments where the seawater composition differs from the bulk characteristics.

There are two types of conductometric procedures commonly used.
Firstly, a Wheatstone Bridge circuit can be set up whereby the ratio of
the resistance of unknown seawater to standard seawater balances the
ratio of a fixed resistor to a variable resistor. The system uses alternating
current to minimise electrode fouling. Alternatively, the conductivity can
be measured by magnetic induction, in which case the sensor consists of a
plastic tube containing sample seawater that links two transformers. An
oscillator establishes a current in one transformer that induces current
flow within the tube, the magnitude of which depends upon the salinity
of the sample. This in turn induces a current in the second transformer,
which can then be measured. This design has been exploited for *in situ*
conductivity measurements.

1.4   Oceanic Circulation

The distribution of components within the ocean is determined by both
transportation and transformation processes. A brief outline of oceanic
circulation is necessary to ascertain the relative influences. Two main
flow systems must be considered. Surface circulation is established by
tides and the prevailing wind patterns, and deep circulation is determined
by gravitational forces. Both are modified by Coriolis force, the acceleration due to the earth's rotation. It acts to deflect moving fluids (*i.e.*
both air and water) to the right in the northern hemisphere and to the left
in the southern hemisphere. The magnitude of the effect is a function of
latitude, being nil at the equator and increasing poleward.

Surface oceanic circulation is depicted in Figure 4. For the most part,
the circulation patterns describe gyres constrained by the continental
boundaries. The prevailing winds acting under the influence of Coriolis
force result in clockwise and counter-clockwise flow in the northern and
southern hemispheres, respectively. The flow fields are non-uniform,
exhibiting faster currents along the western margins. These are manifest,
for example, as the Gulf Stream, Kuroshio Current, and Brazil Current.
Circulation within the Indian Ocean is exceptional in that there are
distinct seasonal variations in accord with the monsoons. The absence of
other continents within the immediate boundary region of Antarctica
gives rise to a circumpolar current within the Southern Ocean.

The surface circulation is restricted to the upper layer influenced by the
wind, typically about 100 m. However, underlying water can be transported up into this zone when horizontal advection is insufficient to
maintain the superimposed flow fields. This process is called upwelling
and is of considerable importance in that biochemical respiration of

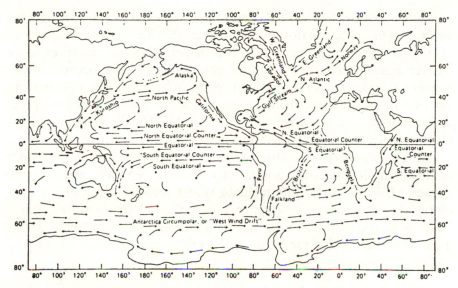

Figure 4   *The surface circulation in the ocean*
(From Stowe[4])

organic material at depth ensures that the ascending water is nutrient-
rich. Upwelling occurs in the eastern oceanic boundaries where long-
shore winds result in the offshore transport of the surface water.
Examples are found off the coasts of Peru and West Africa. Similar
processes cause upwelling off Arabia, but this is seasonal due to the
monsoon effect. A divergence is a zone in which the flow fields separate.
In such a case, upwelling may result as observed in the equatorial Pacific.
It should be noted that a region in which the streamlines come together is
known as a convergence, and water sinks in this zone.

   The density of the water controls the deep circulation. If the density of
a water body increases, it has a tendency to sink. Subsequently it will
spread out over a horizon of uniform σ_θ. As the density can be raised by
either an increase in the salinity or a decrease in the temperature, the
deep water circulatory system is also known as thermohaline circulation.
The densest waters are formed in polar regions due to the relatively low
temperatures and the salinity increase that results from ice formation.
Antarctic Bottom Water (ABW) is generated in the Weddell Sea and
flows northward into the South Atlantic. North Atlantic Deep Water
(NADW) is formed in the Norwegian Sea and off the southern coast of
Greenland. The flow of the NADW can be traced southwards through
the Atlantic Ocean to Antarctica where it is diverted eastward into the
Southern Indian Ocean and South Pacific. There it heads northwards

[4] K. S. Stowe, 'Ocean Science', John Wiley & Sons, New York, 1979.

and either enters the North Pacific or becomes mixed upward into the surface layer in the equatorial region. The transit time is of the order of 1000 years. As noted previously, the thermocline acts as an effective barrier against mixing of dissolved components in the ocean. Consequently, this deep water formation process in high latitudes is important because it facilitates the relatively rapid transport of material from the surface of the ocean down to great depths. The deep advection of atmospherically derived CO_2 is a pertinent example.

Diverse processes can form intermediate waters within the water column. In the southern South Atlantic, the NADW overrides the denser ABW. Antarctic Intermediate Water results from water sinking along the Antarctic Convergence ($\sim 50\,^\circ$S). Relatively warm, saline water exits the Mediterranean Sea at depth and can be identified as a distinctive layer within the North and South Atlantic.

2 SEAWATER COMPOSITION AND CHEMISTRY

2.1 Major Constituents

The major constituents in seawater are conventionally taken to be those elements present in typical oceanic water of salinity 35 that have a concentration greater than 1 mg kg^{-1}, excluding Si which is an important nutrient in the marine environment. The concentrations and main species of these elements are presented in Table 1. One of the most significant observations from the *Challenger* expedition of 1872–76 was that these major components existed in constant relative amounts. As already explained, this feature was exploited for salinity determinations. Inter-element ratios are generally constant, and often expressed as a ratio to Cl‰ as shown in Table 1. This implies conservative behaviour, with concentrations depending solely upon mixing processes, and indeed, salinity itself is a conservative index.

Because of this behaviour, individual seawater constituents can be utilized for source apportionment studies in non-marine environments. For instance, an enrichment factor (*EF*) for a substance X is defined as:

$$EF_X = \frac{(X/Na^+)_{sample}}{(X/Na^+)_{seawater}}$$

An enrichment factor of 1 indicates that the substance exists in comparable relative amounts in the sample and in seawater, thereby giving a good indication of a marine origin. If $EF_X > 1$, then it is enriched with respect to seawater. Conversely, depletion is signified when values $EF_X < 1$. Another example of the application of inter-element ratios

Table 1 *Chemical species and concentrations of the major elements in seawater*

Element	Chemical species	Concentration for S = 35 (mol l^{-1})	(g kg^{-1})	Ratio to chlorinity (Cl = 19.374‰)
Na	Na$^+$	4.79×10^{-1}	10.77	5.56×10^{-1}
Mg	Mg^{2+}	5.44×10^{-2}	1.29	6.66×10^{-2}
Ca	Ca^{2+}	1.05×10^{-2}	0.4123	2.13×10^{-2}
K	K$^+$	1.05×10^{-2}	0.3991	2.06×10^{-2}
Sr	Sr^{2+}	9.51×10^{-5}	0.00814	4.20×10^{-4}
Cl	Cl$^-$	5.59×10^{-1}	19.353	9.99×10^{-1}
S	SO$_4^{2-}$, NaSO$_4^-$	2.89×10^{-2}	0.905	4.67×10^{-2}
C (inorganic)	HCO$_3^-$, CO$_3^{2-}$	2.35×10^{-3}	0.276	1.42×10^{-2}
Br	Br$^-$	8.62×10^{-4}	0.673	3.47×10^{-3}
B	B(OH)$_3$, B(OH)$_4^-$	4.21×10^{-4}	0.0445	2.30×10^{-3}
F	F$^-$, MgF$^+$	7.51×10^{-5}	0.00139	7.17×10^{-5}

Based on Dyrssen and Wedborg[5]

can be found in examining the geochemical cycle of sulfur. Concentrations of SO$_4^{2-}$ and Na$^+$ in ice cores and marine aerosols exhibit a SO$_4^{2-}$: Na$^+$ greater than that observed in seawater. This excess can be readily calculated and is known as non-sea salt sulfate (NSSS). Contributions to NSSS include SO$_2$ derived from both volcanic and anthropogenic sources, together with dimethyl sulfide (DMS) of marine biogenic origin.

Not all the major constituents consistently exhibit conservative behaviour in the ocean. The most notable departures occur in deep waters where Ca^{2+} and HCO$_3^-$ exhibit anomalously high concentrations due to the dissolution of calcite. The concept of relative constant composition does not apply in a number of atypical environments associated with boundary regions. Inter-element ratios for major constituents can be quite different in estuaries and near hydrothermal vents. Obviously, these are not solutions of sea salt (with the implication that accuracy of salinity measurements by chemical and conductometric means is limited).

The residence times for some elements are presented in Table 2. The major constituents normally have long residence times. The residence time is a crude measure of a constituent's reactivity in the reservoir. The aqueous behaviour and rank ordering can be appreciated simply in terms of the ionic potential given by the ratio of electronic charge to ionic radius (Z/r). Elements with $Z/r < 3$ are strongly cationic. The positive charge density is relatively diffuse, but sufficient to attract an envelope of water molecules forming a hydrated cation. As the ionic potential

[5] D. Dyrssen and M. Wedborg, 'The Sea', ed. E. Goldberg, John Wiley & Sons, New York, 1974, p. 181.

Table 2   *The residence time and speciation of some elements in the ocean*

Element	Principal species	Concentration (mol l^{-1})	Residence time (yr)
Li	Li$^+$	2.6×10^{-5}	2.3×10^6
B	B(OH)$_3$, B(OH)$_4^-$	4.1×10^{-4}	1.3×10^7
F	F$^-$, MgF$^+$	6.8×10^{-5}	5.2×10^5
Na	Na$^+$	4.68×10^{-1}	6.8×10^7
Mg	Mg^{2+}	5.32×10^{-2}	1.2×10^7
Al	Al(OH)$_4^-$, Al(OH)$_3$	7.4×10^{-8}	1.0×10^2
Si	Si(OH)$_4$	7.1×10^{-5}	1.8×10^4
P	HPO$_4^{2-}$, PO$_4^{3-}$, MgHPO$_4$	2×10^{-6}	1.8×10^5
Cl	Cl$^-$	5.46×10^{-1}	1×10^8
K	K$^+$	1.02×10^{-2}	7×10^6
Ca	Ca^{2+}	1.02×10^{-2}	1×10^6
Sc	Sc(OH)$_3$	1.3×10^{-11}	4×10^4
Ti	Ti(OH)$_4$	2×10^{-8}	1.3×10^4
V	H$_2$VO$_4^-$, HVO$_4^{2-}$, NaVO$_4^-$	5×10^{-8}	8×10^4
Cr	CrO$_4^{2-}$, NaCrO$_4^-$	5.7×10^{-9}	6×10^3
Mn	Mn^{2+}, MnCl$^+$	3.6×10^{-9}	1×10^4
Fe	Fe(OH)$_3$	3.5×10^{-8}	2×10^2
Co	Co^{2+}, CoCO$_3$, CoCl$^+$	8×10^{-10}	3×10^4
Ni	Ni^{2+}, NiCO$_3$, NiCl$^+$	2.8×10^{-8}	9×10^4
Cu	CuCO$_3$, CuOH$^+$, Cu^{2+}	8×10^{-9}	2×10^4
Zn	ZnOH$^+$, Zn^{2+}, ZnCO$_3$	7.6×10^{-8}	2×10^4
Br	Br$^-$	8.4×10^{-4}	1×10^8
Sr	Sr^{2+}	9.1×10^{-5}	4×10^6
Ba	Ba^{2+}	1.5×10^{-7}	4×10^4
La	La^{3+}, LaCO$_3^+$, LaCl^{2+}	2×10^{-11}	6×10^2
Hg	HgCl$_4^{2-}$	1.5×10^{-10}	8×10^4
Pb	PbCO$_3$, Pb(CO$_3$)$_2^{2-}$, PbCl$^+$	2×10^{-10}	4×10^2
Th	Th(OH)$_4$	4×10^{-11}	2×10^2
U	UO$_2$(CO$_3$)$_2^{4-}$	1.4×10^{-8}	3×10^6

Based on Brewer[6] and Bruland.[7]

increases, the force of attraction towards the water similarly rises to the extent that one oxygen–hydrogen bond in the molecule breaks. This causes the solution pH to fall and metal hydroxides to form. Neutral hydroxides tend to be relatively insoluble and so precipitate. However, in the more extreme case for which $Z/r > 12$, the attraction toward the oxygen is so great that both bonds in the associated water molecules are broken. The reaction product is an oxyanion, usually quite soluble because of the associated anionic charge. Thus in seawater, those elements (Al, Fe) having a tendency to form insoluble hydroxides have short residence times. This is also true for elements that exist preferen-

[6] P. Brewer, 'Chemical Oceanography', ed. J. P. Riley and G. Skirrow, Academic Press, London, 2nd Edn., 1975, Vol. 1, p. 415.
[7] K. W. Bruland, 'Chemical Oceanography', ed. J. P. Riley and R. Chester, Academic Press, London, 1983, Vol. 8, p. 157.

tially as neutral oxides (Mn, Ti). Hydrated cations (Na^+, Ca^{2+}) and strongly anionic species (Cl^-, Br^-, $UO_2(CO_3)_2^{4-}$) have long residence times. This treatment is, of course, somewhat of an oversimplification, ignoring the rather significant role that biological organisms play in nutrient and trace element chemistry.

2.2 Dissolved Gases

2.2.1 Gas Solubility and Air–Sea Exchange Processes. The ocean contains a vast array of dissolved gases. Some of the gases such as Ar and chlorofluorocarbons behave conservatively and can be utilized as tracers for water mass movements and ventilation rates. Equilibrium processes at the air–sea interface generally lead to saturation, and then the concentration remains unchanged once the water sinks. Thus, the gas concentration is characteristic of the lost contact with the atmosphere. Deep waters usually contain no CFCs as such anthropogenic compounds have only a recent history of use. There are several important non-conservative gases, which exhibit wide variations in concentration due to biological activity. O_2 determines the redox potential in seawater and CO_2 buffers the ocean at pH 8. Ocean–atmosphere exchange processes for gases such as CO_2 and dimethyl sulfide may play an important role in climate change.

Both temperature and salinity affect the solubility of gases in water. Empirical relationships can be found elsewhere.[8,9] The trends are such that gas solubility increases with a decrease in temperature or an increase in salinity. The changes in solubility are non-linear and differ dramatically for various gases. Figure 5 depicts the solubility of several gases as a function of temperature.

At the ocean–atmosphere interface, exchange of gases occurs to achieve equilibrium between the two systems and, consequently, gases become saturated. However, supersaturation can be achieved by several mechanisms. Firstly, bubbles that form from white cap activity can be entrained and dissolved at depth. The slight but significantly different pressure relative to the surface favours gas dissolution and results in a higher equilibrium concentration. Secondly as evident from Figure 5, if two water masses that have been equilibrated at different temperatures are mixed, then the resulting water body would be supersaturated. Thirdly, gases that are produced *in situ* by biological activity may become supersaturated, particularly when evasion to the atmosphere is hindered.

[8] R. F. Weiss, *Deep-Sea Res.*, 1970, **17**, 721.
[9] D. Kester, 'Chemical Oceanography', ed. J. P. Riley and G. Skirrow, Academic Press, London, 2nd Edn., 1975, Vol. 1, p. 497.

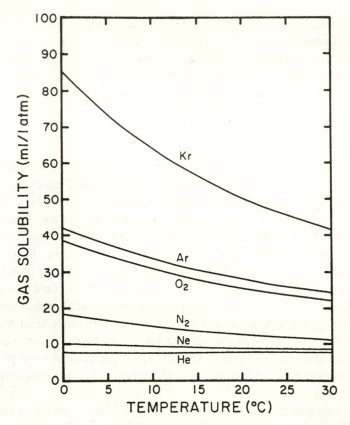

Figure 5   *The solubility of various gases in seawater as a function of temperature*
(From Broecker and Peng[10])

The gas solubility for a water body in equilibrium with the overlying air mass can be expressed in several ways. It is convenient to consider Henry's Law that states:

$$H = c_a c_w^{-1}$$

where H is the Henry's law constant and c_a and c_w refer to the concentration of a gas in air and water, respectively. As discussed by Liss (1983)[11], air–sea exchange occurs when a concentration gradient exists (*i.e.* $\Delta C = c_a H^{-1} - c_w$) and the magnitude of the consequential flux, F, is given as:

$$F = K\Delta C$$

[10] W. S. Broecker and T. H. Peng, 'Tracers in the Sea', Lamont-Doherty Geological Observatory, Palisades, 1982, p. 690.
[11] P. S. Liss, 'Air–Sea Exchange of Gases and Particles', ed. P. S. Liss and W. G. N. Slinn, Reidel, Dordrecht, 1983, p. 241.

Table 3 *The net global fluxes of some trace gases across the air/sea interface*

Gas	Global air–sea direction[a]	Flux magnitude[b]
CH_4	+	10^{12}–10^{13}
CO_2	−	6×10^{15}
N_2O	+	6×10^{12}
$\lbrace CCl_4$	−	10^{10}
CCl_4	=	~ 0
$\lbrace CCl_3F$	−	5×10^9
CCl_3F	=	~ 0
CH_3I	+	$3\text{–}13 \times 10^{11}$
CO	+	$100 \pm 90 \times 10^{12}$
H_2	+	$4 \pm 2 \times 10^{12}$
Hg	+	$\sim 2 \times 10^9$

From Chester.[1]
[a] + sea to air, − air to sea, = no net flux.
[b] $g\,y^{-1}$.

where the proportionality constant, K, has dimensions of velocity and so is frequently referred to as the transfer velocity (see also Chapter 6).

Air–sea exchange processes are consequently dependent upon the concentration gradient and the transfer velocity. The transfer velocity is not a constant, but rather depends upon several physical parameters such as temperature, wind speed, and wave state. The exchange can also be attenuated by the presence of a surface film or slick. Alternatively, the exchange can be facilitated by bubble formation. The concentration gradient determines the direction of the flux, into or out of the ocean. Net global fluxes for some gases are presented in Table 3. The atmosphere serves as the source of material for conservative gases, especially those of anthropogenic origin, but several gases produced *in situ* by biological activity evade from the ocean.

2.2.2 Oxygen. Oxygen is a non-conservative gas and a typical oceanic profile is shown in Figure 6. The concentration varies throughout the water column, its distribution being greatly influenced by biological activity. The generalized chemical equation for carbon fixation is often given as:

$$nCO_2 + nH_2O \rightleftharpoons (CH_2O)_n + nO_2$$

During photosynthesis this reaction proceeds to the right, thereby producing organic material, designated by $(CH_2O)_n$, and O_2. The surface waters become equilibrated with respect to atmospheric O_2, but they can get supersaturated during periods of intense photosynthetic activity. Respiration occurs as the above reaction proceeds to the left and

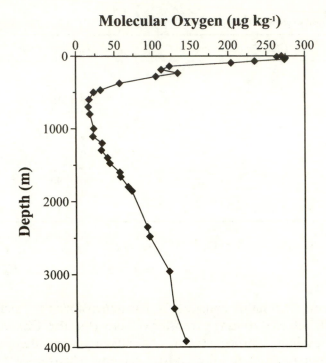

Figure 6   *A profile of molecular oxygen in the North Pacific Ocean*
(Data from Bruland[12])

O_2 is consumed. Photosynthesis is obviously restricted to the upper ocean (in the photic zone) and ordinarily exceeds respiration. However, the relative importance of the two processes changes with depth. The oxygen compensation depth is the horizon in the water column at which the rate of O_2 production by photosynthesis equals the rate of respiratory O_2 oxidation.

Below the photic zone, O_2 is utilized in chemical and biochemical oxidation reactions. As evident in Figure 6, the concentration diminishes with depth to develop an oxygen minimum zone. Thereafter, the O_2 concentration in deeper waters begins to increase because these waters originated from polar regions. They were cold and in equilibrium with atmospheric gases at the time of sinking, but subsequently lost little of the dissolved O_2 because the flux of organic material to deep waters is relatively small.

The dissolved O_2 content of seawater has a significant control on the redox potential, often designated in environmental chemistry by pε. This

---

[12] K. W. Bruland, *Earth Planet. Sci. Lett.*, 1980, **47**, 176.

is defined with reference to electron activity in an analogous fashion to pH and thus:

$$p\varepsilon = -\log\{e^-\}$$

The relationship between $p\varepsilon$ and the more familiar electrode potential E or E_H is:

$$p\varepsilon = \frac{F}{2.303RT}E$$

and for the standard state:

$$p\varepsilon^\ominus = \frac{F}{2.303RT}E^\ominus$$

where F is Faraday's constant, R is the universal gas constant, and T is the absolute temperature in Kelvin. Whereas a high value of $p\varepsilon$ indicates oxidizing conditions, a low value signifies reducing conditions. Oxygen plays a role via the reaction:

$$O_2 + 4H^+ + 4e^- \rightleftharpoons 2H_2O$$

At $20\,°C$, $K = 10^{83.1}$ and so water of pH $= 8.1$ in equilibrium with atmospheric O_2 ($p_{O_2} = 0.21$ atm) has $p\varepsilon = 12.5$. This conforms to surface conditions but the $p\varepsilon$ decreases as the O_2 content diminishes with depth. The oxygen minimum is particularly well developed beneath the highly productive surface waters of the eastern tropical Pacific Ocean where there is a large flux of organic material to depth and subsequently considerable oxidation. The O_2 becomes sufficiently depleted that the resulting low redox conditions causes NO_3^- to be reduced to NO_2^-.

When circulation is restricted vertically due to thermal or saline stratification and horizontally by topographic boundaries, the water becomes stagnant and the oxygen may be completely utilized producing anoxic conditions. Such regions represent atypical marine environments where reducing conditions prevail. Well known examples include the Black Sea, which is permanently anoxic below 200 m, and the Cariaco Trench, a depression in the Venezuelan continental shelf. Some fjords, such as Saanich Inlet in western Canada and Dramsfjord of Norway, may be intermittently anoxic. Periodic flushing of these inlets by dense, oxygenated waters displaces deep anoxic water to the surface causing massive fish mortality.

O_2 can be used as a tracer to help identify the origin of water masses. The warm, saline intrusion into the Atlantic Ocean from the Mediterranean Sea is relatively O_2 deficient. Alternatively, the waters downwelling from polar regions have elevated O_2 concentrations.

2.2.3   Carbon Dioxide and Alkalinity.   Marine chemists sometimes adopt activity conventions quite different from those traditionally used in chemistry. It is useful to preface a discussion about the carbon dioxide–calcium carbonate system in the oceans with a brief outline of pH scales. Although originally introduced in terms of ion concentration, today the definition of pH is based on hydrogen ion activity and is:

$$pH = -\log a_H$$

where a_H refers to the relative hydrogen ion activity (*i.e.* dimensionless, as is pH). Defined using concentration scales, the pH can be:

$$pH = -\log(c_H \gamma_H / c^\circ)$$

or

$$pH = -\log(m_H \gamma_H / m^\circ)$$

where c_H and m_H represent molar and molal concentrations, c° and m° are the respective standard state conditions (1 mol l^{-1} and 1 mol kg^{-1}), and γ_H is the appropriate activity coefficient. Obviously γ_H differs in these two expressions as $c^\circ \neq m^\circ$. However, different activity scales may also be used. In the *infinite dilution activity scale*, $\gamma_H \to 1$ as the concentration of hydrogen ions and all other ions approach 0. For analyses, pH meters are calibrated using dilute buffers prepared in pure water. Alternatively, in the *constant ionic medium activity scale*, $\gamma_H \to 1$ as the concentration of hydrogen ions approaches 0 while all other components are maintained at some constant level. Calibrant buffers are prepared in solutions of constant ionic composition, and in marine chemistry this is often a solution of synthetic seawater. While these two methodologies are equally justifiable from a thermodynamic point of view, it is important to appreciate that pH scales so defined are quite different. As a further consequence, the absolute values for dissociation constants also differ.

The biogeochemical cycle of inorganic carbon in the ocean is extremely complicated. It involves the transfer of gaseous carbon dioxide from the atmosphere into solution. Not only is this a reactive gas that readily undergoes hydration in the ocean, but also it is fixed as organic material by marine phytoplankton. Inorganic carbon can be regenerated either by photochemical oxidation in the photic zone or via respiratory oxidation of organic material at depth. Surface waters are supersaturated with respect to aragonite and calcite, forms of $CaCO_3$, but precipitation is limited to coastal lagoons such as found in the Bahamas. However, several marine organisms utilize calcium carbonate to form shells. Sinking shells can remove inorganic carbon from surface waters which is then regenerated following dissolution in the under-saturated waters found at depth. Nonetheless, calcitic oozes of biogenic origin constitute a major component in marine sediments. Finally, the inorganic carbon equilibrium is responsible for buffering seawater at a pH near 8 on time scales of centuries of millennia.

There are several equilibria to be considered. Firstly, CO_2 is exchanged across the air–sea interface:

$$CO_{2(g)} \rightleftharpoons CO_{2(aq)}$$

The equilibrium process obeys Henry's Law, but the dissolved CO_2 reacts rapidly with water to become hydrated as:

$$CO_{2(aq)} + H_2O \rightleftharpoons H_2CO_3$$

Relative to the exchange process, the hydration reaction forming carbonic acid occurs quite quickly. This means that the concentration of dissolved CO_2 is extremely low. The two processes can be considered together as:

$$CO_{2(g)} + H_2O \rightleftharpoons H_2CO_3$$

The equilibrium constant is then:

$$K_{CO_2} = \frac{\{H_2CO_3\}}{p_{CO_2}\{H_2O\}}$$

where p_{CO_2} is the partial pressure of CO_2 in the marine troposphere. Carbonic acid undergoes dissociation:

$$H_2CO_3 + H_2O \rightleftharpoons H_3O^+ + HCO_3^-$$

$$HCO_3^- + H_2O \rightleftharpoons H_3O^+ + CO_3^{2-}$$

for which the first and second dissociation constants (using $\{H^+\}$ rather than $\{H_3O^+\}$) are:

$$K_1 = \frac{\{H^+\}\{HCO_3^-\}}{\{H_2CO_3\}}$$

$$K_2 = \frac{\{H^+\}\{CO_3^{2-}\}}{\{HCO_3^-\}}$$

The hydrogen ion activity can be established with a pH meter. However, as discussed above, this measurement must be operationally defined. On the other hand, the individual ion activities of bicarbonate and carbonate ions cannot be measured. Instead, ion concentrations are determined, as outlined below, by titration. Accordingly, the equilibrium constants are redefined in terms of concentrations. These are then known as apparent rather than true equilibrium constants and distinguished using a prime notation. It must be appreciated that apparent equilibrium constants are not invariant, but rather are affected by temperature, pressure, salinity, and, as outlined previously, the pH scale adopted. The apparent dissociation constants are:

$$K_1' = \frac{\{H^+\}\{HCO_3^-\}}{\{H_2CO_3\}}$$

$$K_2' = \frac{\{H^+\}\{CO_3^{2-}\}}{\{HCO_3^-\}}$$

It should be noted that whereas ion activities are denoted by curly brackets { }, concentrations are designated by square brackets [ ]. Analogous to the pH, pK conventionally refers to $-\log K$. Numerical values for the constants pK_{CO_2}, pK_1', and pK_2' based on a constant ionic medium scale (*i.e.* seawater with chlorinity = 19‰) are given in Table 4. This provides sufficient information to calculate the speciation of carbonic acid in seawater at a given temperature as a function of pH.

Table 4 *Equilibrium constants for the carbonate system*

T (°C)	pK_{CO_2}*	pK_1'	pK_2'
0	1.19	6.15	9.40
5	1.27	6.11	9.34
10	1.34	6.08	9.28
15	1.41	6.05	9.23
20	1.47	6.02	9.17
25	1.53	6.00	9.10
30	1.58	5.98	9.02

Adapted from Stumm and Morgan.[13]

[13] W. Stumm and J. J. Morgan, 'Aquatic Chemistry', John Wiley & Sons, New York, 3rd Edn., 1996, p. 1022.

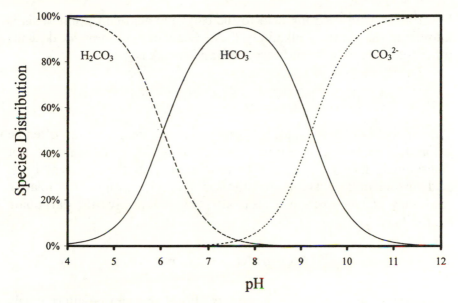

Figure 7 *The distribution of carbonic acid species in seawater of 35 psu at 15 °C as a function of pH*

This is shown for carbonic acid at 15 °C in seawater equilibrated with atmospheric carbon dioxide ($\sim 3.5 \times 10^{-4}$ atm) in Figure 7. While there are several confounding features, the pH of seawater can be considered to be buffered by the bicarbonate:carbonate pair. The pH is generally about 8, but is sensitive to the concentration ratio $[HCO_3]:[CO_3^{2-}]$ as evident from rearranging the expression for K_2' to become:

$$\{H^+\} = K_2' \frac{[HCO_3^-]}{[CO_3^{2-}]}$$

To understand the response of the oceanic CO_2 system to *in situ* biological activity or enhanced CO_2 concentrations in the atmosphere, it is necessary to consider in more detail the factors influencing the inorganic carbon cycle. Two useful parameters can be introduced. Firstly, the total concentration of inorganic carbon, ΣCO_2, in seawater is:

$$\Sigma CO_2 = [CO_2] + [H_2CO_3] + [HCO_3^-] + [CO_3^{2-}]$$

The first term is negligible and as evident in Figure 7, the major species at pH 8 are HCO_3^- and CO_3^{2-}.

Alkalinity is defined as a measure of the proton deficit in solution and should not be confused with basicity. Alkalinity is operationally defined by titration with a strong acid to the carbonic acid end point. This is known as the titration alkalinity (TA). Seawater contains weak acids other than bicarbonate and carbonate and so TA is given as:

$$TA = [HCO_3^-] + 2[CO_3^{2-}] + [B(OH)_4^-] + [OH^-] - [H^+]$$

The influence of $[OH^-]$ and $[H^+]$ on the TA are small and can often be ignored. The borate contributes about 3% of the TA and, if not determined independently, can be estimated from the apparent boric acid dissociation constants and the salinity, relying upon the relative constancy of composition of sea salt. This would give the carbonate alkalinity (CA):

$$CA = [HCO_3^-] + 2[CO_3^{2-}]$$

Considering the dissociation constants above, this can be alternatively expressed as:

$$CA = \frac{K_1'[H_2CO_3]}{\{H^+\}} + \frac{2K_1'K_2'[H_2CO_3]}{\{H^+\}^2}$$

or

$$CA = \frac{K_1'K_{CO_2}p_{CO_2}}{\{H^+\}} + \frac{2K_1'K_2'K_{CO_2}p_{CO_2}}{\{H^+\}^2}$$

This equation can be rearranged to give the following quadratic expression that can be solved for the pH:

$$CA\{H^+\}^2 - K_1'K_{CO_2}p_{CO_2}\{H^+\} - 2K_1'K_2'K_{CO_2}p_{CO_2} = 0$$

Worked Example 1

For seawater (35 psu, 15 °C) with an alkalinity of 2.30 meq l^{-1} and in equilibrium with atmospheric $CO_2 = 3.65 \times 10^{-4}$ atm, calculate (i) the pH and (ii) the speciation of carbonic acid.

(i) pH calculation

Data, including constants from Table 4, are: $K_{CO_2} = 10^{-1.41}$, $K_1' = 10^{-6.05}$, $K_2' = 10^{-9.23}$, CA $= 2.30 \times 10^{-3}$, $p_{CO_2} = 3.65 \times 10^{-4}$

The pH is obtained from the calculation of $\{H^+\}$ using the above quadratic equation. Thus,

$$\{H^+\} = -\frac{b + \sqrt{b^2 - 4ac}}{2a}$$

where

$a = CA$ $\qquad\qquad\qquad\qquad\qquad\qquad\qquad\qquad\qquad = 2.30 \times 10^{-3}$
$b = -K_1'K_{CO_2}p_{CO_2}$ $\quad = -10^{-6.05} \times 10^{-1.41} \times 10^{-3.46}$ $\qquad = -1.20 \times 10^{-11}$
$c = -2K_1'K_2'K_{CO_2}p_{CO_2}$ $= -2 \times 10^{-6.05} \times 10^{-9.23} \times 10^{-1.41} - 10^{-3.46}$ $= -1.42 \times 10^{-20}$

Giving $\{H^+\} = 6.22 \times 10^{-9}$ and pH $= 8.21$

(ii) Carbonic acid speciation calculations
Knowing the pH, each of the three major species (H_2CO_3, HCO_3^-, CO_3^{2-}) can be calculated as a fraction (or percentage) of the ΣCO_2. Note that the negligible contribution due to dissolved CO_2 is ignored. The necessary expressions are derived from the definitions of the K_1' and K_2' given previously.

$$[HCO_3^-] = \frac{K_1'}{\{H^+\}}[H_2CO_3]$$

$$[CO_3^{2-}] = \frac{K_2'}{\{H^+\}}[HCO_3^-] = \frac{K_1'K_2'}{\{H^+\}^2}[H_2CO_3]$$

Substituting into the expression for the summation of all carbonic species gives:

$$\Sigma CO_2 = [H_2CO_3] + [HCO_3^-] + [CO_3^{2-}]$$
$$= [H_2CO_3] + K_1'\{H^+\}^{-1}[H_2CO_3] + K_1'K_2'\{H^+\}^{-2}[H_2CO_3]$$
$$= [H_2CO_3](1 + K_1'\{H^+\}^{-1} + K_1'K_2'\{H^+\}^{-2})$$

Thereafter, the fractional contribution of each species to the total can be calculated using:

$$\frac{[H_2CO_3]}{\Sigma[CO_2]} = (1 + K_1'\{H^+\}^{-1} + K_1'K_2'\{H^+\}^{-2})^{-1}$$

$$\frac{[HCO_3^-]}{\Sigma[CO_2]} = (K_1'\{H^+\})(1 + K_1'\{H^+\}^{-1} + K_1'K_2'\{H^+\}^{-2})^{-1}$$

$$\frac{[CO_3^{2-}]}{\Sigma[CO_2]} = (K_1'K_2'\{H^+\}^{-2})(1 + K_1'\{H^+\}^{-1} + K_1'K_2'\{H^+\}^{-2})^{-1}$$

Substituting the values for K_1' and K_2' and using the previously calculated pH of 8.21, the fractional contribution of each species is 0.006, 0.907, and 0.087 for H_2CO_3, HCO_3^-, and CO_3^{2-}, respectively.

___

Consider now the effect of altering the p_{CO_2} in the water. The alkalinity should not change in response to variations in CO_2 alone because the hydration and dissociation reactions give rise to equivalent amounts of H^+ and anions. CO_2 can be lost by evasion to the atmosphere (a process usually confined to equatorial regions) or by photosynthesis. This causes the ΣCO_2 to diminish and the pH to rise, an effect that can be quite dramatic in tidal rock pools in which pH may then rise to 9. Conversely, an increase in p_{CO_2}, either by invasion from the atmosphere or release following respiration, prompts an increase in ΣCO_2 and a fall in pH. Thus, the depth profiles of pH would mimic that of O_2 but the ΣCO_2 would exhibit a maximum at the oxygen minimum.

There are further confounding influences, in particular concerning $CaCO_3$. $CaCO_3$ in the form of aragonite or calcite is used by many organisms to form calcareous shells (tests). The shells sink and dissolve when the organism dies. The solubility is governed by:

$$Ca^{2+} + CO_3^{2-} \rightleftharpoons CaCO_{3(s)}$$

Surface waters are supersaturated with respect to $CaCO_3$, but precipitation rarely occurs, possibly due to an inhibitory effect by Mg^{2+} forming ion pairs with CO_3^{2-}. The solubility of $CaCO_3$ increases with depth, due to both a pressure effect and the decrease in pH following respiratory release of CO_2, with the result that the shells dissolve. This behaviour not only increases the alkalinity but also accounts for the non-conservative nature of Ca^{2+} and inorganic carbon in deep waters. The depth at which appreciable dissolution begins is known as the lysocline. At a greater depth, designated as the carbonate compensation depth (CCD), no calcareous material is preserved in the sediments. The depths of the lysocline and CCD are influenced by the flux of organic material and shells, and tend to be deeper under high productivity zones.

The CO_2 and $CaCO_3$ systems are coupled in that the pH buffering in the ocean is due to the reaction:

$$CO_2 + H_2O + CaCO_3 \rightleftharpoons Ca^{2+} + 2HCO_3^-$$

In addition to the effects noted previously, an input of CO_2 promotes the dissolution of $CaCO_3$. The reaction does not proceed to the right without constraint, but rather meets a resistance given by the Revelle factor, R:

$$R = \frac{\mathrm{d}p_{CO_2}/p_{CO_2}}{\mathrm{d}\Sigma CO_2/\Sigma CO_2}$$

This value is approximately 10, indicating that the ocean is relatively well buffered against changes in ΣCO_2 in response to variations in atmospheric p_{CO_2}. Although the ocean does respond to an increase in the atmospheric burden of CO_2, the time scales involved are quite considerable. The surface layer can become equilibrated on the order of decades, but as the thermocline inhibits exchange into deep waters, the equilibration of the ocean as a whole with the atmosphere proceeds on the order of centuries. The ventilation of deep water by downwelling water masses in polar latitudes only partly accelerates the overall process.

2.2.4 Dimethyl Sulfide and Climatic Implications. The Gaia hypothesis of Lovelock[14] states that the biosphere regulates the global environment for self-interest. This presupposes that controls, perhaps poorly understood or unknown, serve to maintain the present *status quo*. Charlson and co-workers have made use of this hypothesis to suggest that biogenic production of dimethyl sulfide (DMS) and the consequent formation of atmospheric cloud condensation nuclei (CCN, *i.e.* small particles onto which water can condense) acts as a feedback mechanism to counteract the global warming resulting from elevated greenhouse gas concentrations in the atmosphere.[15] The cycle is illustrated in Figure 8. Global warming, with concurrent warming of the ocean surface, leads to enhanced phytoplankton productivity. This promotes the production and evasion to the atmosphere of DMS. The DMS undergoes oxidation to form CCN which promote cloud formation and increase the planetary albedo (*i.e.* reflectivity with respect to sunlight) thereby causing a cooling effect. From a biogeochemical perspective, the two key features are the controls on the biogenic production of DMS and the formation of CCN following aerial oxidation of DMS. These will be considered below in more detail. With respect to the physics, the most important aspects of the proposed climate control mechanism are that the enhancement of the albedo is due to an increase in the number and type of CCN, and that this CCN production occurs in the marine boundary layer. The albedo of calm seawater is very low ($\sim 2\%$) in comparison to vegetated regions (10–25%), deserts ($\sim 35\%$), and snow covered surfaces ($\sim 90\%$).

That biological processes within the oceans act as a major source of reduced sulfur gases is well established.[16] Of particular importance is the generation of DMS. Surface concentrations, approximately in the range

[14] J. Lovelock, 'Gaia. A New Look at Life on Earth', Oxford University Press, Oxford, 1979.
[15] R. J. Charlson, J. E. Lovelock, M. O. Andreae, and S. G. Warren, *Nature*, 1987, **326**, 655.
[16] M. O. Andreae, 'The Role of Air–Sea Exchange in Geochemical Cycling', ed. P. Buat-Ménard, Reidel, Dordrecht, 1986, p. 331.

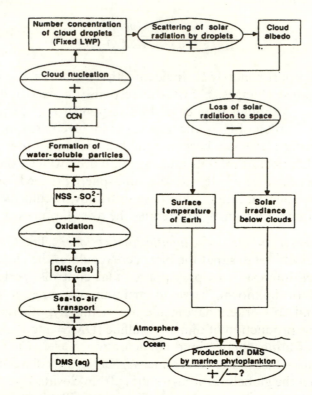

Figure 8   *The possible climatic influence of dimethyl sulfide of marine biogenic origin (From Charlson et al.*[15])

$0.7–17.8$ nmol 1^{-1}, exhibit large temporal and geographic variations. Oceanic distributions indicate that DMS is produced within the photic zone, which is consistent with a phytoplankton source, but DMS concentrations are poorly correlated with normal indicators of primary productivity. While *Phaeocystis* and *Coccolithoporidae* have been identified as important DMS producers, there is still uncertainty as to the full potential for biological DMS formation. With respect to climate modification, questions remain as to the biological response to global warming. For the model of Charlson and co-workers[15] to hold, either organisms might increase DMS formation or biological succession could change in such a way as to favour DMS producers. Thus, marine biogenic source strengths and the controlling factors remain important unresolved issues in sulfur biogeochemistry.

DMS concentrations in the remote marine troposphere vary in the range of $0.03–32$ nmol m^{-3}. Not surprisingly, and as with the seawater concentrations, considerable temporal and geographic disparities occur. Furthermore, atmospheric DMS concentrations exhibit diurnal variations, with a night-time maximum and an afternoon minimum consistent with a photochemical sink. Whereas oxidation involves HO• free radicals during the day, a reaction with NO_3• may be important at night. Relatively low levels are associated with air masses derived from continental areas, owing to the enhanced concentrations of oxidants. While oceanic venting rates are dependent upon a number of meteorological and oceanographic conditions, there is no question that the marine photic zone acts as the major source of DMS to the overlying troposphere.

The oxidation of DMS in the atmosphere could yield several products, namely dimethyl sulfoxide (DMSO), methanesulfonate (MSA) or sulfate. Insofar as aerosol formation is concerned, the two key products are MSA and SO_4^{2-}. Atmospheric particles in the sub-micron size range exert a significant influence on the earth's climate. The effect can be manifested via three mechanisms. Firstly, the particles themselves may enhance backscatter of solar radiation. Secondly, they act as cloud condensation nuclei promoting cloud formation and so increasing the earth's albedo. Thirdly, such clouds affect the hydrological cycle. The evidence for such a biofeedback mechanism limiting global warming remains circumstantial.

2.3 Nutrients

Although several elements are necessary to sustain life, traditionally in oceanography the term 'nutrients' has referred to nitrogen (usually nitrate but also ammonia), phosphate, and silicate. The rationale for this classification was that analytical techniques had long been available that allowed the precise determination of these constituents despite their relatively low concentrations. They were observed to behave in a consistent and explicable manner, but quite differently from the major constituents in seawater.

The distributions of these three nutrients are determined by biological activity. Nitrate and phosphate become incorporated into the soft parts of organisms. As evident in the modified carbon fixation equation of Redfield given below:[17]

$$106CO_2 + 16NO_3^- + HPO_4^{3-} + 122H_2O + 18H^+ \rightleftharpoons$$

$$C_{106}H_{263}O_{110}N_{16}P + 138O_2$$

the uptake of these nutrients into tissues occurs in constant relative amounts. The ratio (*i.e.* Redfield ratio) for C:N:P is 106:16:1. Silicate is utilized by some organisms, particularly diatoms (phytoplankton) and radiolaria (zooplankton), to form siliceous skeletons. Such skeletons

[17] A. C. Redfield, *Am. J. Sci.*, 1958, **46**, 205.

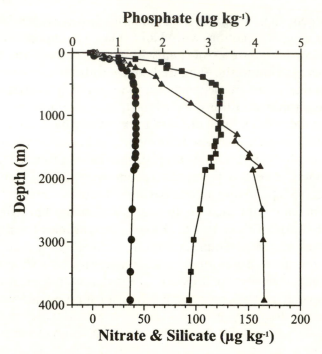

Phosphate (μg kg⁻¹)

Figure 9   *The depth distribution of nitrate (●), phosphate (■), and silicate (▲) in the*
North Pacific Ocean
(Date from Bruland[12])

consist of an amorphous, hydrated silicate, $SiO_2 \cdot nH_2O$, often called
opaline silica.

A depth profile of nitrate, phosphate, and silicate in the North Pacific
Ocean is presented in Figure 9. Nutrients behave much like ΣCO_2 and
are removed in the surface layer, especially in the photic zone. Thus,
concentrations can become quite low, and indeed sufficiently low to limit
further photosynthetic carbon fixation. The organisms sink following
death. The highest concentrations occur where respiration and bacterial
decomposition of the falling organic material are greatest, that is at the
oxygen minimum. The nutrients, including silica, are consequently
regenerated and their concentrations in deep waters are much greater
than those observed in the surface waters, thereby accounting for the
fertilizing effect of upwelling. It should be noted that the siliceous
remains behave differently from the calcareous shells discussed previously. The oceans everywhere are undersaturated with respect to
silica. Its solubility exhibits no pronounced variation with depth and
there is no horizon analogous to the CCD (see Section 2.2.3). Silica is
preserved to any great extent only in deep sea sediments associated with
the highly productive upwelling zones in the ocean.

2.4 Trace Elements

Trace elements in seawater are taken to be those that are present in quantities less than 1 mg l^{-1}, excluding the nutrient constituents. The distribution and behaviour of minor elements have been reviewed in the light of data that conform to an oceanographically consistent manner.[1,7] Analytical difficulties are readily comprehensible when it is appreciated that the concentration for some of these elements can be extremely low, *i.e.* a few pg l^{-1} for platinum group metals.[18] Some trace elements, such as Cs^+, behave conservatively and therefore absolute concentrations depend upon salinity. More often, the elements are non-conservative and their distributions in both surface waters and the water column vary greatly, reflecting the differing source strengths and removal processes in operation. Generalizations regarding residence times cannot be made, as biologically active elements are removed from seawater relatively rapidly but conservative constituents and platinum group metals have rather long residence times of the order of 10^5 years.

Considering firstly the distribution in surface waters, several elements exhibit high concentrations in coastal waters in comparison to levels in the centres of oceanic gyres. Typically, this distribution arises because the elements originate predominantly from riverine inputs or through diffusion from coastal sediments. However, as they are effectively removed from the surface waters in the coastal regions, little material is advected horizontally to the open sea. Examples of elements that behave in this way are Cd, Cu, and Ni. In contrast, the concentration of Pb, including ^{210}Pb, is greater in the gyres. This results from a strong widespread aeolian (wind-borne) input coupled with less effective removal from surface waters in the gyres.

Clearly the removal mechanisms have an appreciable effect on dissolved elemental abundances. The two major processes in operation are uptake by biota and scavenging by suspended particulate material. In the first instance, the constituent mimics the behaviour of nutrients. This is evident in the metal:nutrient correlation for Cd:P and Zn:Si (Figure 10).

No consistent pattern for depth profiles of trace elements exists. Conservative elements trend with salinity variations, provided they have no significant submarine sources. Non-conservative elements may exhibit peak concentrations at different depths in oxygenated waters as:

1. surface enrichment;
2. maximum at the O_2 minimum;

---

[18] E. Goldberg, M. Koide, J. S. Yang, and K. K. Bertine, 'Metal Speciation: Theory, Analysis and Applications', ed. J. R. Kramer and H. E. Allen, Lewis Publishers, Chelsea, 1988, p. 201.

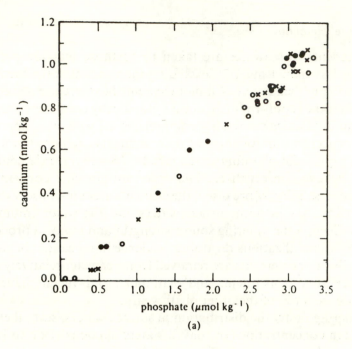

(a)

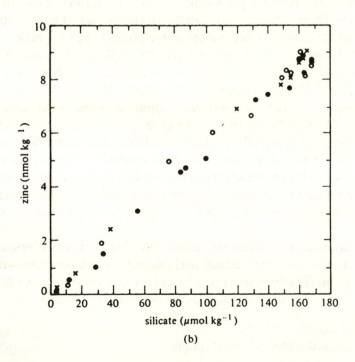

(b)

Figure 10   *Metal : nutrient correlations in the North Pacific for Cd : P and Zn : Si* (From Bruland[12])

3. mid-depth maximum not associated with the O_2 minimum;
4. bottom enrichment.

The criteria for an element, such as Pb, to exhibit a maximum concentration in surface waters are that the only significant input must be at the surface (aeolian supply) and it must be effectively removed from the water column. Constituents such as As, Ba, Cd, Ni, and Zn exhibit nutrient type behaviour. Those elements (Cd) associated with the soft parts of the organism are strongly correlated with phosphate and are regenerated at the O_2 minimum. Elements (*e.g.* Zn) associated with the skeletal material may exhibit a smooth increasing trend with depth. The third case pertains to elements, notably Mn, that have a substantial input from hydrothermal waters. These are released into oceanic waters from spreading ridges. Ocean topography is such once these waters are advected away from such regions towards the abyssal plains, they are then found at some intermediate depth. Bottom enrichment is observed for elements (Mn) that are remobilized from marine sediments. The behaviour of Al combines features outlined above, resulting in a mid-depth minimum concentration. Surface enrichment evident in mid (41 °N) but not high ($\sim$60 °N) latitudes in the North Atlantic results from the solubilization of aeolian material. Removal occurs via scavenging and incorporation into siliceous skeletal material. Subsequent regeneration by shell dissolution increases deep water Al levels.

2.5   Physico-chemical Speciation

Physico-chemical speciation refers to the various physical and chemical forms in which an element may exist in the system. In oceanic waters, it is difficult to determine speciation directly. Whereas some individual species can be analysed, others can only be inferred from thermodynamic equilibrium models as exemplified by the speciation of carbonic acid in Figure 7. Often an element is fractionated into various forms that behave similarly under a given physical (*e.g.* filtration) or chemical (*e.g.* ion exchange) operation. The resulting partition of the element is highly dependent upon the procedure utilized, and so known as *operationally defined*. In the following discussion, speciation will be exemplified with respect to size distribution, complexation characteristics, redox behaviour, and methylation reactions.

Physico-chemical speciation determines the environmental mobility of an element, especially with respect to partitioning between the water and sediment reservoirs. The influence can be manifested through various mechanisms as summarized in Figure 11. Settling velocities, and by implication the residence time, are controlled by the size of the particle.

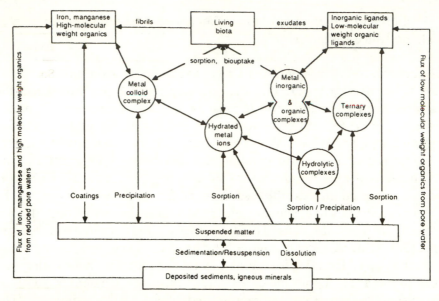

Figure 11 *Speciation of metal ions in seawater and the main controlling mechanisms* (From Öhman and Sjöberg[19])

Thus, dissolved to particulate interactions involving adsorption, precipitation, or biological uptake can effectively remove a constituent from the water column. The redox state can have a comparable influence. Mn and Fe are reductively remobilized in sediments. Following their release from either hydrothermal sources or interstitial waters, they rapidly undergo oxidation to form colloids, which are then quickly removed from the water column. Speciation also determines the bioavailability of an element for marine organisms. It is generally accepted that the uptake of trace elements is limited to free ions and some types of lipid-soluble organic complexes. This is important in that some elements may be essential, but toxic at elevated concentrations. Via complexation reactions, organisms may be able to modify the free ion concentration.

An element can exist in natural waters in a range of forms that exhibit a size distribution as indicated in Table 5. Entities in true solution include ions, ion pairs, complexes, and a wide range of organic molecules that can span several size categories. At the smallest extreme are ions, which exist in solution with a co-ordinated sphere of water molecules as discussed previously. Na^+, K^+, and Cl^- exist predominantly as free hydrated ions. Collisions of oppositely charged ions occur due to electrostastic attraction, and can produce an ion pair. An ion pair is the

[19] L. Öhman and S. Sjöberg, 'Metal Speciation: Theory, Analysis and Applications', ed. J. R. Kramer and H. E. Allen, Lewis Publishers, Chelsea, 1988, p. 1.

Table 5 *The size distribution of trace metal species in natural waters*

Size range	Metal species	Examples	Phase state
<1 nm	Free metal ions	Mn^{2+}, Cd^{2+}	Soluble
1–10 nm	Inorganic ion pairs, inorganic complexes, low molecular mass organic complexes	$NiCl^+$, $HgCl_4^{2-}$, Zn-fulvates	Soluble
10–100 nm	High molecular mass organic complexes	Pb-humates	Colloidal
100–1000 nm	Metal species adsorbed onto inorganic colloids, metals associated with detritus	$Co-MnO_2$, $Pb-Fe(OH)_3$	Particulate
>1000 nm	Metals adsorbed into living cells, metals adsorbed onto, or incorporated into mineral solids and precipitates	Cu-clays, $PbCO_{3(s)}$	Particulate

From de Mora and Harrison.[20]

transient coupling of a cation and anion during which each retains its co-ordinated water envelope. While impossible to measure directly, concentrations can be calculated with knowledge of the ion activities and stability constant. For the formation of the ion pair $NaSO_4^-$ via:

$$Na^+ + SO_4^{2-} \rightleftharpoons NaSO_4^-$$

the stability constant is defined as:

$$K = \frac{\{NaSO_4^-\}}{\{Na^+\}\{SO_4^{2-}\}}$$

Ion pair formation is important for Ca^{2+}, Mg^{2+}, SO_4^{2-}, and HCO_3^-.

If the attraction is sufficiently great, a dehydration reaction can occur leading to covalent bonding. A complex consists of a central metal ion sharing a pair of electrons donated by another constituent, termed a ligand, acting as a Lewis base. The metal ion and ligand share a single water envelope. Ligands can be neutral (*e.g.* H_2O) or anionic (*e.g.* Cl^-, HCO_3^-) species. A metal ion can co-ordinate with one or more ligands, which need not be the same chemical entity. Alternatively, the cation can share more than one electron pair with a given ligand thereby forming a ring structure. This type of complex, known as a chelate, exhibits

[20] S. J. de Mora and R. M. Harrison, 'Hazard Assessment of Chemicals, Current Developments', ed. J. Saxena, Academic Press, London, Vol. 3, 1984, p. 1.

enhanced stability largely due to the entropy effect of releasing large numbers of molecules from the water envelopes.

Complex formation is an equilibrium process. Ignoring charges, for the general case of a metal, M, and ligand, L, complex formation occurs as:

$$M + L \rightleftharpoons ML$$

for which the formation constant, K_1, is given by:

$$K_1 = \frac{\{ML\}}{\{M\}\{L\}}$$

A second ligand may then be co-ordinated as:

$$ML + L \rightleftharpoons ML_2$$

$$K_2 = \frac{\{ML_2\}}{\{M\}\{L\}}$$

The equilibrium constant for ML_2 can be expressed solely in terms of the activities of the M and L:

$$\beta_2 = \frac{\{ML_2\}}{\{M\}\{L\}^2}$$

where β_2 is the product $K_1 K_2$ and is known as the stability constant. The case can be extended to include n ligands as:

$$\beta_n = \frac{\{ML_n\}}{\{M\}\{L\}^n}$$

Worked Example 2

Assuming that $\gamma = 1$ for all species, calculate the speciation of mercury in typical seawater (35 psu at 25 °C) given the following values for stepwise stability constants for successive chlorocomplexes ($K_1 = 10^{6.74}$, $K_2 = 10^{6.48}$, $K_3 = 10^{0.85}$, $K_4 = 10^{1.00}$). Note that [Cl$^-$] is 0.559 mmol l^{-1} and that it is not necessary to know the mercury concentration in seawater.

The total mercury concentration is given as the sum of all contributing species. Thus

$$[Hg]_T = [Hg^{2+}] + [Hg\,Cl^+] + [Hg\,Cl_2] + [Hg\,Cl_3^-] + [Hg\,Cl_4^{2-}]$$

From the definition of the stability constants, we know that

$$[\text{Hg Cl}^+] = K_1[\text{Hg}^{2+}][\text{Cl}^-] \text{ and } [\text{Hg Cl}_2] = \beta_2[\text{Hg}^{2+}][\text{Cl}^-]^2, \text{ etc.}$$

Thus,

$$[\text{Hg}]_T = [\text{Hg}^{2+}](1 + K_1[\text{Cl}^-] + \beta_2[\text{Cl}^-]^2 + \beta_3[\text{Cl}^-]^3 + \beta_4[\text{Cl}^-]^4).$$

Now let $D = (1 + K_1[\text{Cl}^-] + \beta_2[\text{Cl}^-]^2 + \beta_3[\text{Cl}^-]^3 + \beta_4[\text{Cl}^-]^4)^{-1}$, and note that this is a constant for a stipulated chloride concentration. Thus, at the seawater chloride concentration ($[\text{Cl}^-] = 10^{-0.25}$) this gives

$$D = (1 + 10^{6.74} \times 10^{-0.25} + 10^{6.74} \times 10^{6.48} \times 10^{-0.25} + 10^{6.74} \times 10^{6.48} \times 10^{0.85} \times 10^{-0.25} + 10^{6.74} \times 10^{6.48} \times 10^{0.85} \times 10^{1.00} \times 10^{-0.25})^{-1}$$

$$= 7.27 \times 10^{-15}$$

The fractional (or percentage) contribution of each species can be determined using

Fractional contribution	Expression	Value for $[\text{Cl}^-] = 0.559 \text{ mmol } l^{-1}$
$[\text{Hg}^{2+}]{:}[\text{Hg}^{2+}]_T$	D	7.3×10^{-15}
$[\text{Hg Cl}^+]{:}[\text{Hg}^{2+}]_T$	$D \times K_1[\text{Cl}^-]$	2.2×10^{-8}
$[\text{Hg Cl}_2]{:}[\text{Hg}^{2+}]_T$	$D \times \beta_2[\text{Cl}^-]^2$	3.7×10^{-2} (4%)
$[\text{Hg Cl}_3^-]{:}[\text{Hg}^{2+}]_T$	$D \times \beta_3[\text{Cl}^-]^3$	1.5×10^{-1} (15%)
$[\text{Hg Cl}_4^{2-}]{:}[\text{Hg}^{2+}]_T$	$D \times \beta_4[\text{Cl}^-]^4$	8.2×10^{-1} (82%)

Thus, at the high chloride concentration found in seawater, the mercury speciation is dominated by the tetra- and tri-chloro species.

If the total mercury content for a given seawater sample is known, the concentration of each species can be readily calculated, *e.g.*:

$$[\text{Hg}^{2+}] = f_{\text{Hg}^{2+}} \times [\text{Hg}_T^{2+}]$$

This is a simplified example that has not considered other species (bromides, hydroxides, *etc.*) that might be important in seawater.

The speciation of constituents in solution can be calculated if the individual ion activities and stability constants are known. This information is relatively well known with respect to the major constituents in seawater, but not for all trace elements. Some important confounding variables create considerable difficulties in speciation modelling. Firstly, it is assumed that equilibrium is achieved, meaning that neither biological interference nor kinetic hindrance prevents this state. Secondly, seawater contains appreciable amounts (at least in relation to the trace metals) of organic matter. However, the composition of the organic matrix, the number of available binding sites, and the appropriate stability constants are poorly known. Nevertheless, speciation models can include estimates of these parameters. Organic material can form chelates with relatively high stability constants and dramatically decrease the free ion activity of both necessary and toxic trace elements. Organisms may make use of such chemistry, producing compounds either to sequester metals in limited supply or to detoxify contaminants. Thirdly, surface adsorption of dissolved species onto colloids or suspended particles may remove them from solution. As with organic matter, an exact understanding of the complexation characteristics of the suspended particles is not available, but approximations can also be incorporated into speciation models.

Elements may be present in a variety of phases other than in true solution. Colloidal formation is particularly important for elements such as Fe and Mn, which produce amorphous oxyhydroxides with very great complexation characteristics. Adsorption processes cannot be ignored in biogeochemical cycling. Particles tend to have a much shorter residence time in the water column than do dissolved constituents. Scavenging of trace components by falling particles accelerates deposition to the sediment sink.

Several elements in seawater may undergo alkylation via either chemical or biological mechanisms.[21] Type I mechanisms involve methyl radical or carbonium ion transfer and no formal change in the oxidation state of the acceptor element. The incoming methyl group may be derived for example from methylcobalamin coenzyme, *S*-adenosyl-methionine, betaine, or iodomethane. Elements involved in Type I mechanisms include Pb, Tl, Se, and Hg. Other reaction sequences involve the oxidation of the methylated element. The methyl source can be a carbanion from methylcobalamin coenzyme. Oxidative addition from iodomethane and enzymatic reactions has also been suggested. Some elements that can undergo such methylation processes are As, Sb, Ge, Sn and S. Methylation can enhance the toxicity of some elements, especially for Pb and Hg. The environmental mobility can also be affected. Methylation in the surface waters can enhance volatility and so favour evasion from the sea, as observed for S, Se, and Hg. Methylation within the sediments may facilitate transfer back into overlying waters.

Elements may exhibit multiple oxidation states in seawater. Redox

[21] P. Craig, 'Organometallic Compounds in the Environment', ed. P.S. Craig, Longman, Harlow, 1986, p. 1.

processes can be modelled in an analogous manner to the ion pairing and complexation outlined previously. The information is often presented graphically in the form of a predominance area diagram, that is a plot of pε *versus* pH showing the major species present for the designated conditions. Although a single oxidation state might be anticipated from equilibrium considerations, there are several ways in which multiple oxidation states might arise. Biological activity can produce non-equilibrium species, as evident in the alkylated metals discussed above. Whereas Mn^{IV} and Cu^{II} might be expected by thermodynamic reasoning, photochemical processes in the surface waters can lead to the formation of significant amounts of Mn^{II} and Cu^{I}. Fe^{III} is the favoured redox state of Fe in seawater, but it is relatively insoluble and exists predominantly in a colloidal phase. Photochemical reduction to Fe^{II}, which only slowly oxidizes to Fe^{III}, might act as a very important mechanism rendering Fe bioavailable to marine organisms.

Goldberg and co-workers have presented impressive information (given that seawater concentrations are as low as 1.5 pg l^{-1} for Ir and 2 pg l^{-1} for Ru) on the speciation, including redox state, of platinum group metals as a means of interpreting distributions in seawater and marine sediments.[18] Pt and Pd are stabilized in seawater as tetrachlorodivalent anions. Their relative abundance of 5 Pt:1 Pd agreeing with a factor of five difference in β_4. Pt is enriched in ferromanganese nodules following oxidation to a stable (IV) state, behaviour not observed for Pd. Rh exists predominantly in the heptavalent state, but accumulates in reducing sediments as lower valence sulfides. Au and Ag are present predominantly in solution as monovalent forms. Ag, but not Au, accumulates in anoxic coastal sediments.

3 SUSPENDED PARTICLES AND MARINE SEDIMENTS

3.1 Description of Sediments and Sedimentary Components

The sediments represent the major sink for material in the oceans. The main pathway to the sediments is the deposition of suspended particles. Such particles may be only in transit through the ocean from a continental origin or be formed *in situ* by chemical and biological processes. Sinking particles can scavenge material from solution. Accordingly, this section introduces the components found in marine sediments, but emphasises processes that occur within the water column that lead to the formation and alteration of the deposited material.

Marine sediments cover the ocean floor to a thickness averaging 500 m. The deposition rates vary with topography. The rate may be several mm per year in nearshore shelf regions, but is only 0.2–7.5 mm

Table 6   *The four categories of marine sedimentary components with examples of mineral phases*

Classification	Mineral example	Chemical formula
Lithogenous	Quartz	SiO_2
	Microcline	$KAlSi_3O_8$
	Kaolinite	$Al_4Si_4O_{10}(OH)_8$
	Montmorillonite	$Al_4Si_8O_{20}(OH)_4 \cdot nH_2O$
	Illite	$K_2Al_4(Si, Al)_8O_{20}(OH)_4$
	Chlorite	$(Mg, Fe^{2+})_{10}Al_2(Si, Al)_8O_{20}(OH, F)_{16}$
Hydrogenous	Fe–Mn minerals	$FeO(OH)–MnO_2$
	Carbonate fluoroapatite	$Ca_5(PO_4)_{3-x}(CO_3)_xF_{1+x}$
	Barite	$BaSO_4$
	Pyrite	FeS_2
	Aragonite	$CaCO_3$
	Dolomite	$CaMg(CO_3)_2$
Biogenous	Calcite	$CaCO_3$
	Aragonite	$CaCO_3$
	Opaline silica	$SiO_2 \cdot nH_2O$
	Apatite	$Ca_5(F, Cl)(PO_4)_3$
	Barite	$BaSO_4$
	Organic matter	
Cosmogenous	Cosmic spherules	
	Meteoric dusts	

From Harrison and de Mora.[22]

per 1000 years on the abyssal plains. Oceanic crustal material is formed along spreading ridges and moves outwards eventually to be lost in subduction zones, the major trenches in the ocean. Because of this continual movement, the sediments on the seafloor are no older than Jurassic in age, about 166 million years.

The formation of marine sediments depends upon chemical, biological, geological, and physical influences. There are four distinct processes that can be readily identified. Firstly, the source of the material is obviously important. This is usually the basis on which sediment components are classified and will be considered below in more detail. Secondly, the material and its distribution on the ocean floor are influenced by its transportation history, both to and within the ocean. Thirdly, there is the deposition process that must include particle formation and alteration in the water column. Finally, the sediments may be altered after deposition, a process known as diagenesis. Of particular importance are reactions leading to changes in the redox state of the sediments.

[22] R. M. Harrison and S. J. de Mora, 'Introductory Chemistry for the Environmental Sciences', 2nd Edn., Cambridge University Press, Cambridge, 1996, p. 373.

The components in marine sediments are classified according to origin. Examples are given in Table 6. Lithogenous (or terrigenous) material comes from the continents as a result of weathering processes. The relative contribution of lithogenous material to the sediments will depend upon proximity to the continent and the source strength of material derived elsewhere. The most important components in the lithogenous fraction are quartz and the clay minerals (kaolinite, illite, montmorillonite, and chlorite). The distribution of the clay minerals varies considerably. Illite and montmorillonite tend to be ubiquitous in terrestrial material, but the latter has a secondary origin associated with submarine volcanic activity. Kaolinite typifies intense weathering observed in tropical and desert conditions. Therefore, it is relatively enriched in equatorial regions. On the other hand, chlorite is indicative of the high latitude regimes where little chemical weathering occurs. The lithogenous components tend to be inert in the water column and represent detrital deposition. Nonetheless, the particle surfaces can act as important sites for adsorption of organic material and trace elements.

Hydrogenous components, also known as chemogenous or halmeic material, are those produced abiotically within the water column. This may comprise primary material formed directly from seawater upon exceeding a given solubility product, termed authigenic precipitation. The best known example of authigenic material is ferromanganese nodules found throughout the oceans. Alternatively, secondary material may be formed as components of continental or volcanic origin become altered by low temperature reactions in seawater, a mechanism known as halmyrolysis. Halmyrolysis reactions can occur in the estuarine environment, being essentially an extension of chemical weathering of lithogenous components. Such processes continue at the sediment–water interface. Accordingly, there are considerable overlaps between the terms weathering, halmyrolysis, and diagenesis. Owing to the importance that surface chemistry has on the final composition, authigenic precipitation and halmyrolysis are considered further in Section 3.2.

Biogenous (or biotic) material is produced by the fixation of mineral phases by marine organisms. The most important phases are calcite and opaline silica, although aragonite and magnesian calcite are also deposited. As indicated in Table 7, several plants and animals are involved, but the planktonic organisms are the most important with respect to the world ocean. The source strength depends upon the species composition and productivity of the overlying oceanic waters. For instance, siliceous oozes are found in polar latitudes (diatoms) and along the equator (radiolaria). The relative contribution of biogenous material to the sediments depends upon its dilution by material from other sources and

Table 7   *Quantitatively important plants and animals that secrete calcite,*
aragonite, Mg-calcite, and opaline silica

Mineral	Plants	Animals
Calcite	Coccolithophorids[a]	Foraminifera[a]
		Molluscs
		Bryozoans
Aragonite	Green algae	Molluscs
		Corals
		Pteropods[a]
		Bryozoans
Mg-calcite	Coralline (red) algae	Benthic foraminifera
		Echinoderms
		Serpulids (tubes)
Opaline silica	Diatoms[a]	Radiolaria[a]
		Sponges

From Berner and Berner.[23]
[a]Planktonic organisms.

the extent to which the material can be dissolved in seawater. As noted previously, both calcareous and siliceous skeletons are subject to considerable dissolution in the water column and at the sediment–water interface.

There are two sources that give rise to minor components in the marine sediments. Cosmogenous material is that derived from an extraterrestrial source. Such material tends to comprise small (*i.e.* <0.5 mm) black micrometeorites or cosmic spherules. The composition is either magnetite or a silicate matrix including magnetite. They are ubiquitous but scarce, with relative contributions to the sediments decreasing with an increase in sedimentation rate. Finally, there are anthropogenic components, notably heavy metals and Sn, which can have a significant influence on sediments in coastal environments.

As noted in Section 1.1, the principal modes of transport of particulate material to the ocean are by rivers or via the atmosphere. Within the oceans, distribution is further affected by ice rafting, turbidity currents, organisms, and oceanic currents. Turbidity currents refer to the turbid and turbulent flow of sediment-laden waters along the seafloor caused by sediment slumping. They are especially important in submarine canyons and can transport copious amounts of material, including coarse-grained sediments, to the deep sea. Ice rafting can also transport substances to the deep sea. Although, ice rafting is presently confined to the polar latitudes (40 °N and 55 °S), there have been considerable variations in ice

[23] E. K. Berner and R. A. Berner, 'The Global Water Cycle', Prentice-Hall, Englewood Cliffs, NJ, 1987, p. 397.

limits within the geologic record. Organisms are notable not only for biogenous sedimentation, but also because they can influence fine-grained lithogenous material that becomes incorporated into faecal pellets, consequently accelerating the settling rate. Ocean currents are important for the distribution of material with a long residence time. Major surface currents are zonal and tend to reinforce the pattern of aeolian supply. On the other hand, deep water currents are of little consequence as velocities are slow relative to the settling rates.

3.2 Surface Chemistry of Particles

3.2.1 Surface Charge. Particles in seawater tend to exhibit a negative surface charge. There are several mechanisms by which this might arise. Firstly, the negative charge can result from crystal defects (*i.e.* vacant cation positions) or cation substitution. Clay minerals are layered structures of octahedral AlO_6 and tetrahedral SiO_4. Either substitution of Mg^{II} and Fe^{II} for the Al^{III} in octahedral sites or replacement of Si^{IV} in tetrahedral location by Al^{III} can cause a net negative charge. Secondly, a surface charge can result from the differential dissolution of an electrolytic salt such as barite ($BaSO_4$). A charge will develop whenever the rate of dissolution of cations and anions differs. Thirdly, organic material can be negatively charged due to the dissociation of acidic functional groups.

Adsorption processes can also lead to the development of a negatively charged particle surface. One example is the specific adsorption of anionic organic compounds onto the surfaces of particles. Another mechanism relates to the acid–base behaviour of oxides in suspension. Metal oxides (most commonly Fe, Mn) and clay minerals have frayed edges resulting from broken metal-oxygen bonds. The surfaces can be hydrolysed and exhibit amphoteric behaviour:

$$-X-O^-_{(s)} + H^+_{(aq)} \rightleftharpoons -X-OH_{(s)}$$

$$-X-OH^-_{(s)} + H^+_{(aq)} \rightleftharpoons -X-OH^+_{2(s)}$$

The hydroxide surface exhibits a different charge depending upon the pH. Cations other than H^+ can act as the potential determining ion. The point of zero charge (PZC) is the negative log of the activity at which the surface exhibits no net surface charge. At the PZC:

$$[-X-O^-_{(s)}] \rightleftharpoons [-X-OH^+_{2(s)}]$$

The PZC for some mineral solids found in natural waters are shown in Table 8. Clearly, the extent to which such surfaces can adsorb metal

Table 8   *The point of zero charge (PZC) for some mineral phases*

Mineral	pH_{PZC}
α-Al$_2$O$_3$	9.1
α-Al(OH)$_3$	5.0
γ-AlOOH	8.2
CuO	9.5
Fe$_3$O$_4$	6.5
α-FeOOH	7.8
γ-Fe$_2$O$_3$	6.7
'Fe(OH)$_3$' (amorphous)	8.5
MgO	12.4
δ-MnO$_2$	2.8
β-MnO$_2$	7.2
SiO$_2$	2.0
ZrSiO$_4$	5
Feldspars	2–2.4
Kaolinite	4.6
Montmorillonite	2.5
Albite	2.0
Chrysotile	> 12

From Stumm and Morgan.[13]

cations will be dependent upon the pH of the solution. At the pH typical of seawater, most of the surfaces indicated in Table 8 would be negatively charged and would readily adsorb metal cations.

3.2.2 Adsorption Processes. Physical or non-specific adsorption involves relatively weak attractive forces, such as electrostatic attraction and van der Waals forces. Adsorbed species retain their co-ordinated sphere of water and, hence, cannot approach the surface closer than the radius of the hydrated ion. Adsorption is favoured by ions having a high charge density, *i.e.* trivalent ions in preference to univalent ones. Additionally, an entropy effect promotes the physical adsorption of polymeric species, such as Al and Fe oxides, because a large number of water molecules and monomeric species is displaced.

Chemisorption or specific adsorption involves greater forces of attraction than physical adsorption. As hydrogen bonding or π orbital interactions are utilized, the adsorbed species lose their hydrated spheres and can approach the surface as close as the ionic radius. Whereas multilayer adsorption is possible in physical adsorption, chemisorption is necessarily limited to monolayer coverage.

As outlined previously, hydrated oxide surfaces have sites that are either negatively charged or readily deprotonated. The oxygen atoms tend to be available for bond formation, a favourable process for transition metals. Several mechanisms are possible. An incoming metal ion, M^{z+}, may eliminate an H^+ ion as:

$$-X-O-H + M^{z+} \rightleftharpoons -X-O-M^{(z-1)+} + H^+$$

Alternatively, two or more H^+ ions may be displaced, thereby forming a chelate as:

$$
\begin{array}{c}
-X-O-H \\
\\
-X-O-H
\end{array}
+ M^{z+} \quad \rightleftharpoons \quad
\begin{array}{c}
-X-O \\
\diagdown \\
M^{(z-2)+} + 2H^+ \\
\diagup \\
-X-O
\end{array}
$$

A metal complex, ML_n^{z+}, may be co-ordinated instead of a free ion by displacement of one or more H^+ ions in a manner analogous to the above reaction. In addition, the metal complex might eliminate a hydroxide group giving rise to a metal–metal bond as:

$$-X-O-H + ML_n^{z+} \rightleftharpoons -X-ML_n^{(z-1)+} + OH^-$$

It should be noted that all of these reactions are equilibria for which an appropriate equilibrium constant can be defined and measured. These data can then be incorporated into the speciation models discussed in Section 2.5.

3.2.3 Ion Exchange Reactions. Both mineral particles and particulate organic material can take up cations from solution and release an equivalent amount of another cation into solution. This process is termed cation exchange and the cation exchange capacity (CEC) for a given phase is a measure of the number of exchange sites present per 100 g of material. This is operationally defined by the uptake of ammonium ions from 1 mol l^{-1} ammonium acetate at pH 7. The specific surface area and CEC are given in Table 9 for several sorption-active materials.

There are several factors that influence the affinity of cations towards a given surface. Firstly, the surface coverage will increase as a function of the cation concentration. Secondly, the affinity for the exchange site is enhanced as the oxidation state increases. Finally, the higher the charge density of the hydrated cation, the greater will be its affinity for the exchange site. In order of increasing charge density, the group I and II cations are:

$$Ba < Sr < Ca < Mg < Cs < Rb < K < Na < Li$$

Table 9   *The specific surface area and cation-exchange capacities of several*
sorption-active materials

Material	Specific surface area $(m^2 g^{-1})$	Cation-exchange capacity (meq/100 g)
Calcite ($<2 \mu$m)	12.5	–
Kaolinite	10–50	3–15
Illite	30–80	10–40
Chlorite	–	20–50
Montmorillonite	50–150	80–120
Freshly precipitated $Fe(OH)_3$	300	10–25
Amorphous silicic acid	–	11–34
Humic acids from soils	1900	170–590

From Förstner and Wittman.[24]

3.2.4   Role of Surface Chemistry in Biogeochemical Cycling.   Reactions
at the aqueous–particle interface have several consequences for material
in the marine environment, from estuaries to the deep sea sediment–
water boundary. Within estuarine waters, suspended particles experience
a dramatic change in the composition and concentration of dissolved
salts. A number of halmyrolysates can be formed. Clay minerals undergo
cation exchange as Mg^{2+} and Na^+ replace Ca^{2+} and K^+. Alternatively,
montmorillonite may take up K^+ becoming transformed to illite.
Hydrogenous components, in particular Mn and Fe oxides, may be
precipitated onto the surfaces of suspended particles. The particulate
material generally accumulates organic coatings within estuaries, which,
together with an increase in the ionic strength of the surrounding
solution, leads to the formation of stable colloids. Both the oxide and
organic coatings can subsequently scavenge other elements in the
estuary.

Within the ocean, the exchange of material from the dissolved to the
suspended particulate state influences the distribution of several elements. This scavenging process removes dissolved metals from solution
and accelerates their deposition. The effectiveness of this process is
obvious in the depth profiles of metals, especially those of the surface
enrichment type. Furthermore, the removal can be expressed in terms of
a deep water scavenging residence time as indicated in Table 10.

The scavenging mechanism can be particularly effective in the water–
sediment boundary region. The resuspension of fine sediments generates
a very large surface area for adsorption and ion exchange processes.
Within the immediate vicinity of hydrothermal springs, reduced species

[24] U. Förstner and G. T. W. Wittman, 'Metal Pollution in the Aquatic Environment', Springer-
Verlag, Berlin, 2nd Edn., 1981, p. 486.

Table 10   *The deep water scavenging residence times of some trace elements in the oceans*

Element	Scavenging residence time (yr)	Element	Scavenging residence time (yr)
Sn	10	Mn	51–65
Th	22–33	Al	50–150
Fe	40–77	Sc	230
Co	40	Cu	385–650
Po	27–40	Be	3700
Ce	50	Ni	15 850
Pa	31–67	Cd	117 800
Pb	47–54	Particles	0.365

From Chester.[1]

of Mn and Fe are released and subsequently are oxidized to produce colloidal oxyhydroxides, which have a large surface area and very great sorptive characteristics. Finally, ferromanganese nodules form at the sediment–water interface and become considerably enriched in a number of trace metals via surface reactions.

Ferromanganese nodules result from the authigenic precipitation of Fe and Mn oxides at the seafloor. Two morphological types are recognized, depending upon the growth mechanism. Firstly, spherical encrustations produced atop oxic deep sea sediments grow slowly, accumulating material from seawater. These seawater nodules exhibit a relatively low Mn:Fe ratio and are especially enriched with respect to Co, Fe, and Pb. Secondly, discoid shaped nodules develop in nearshore environments deriving material via diffusion from the underlying anoxic sediments. Such diagenetic nodules grow faster than deep sea varieties, and metals tend to be in lower oxidation states. They have a high Mn:Fe ratio and enhanced content of Cu, Mn, Ni, and Zn. Ferromanganese nodules have concentric light and dark bands in cross-section, related to Fe and Mn oxides, respectively. Patterns of trace element enrichment in the nodules are determined by mineralogy, the controlling mechanisms being related to cation substitution in the crystal structure. Mn phases preferentially accumulate Cu, Ni, Mo, and Zn. Alternatively, Co, Pb, Sn, Ti, and V are enriched in Fe phases.

3.3   Diagenesis

Diagenesis refers to the collection of processes that alters the sediments following deposition. These mechanisms may be physical (compaction), chemical (cementation, mineral segregation, ion exchange reactions), or biological (respiration). The latter are of particular importance as the bacteria control pH and pε in the interstitial waters, master variables that affect a wide range of equilibria. They influence the composition of the interstitial water, which in turn can exert a feedback effect on the overlying seawater. Also, they can ultimately control the mineralogical phases that are lost to the sediment sink.

Organic material accumulates with other sedimentary components at the time of deposition. High biological activity in surface waters and rapid sedimentation ensures that most nearshore and continental margin sediments contain significant amounts of organic matter. Biochemical oxidation of this material exhausts the available O_2, creating anoxic conditions. The oxic/anoxic boundary occurs at the horizon where the respiratory consumption of O_2 balances its downward diffusion. Upon depleting the O_2, other constituents are used as oxidants leading to the stepwise depletion of NO_3^-, NO_2^-, and SO_4^{2-}. Thereafter, organic matter itself may be utilized with the concurrent production of CH_4.

This series of reactions causes progressively greater reducing conditions, with consequent influences on the chemistry of several elements. Metals are reduced and so are present in lower oxidation states. In particular, Mn undergoes reductive dissolution from $MnO_{2(s)}$ to $Mn_{(aq)}^{2+}$. As the divalent state is much more soluble, Mn is effectively remobilized under anoxic conditions and can be released back into overlying seawater. As seen in the previous section, this is one pathway to ferromanganese nodule formation. This can also be true for other elements that had been deposited following incorporation into the Fe and Mn oxide phases. In contrast, some elements can be preserved very effectively in anoxic sediments. Interstitial waters in marine sediments, in contrast to freshwater deposits, have high initial concentrations of SO_4^{2-}. Bacterial sulfate reduction proceeds via the reaction:

$$2CH_2O + SO_4^{2-} \rightleftharpoons H_2S + 2HCO_3^-$$

Thus, sulfide levels in interstitial waters increase. A number of elements form insoluble sulfides, which under these anoxic conditions are precipitated and retained within the sediments. A notable example is the accumulation of pyrite, FeS_2, but also Ag, Cu, Pb, and Zn are enriched in anoxic sediments in comparison with oxic ones.

4 PHYSICAL AND CHEMICAL PROCESSES IN ESTUARIES

Rivers transport material in several phases: dissolved, suspended particulate, and bedload. Physical and chemical processes within an estuary influence the transportation and transformation of this material, thereby affecting the net supply of material to the oceans. Several definitions and geomorphologic classifications of estuaries have been reviewed by Perillo.[25] From a chemical perspective, an estuary is most simply described as the mixing zone between river water and seawater characterized by sharp gradients in the ionic strength and chemical composition. Geographic distinctions can be made between drowned river valleys, fjords and bar built estuaries. They can alternatively be classified in terms of the hydrodynamic regime as:

1. salt wedge;
2. highly stratified;
3. partially stratified;
4. vertically well mixed.

The aqueous inputs into the system are river flow and the tidal prism. The series above is ranked according to the diminishing importance of the riverine flow and the increasing marine contribution. Thus, a salt wedge estuary represents the extreme case, dominated by river flow, in which very little mixing occurs. A fresh, buoyant layer flows outward over denser, saline waters. In contrast, the vertically well mixed estuary is one dominated by the tidal prism. The inflowing river water mixes thoroughly with and dilutes the seawater, but the effective dilution diminishes with distance along the mixing zone.

The position of the mixing zone in the estuary exhibits considerable temporal variations. There can be a strong seasonal effect, largely due to non-uniform river discharge. High winter rainfall leads to a winter-time discharge maximum. However, winter precipitation as snow creates a storage reservoir, such that the river flow maximum occurs following snow melt in spring or even early summer if the catchment area is of high elevation as for the Fraser River in western Canada. On shorter time scales, the mixing zone is influenced by tidal cycles. Thus, the penetration of seawater into the estuary depends upon the spring–neap tidal cycle and the diurnal nature of the tides. Together these influences determine the geographic extent of sediments experiencing a variable salinity regime. Variations in the river discharge affects the mass loading of the discharge, both in terms of suspended sediment and bedload material. The hydrodynamic regime in the estuary influences the deposition of the riverine sedimentary material and the mixing of dissolved material.

The estuary is a mixing zone for river water and seawater, the characteristics of which differ considerably. River water is slightly

[25] G. M. E. Perillo, 'Geomorphology and Sedimentology of Estuaries', ed. G. M. E. Perillo, Elsevier Science, Amsterdam, 1995, p. 17.

acidic and of low ionic strength with a salt matrix predominantly of $Ca(HCO_3)_2$ (~ 120 mg l^{-1}). In contrast, seawater has a higher pH (~ 8), higher ionic strength (~ 0.7) and consists primarily of NaCl (~ 35 g l^{-1}). As a consequence, the salt matrix within the estuary is dominated by the sea-salt end member throughout the mixing zone except for a small proportion at the dilute extreme. Salinity can be used as a conservative index, although conductivity is better, not being subject to systematic conversion errors in the initial mixing region.

In a plot of concentration *versus* some conservative index (*i.e.* S‰, Cl‰, or conductivity), the theoretical dilution curve would comprise a straight line between the river and seawater end members. A dissolved constituent that exhibits such a distribution is said to behave conservatively in the estuary. Whereas a negative slope shows that the riverine end member is progressively diluted during mixing with seawater, a positive slope indicates that the seawater end member has the greater concentration. Conservative behaviour is exhibited, for example, by Na^+, K^+, and SO_4^{2-}. Reactive silica may at times behave conservatively. Non-conservative behaviour can result from an additional supply of material (causing positive deviations from the theoretical dilution curve). Elements that may show a maximum concentration at some intermediate salinity are Mn and Ba. Alternatively, the removal of dissolved material during mixing (*i.e.* negative deviation from the theoretical dilution curve) can be caused by biological activity or by dissolved to particulate transformations. Biological activity can cause non-conservative behaviour for nutrient elements, including reactive silica. Dissolved constituents typically transformed to the particulate phase include Al, Mn, and Fe (see Figure 12) in some estuaries. The pH distribution is usually characterized by a pH minimum in the initial mixing zone, resulting from the non-linear salinity dependence of the first and second dissociation constants of H_2CO_3. Notwithstanding the obvious utility of component–conservative index plots, they can be applied and interpreted only with caution. Often it is difficult to define the exact composition of the end members. Hence, a plot that apparently denotes non-conservative behaviour could arise if temporal fluctuations in the concentration of the component of interest occur on the same relative time scales as estuarine flushing.

A component can undergo considerable physico-chemical speciation alterations in an estuary. With respect to dissolved constituents, the composition and concentration of available ligands changes. Depending upon the initial pH of the riverine water, OH^- may become markedly more important down the estuary. Similarly, chlorocomplexes for metals such as Cd, Hg, and Zn become more prevalent as the salinity increases. Conversely, the competitive influence of seawater-derived Ca and Mg for

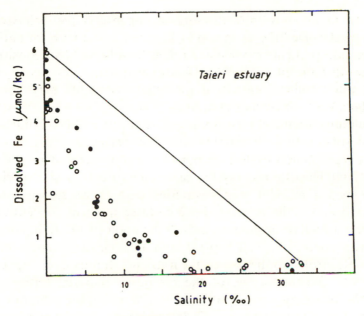

Figure 12   *The distribution of dissolved Fe versus salinity in the Taieri Estuary, New Zealand;* ● *21 October 1980,* ○ *4 December 1980* (From Hunter[26])

organic material decreases the relative importance of humic complexation for Mn and Zn.

Estuaries are particularly well known for dissolved–particulate interactions. Phase changes come about via several mechanisms. Firstly, dissolution—precipitation processes may occur. This is especially important for the authigenic precipitation of Fe and Mn oxyhydroxides. Secondly, components may experience adsorption–desorption reactions. Desorption can occur in the initial mixing zone, partly in response to the pH minimum. Adsorption, particularly in association with the Fe and Mn phases, can accumulate material within the suspended sediments. Thirdly, flocculation and aggregation processes can remove material from solution. This occurs as particulate material with negatively charged surfaces adsorb cations in the estuary. The surface charge diminishes and as the ionic strength increases, the particles experience less electrostatic repulsion. Eventually the situation arises whereby particle collisions lead to aggregation due to weak bonding. This process can be facilitated if particle surfaces are coated with organic material (and is then known as flocculation rather than aggregation). The three types of processes outlined above often happen simultaneously

[26] K. Hunter, *Geochim. Cosmochim. Acta*, 1983, **47**, 467.

in the estuarine environment. Such non-conservative behaviour is typified by dissolved Fe, as shown in Figure 12. The transformation of dissolved into particulate phases can then be followed by deposition to the estuarine sediments. Thus, the flux of material to the ocean can be considerably modified, particularly as such sediments may be transported landward rather than seaward.

Biological activity in the estuarine environment can also influence the speciation of constituents, notably dissolved–particulate partitioning. A complex regeneration cycle determines distributions and modifies the riverine flux. This is especially the case for nutrients, and estuaries are often termed a nutrient trap. Estuaries tend to be regions of high biological productivity as rivers have elevated nutrient concentrations. Moreover, several freshwater organisms die upon encountering brackish water with consequent cell rupture and the release of contents into solution. Regardless of the source, the nutrients stimulate phytoplankton productivity. Whereas some of the biological material (either transported via the freshwater or produced *in situ* within the estuarine zone) can be flushed out to sea, the remainder settles out of the surface water to be deposited onto the floor of the estuary. Respiration of this debris by benthic organisms regenerates nutrients and any contaminants that have been accumulated, releasing them into the bottom water. In estuaries where a two-layer flow is well defined, these nutrients are transported upriver in the salt wedge and entrained into the exiting river water, thereby adding to the available nutrient pool. Polluted rivers having a high nutrient loading are subject to eutrophication due to overstimulation of biological activity. Anoxic conditions within the bottom waters and/or underlying sediments can result, depending upon the organic loading. As mentioned previously (Section 2.2.2), some fjords develop anoxic conditions when bottom waters stagnate due to limited mixing.

5 MARINE CONTAMINATION AND POLLUTION

Both contamination and pollution entail the perturbation of the natural state of the environment by anthropogenic activity. The two terms are distinguishable in terms of the severity of the effect, whereby pollution induces the loss of potential resources.[27] Additionally, a clear cause–effect relationship must be established for a substance to be classified as a pollutant towards a particular organism. The human-induced disturbances take many forms but the greatest effects tend to be in coastal environments due to the source strengths and pathways. Waters and sediments in coastal regions bear the brunt of industrial and sewage

[27] E. Goldberg, *Mar. Pollut. Bull.*, 1992, **25**, 45.

discharges, and are subject to dredging and spoil dumping. Agricultural runoff may contain pesticide residues and elevated nutrients, the latter of which may overstimulate biological activity producing eutrophication and anoxic conditions. The deep sea has not escaped contamination. The most obvious manifestations comprise crude oil, petroleum products, and plastic pollutants, but includes the long-range transport of long-lived radionuclides from coastal sources. Additionally, the aeolian transport of heavy metals has enhanced natural fluxes of some elements, particularly lead. Three case studies are introduced below to illustrate diverse aspects of marine contamination and pollution.

5.1 Oil Slicks

Major releases of oil have been caused by the grounding of tankers (*e.g. Torrey Canyon*, South-West England, 1967; *Argo Merchant*, Nantucket Shoals, USA, 1976; *Amoco Cadiz*, North-West France, 1978; *Exxon Valdez*, Alaska, 1989) or by the accidental discharge from offshore platforms (*e.g. Chevron MP-41C*, Mississippi Delta, 1970; *Ixtox I*, Gulf of Mexico, 1979). Because oil spills receive considerable public attention and provoke substantial anxiety, oil pollution must be put into perspective. Crude oil has been habitually introduced into the marine environment from natural seeps at a rate of approximately $340 \times 10^6 \, l \, y^{-1}$. Anthropogenic activity has recently augmented this supply by an order of magnitude; however, most of this additional oil has originated from relatively diffuse sources relating to municipal runoff and standard shipping operations. Exceptional episodes of pollution occurred in the Persian Gulf in 1991 ($910 \times 10^6 \, l \, y^{-1}$) and due to the *Ixtox I* well in the Gulf of Mexico in 1979 ($530 \times 10^6 \, l \, y^{-1}$). In contrast to such mishaps, the *Amoco Cadiz* discharged only $250 \times 10^6 \, l \, y^{-1}$ of oil in 1978 accounting for the largest spill from a tanker. The cumulative pollution from tanker accidents on an annual basis matches that emanating from natural seepage. Nevertheless, the impacts can be severe when the subsequent slick impinges on coastal ecosystems.

Regardless of the source, the resultant oil slicks are essentially surface phenomena that are affected by several transportation and transformation processes.[28] With respect to transportation, the principal agent for the movement of slicks is the wind, but length scales are important. Whereas small (*i.e.* relative to the slick size) weather systems, such as thunderstorms, tend to disperse the slick, cyclonic systems can move the slick essentially intact. Advection of a slick is also affected by waves and

---

[28] S. Murray, 'Pollutant Transfer and Transport in the Sea', ed. G. Kullenberg, CRC Press, Boca Raton, 1982, Vol. 2, p. 169.

currents. To a more limited extent, diffusion can also act to transport the oil.

Transformation of the oil involves phase changes and/or degradation. Several physical processes can invoke phase changes. Evaporation of the more volatile components is a significant loss mechanism, especially for light crude oil. Oil slicks spread as a buoyant lens under the influence of gravitational forces, but generally separate into distinctive thick and thin regions. Such pancake formation is due to the fractionation of the components within the oil mixture. Sedimentation can play a role in coastal waters when rough seas bring dispersed oil droplets into contact with suspended particulate material and the density of the resulting aggregate exceeds the specific density of seawater. Colloidal suspensions can consist of either water-in-oil or oil-in-water emulsions, which behave distinctly differently. Water-in-oil emulsification creates a thick, stable colloid that can persist at the surface for months. The volume of the slick increases and it aggregates into large lumps known as 'mousse', thereby acting to retard weathering. Conversely, oil-in-water emulsions comprise small droplets of oil in seawater. This aids dispersion and increases the surface area of the slick, which can accelerate weathering processes.

Chemical transformations of oil are evoked through photochemical oxidation and microbial biodegradation. Not only is the latter more important in nature, but strategies can be adopted to stimulate biological degradation, consequently termed bioremediation. All marine environments contain micro-organisms capable of degrading crude oil. Furthermore, most of the molecules in crude oils are susceptible to microbial consumption. Oil contains little nitrogen or phosphorus so microbial degradation of oil tends to be nutrient limited. Bioremediation often depends upon the controlled and gradual delivery of these nutrients, while taking care to limit the concurrent stimulation of phytoplankton activity. Approaches that have been adopted are the utilization of slow-release fertilizers, oleophilic nutrients and a urea-foam polymer fertilizer incorporating oil-degrading bacteria. Bioremediation techniques were successfully applied in the clean-up of Prince William Sound and the Gulf of Alaska following the *Exxon Valdez* accident. Alternative bioremediation procedures relying on the addition of exogenous bacteria have still to be proved. Similarly, successful bioremediation of floating oil spills has yet to be demonstrated.

Source apportionment of crude oil in seawater and monitoring the extent of weathering and biodegradation constitute important challenges in environmental analytical chemistry. As the concentration of individual compounds varies from one sample of crude oil to another, the relative amounts define a signature characteristic of the source. Compounds that degrade at the same rate stay at fixed relative amounts throughout the lifetime of an oil slick. Hence, a 'source ratio', which represents the concentration ratio for a pair of compounds exhibiting such behaviour remains constant. Conversely, a 'weathering ratio' reflects the concentration ratio for two compounds that degrade at different rates and consequently continually changes. Oil spill monitoring programmes conventionally determine four fractions:[29]

1. volatile hydrocarbons;
2. alkanes;
3. total petroleum hydrocarbons;
4. polycyclic aromatic hydrocarbons (PAHs).

The volatile hydrocarbons, albeit comparatively toxic to marine organisms, evaporate relatively quickly and hence serve little purpose as diagnostic aids. The alkanes and total petroleum hydrocarbons make up the bulk of the crude oil. They can be used to some extent for source identification and monitoring weathering progress. The final fraction, the PAHs, comprises only about 2% of the total content of crude oil but includes compounds that are toxic. Moreover, these components exhibit marked disparities in weathering behaviour due to differences in water solubility, volatility and susceptibility towards biodegradation. As demonstrated in Figure 13, both a source ratio (C3-dibenzothiophenes: C3-phenanthrenes) and a weathering ratio (C3-dibenzothiophenes: C3-chrysenes) have been defined from amongst such compounds that enable the extent of crude oil degradation to be estimated in the marine environment, as well as for subtidal sediments and soils.[29]

5.2 Plastic Debris

The increasing accumulation of litter and debris along shorelines epitomizes a general deterioration of environmental quality on the high seas. The material originates not only from coastal sources, but also arises from the ancient custom of dumping garbage from ships. Drilling rigs and offshore production platforms have similarly acted as sources of contamination. Some degree of protection in recent years has accrued from both the London Dumping Convention (LDC) and the International Convention for the Prevention of Pollution from Ships (MARPOL) which outlaw such practices. However, the problem of seaborne litter remains global in extent and not even Antarctica has been left unaffected.[30]

[29] G. S. Douglas, A. E. Bence, R. C. Prince, S. J. McMillen and E. L. Butler, *Environ. Sci. Technol.*, 1996, **30**, 2332.
[30] M. R. Gregory and P. G. Ryan, 'Marine Debris: sources, impacts and solutions', ed. J. M. Coe and D. B. Rogers, Springer, New York, 1996, p. 49.

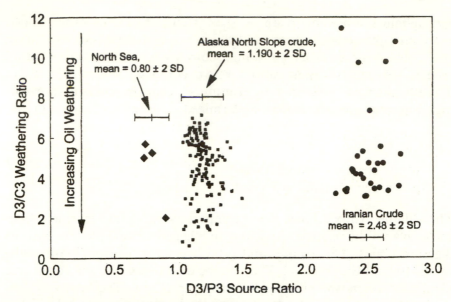

Figure 13   *Plot of weathering ratio (C3-dibenzothiophenes:C3-chrysenes) versus source ratio (C3-dibenzothiophenes:C3-phenanthrenes) for fresh and degraded oil samples from three different crude oil spills (From Douglas et al.[29])*

The debris consists of many different materials, which tend to be non-degradable and endure in the marine environment for many years. The most notorious are the plastics (*e.g.* bottles, sheets, fishing gear, packaging materials, and small pellets), along with glass bottles, tin cans, and lumber. This litter constitutes an aesthetic eyesore on beaches, but more importantly can be potentially lethal to marine organisms. Deleterious impacts on marine birds and mammals result from entanglement and ingestion. Lost or discarded plastic fishing nets remain functional and can continue 'ghost fishing' for several years. This is similarly true for traps and pots that go astray. Plastic debris settling on soft and hard bottoms can smother benthos and limit gas exchange with pore waters. Despite the negative effects of seaborne plastic debris, this material can have positive consequences, serving as new habitats for opportunistic colonizers.

5.3   Tributyltin

Tributyltin (TBT) provides an interesting case study of a pollutant in the marine environment.[31] Because TBT compounds are extremely poison-

[31] 'Tributyltin: case study of an environmental contaminant', ed. S. J. de Mora, Cambridge University Press, Cambridge, 1996, p. 301.

ous and exhibit broad-spectrum biocidal properties, they have been utilized as the active ingredient in marine anti-fouling paint formulations. Its potency and longevity ensures good fuel efficiencies for ship operations and guarantees a long lifetime between repaintings. TBT-based paints have been used on boats of all sizes, from small yachts to supertankers, ensuring the global dispersion of TBT throughout the marine environment, from the coastal zone to the open ocean.

Notwithstanding the benefits, TBT's extreme toxicity and environmental persistence has resulted in a wide range of deleterious biological effects on non-target organisms. TBT is lethal to some shellfish at concentrations as low as 0.02 μg TBT-Sn l^{-1}. Lower concentrations result in sub-lethal effects, such as poor growth rates and reduced recruitment leading to the decline of shellfisheries. The most obvious manifestations of TBT contamination have been shell deformation in Pacific oysters (*Crassostrea gigas*), and the development of imposex (*i.e.* the imposition of male sex organs on females) in marine gastropods. The latter effect has caused dramatic population decline of gastropods at locations throughout the world. Laboratory experiments and field observations of deformed oysters and imposex indicate that adverse biological effects occur at concentrations below those detectable. Thus, a 'no-effect' concentration has yet to be demonstrated. TBT has been observed to accumulate in fish and various marine birds and mammals, with as yet unknown consequences. Although it has not been shown to pose a public health risk, one study reported measurable butyltin concentrations in human liver.[32]

The economic consequences of the shellfisheries' decline led to a rapid political response globally. The first publication suggesting TBT to be the causative agent only appeared in 1982,[33] but already the use of TBT-based paints has been banned in some countries, including New Zealand. Other nations have imposed partial restrictions, its use being permitted only on vessels >25 m in length or on those with aluminium hulls and outdrives. This has certainly had the effect to decrease the TBT flux to the marine environment as manifested in sedimentary TBT profiles. Oyster aquaculture in Arcachon Bay benefitted immediately, with a notable decline in shell deformations and TBT body burdens and the complete recovery of production within two years (Figure 14). Comparable improvements in oyster conditions have been reported for Great Britain and Australia. Similarly, there have been many reported instances of restoration of gastropod populations at previously impacted locations. However, large ships continue to act as a source of TBT to the

[32] K. Kannan and J. Falandysz, *Mar. Pollut. Bull.*, 1997, **34**, 203.
[33] C. Alzieu, M. Heral, Y. Thibaud, M. J. Dardignac, and M. Feuillet, *Rev. Trav. Inst. Marit.*, 1982, **45**, 100.

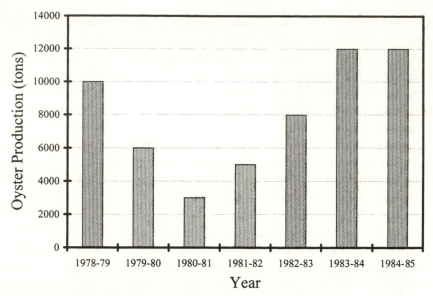

Figure 14   *Annual oyster* (Crassostrea gigas) *production in Arcachon Bay, 1978–85;*
restrictions on TBT use were applied starting January 1982
(Data from Alzieu[34])

marine environment. It should be of concern that imposexed gastropods
have been observed at sites (*e.g.* North Sea and Strait of Malacca) where
the source of TBT can only be attributed to shipping.

TBT exists in solution as a large univalent cation and forms a neutral
complex with Cl^- or OH^-. It is extremely surface active and so is readily
adsorbed onto suspended particulate material. Such adsorption and
deposition to the sediments limits its lifetime in the water column.
Degradation, via photochemical reactions or microbially mediated pathways, obeys first order kinetics. Several marine organisms, as diverse as
phytoplankton to starfish, debutylate TBT. Stepwise debutylation produces di- and mono-butyltin, which are much less toxic in the marine
environment than is TBT. As degradation rates in the water column are
of the order of days to weeks, they are slow relative to sedimentation.
TBT accumulates in the sediments where degradation rates are much
slower, with the half-life being of the order of years.[35] Furthermore,
concentrations are highest in those areas, such as marinas and harbours,
which are most likely to undergo dredging. The intrinsic toxicity of TBT,
its persistency in the sediments and its periodic remobilization by
anthropogenic activity are likely to retard the long-term recovery of the
marine ecosystem.

[34] C. Alzieu, *Mar. Environ. Res.*, 1991, **32**, 7.
[35] C. Stewart and S. J. de Mora, *Environ. Technol.*, 1990, **11**, 565.

Recapitulating, the unrestricted use of TBT has ended in many parts of the world but significant challenges remain. For the most part, the coastal tropical ecosystems remain unprotected and the sensitivity of its indigenous organisms relatively poorly evaluated. TBT endures in sediments globally, with concentrations usually greatest in environments most likely to be perturbed. The widespread introduction of TBT into seawater continues from vessels not subject to legislation. Organisms in regions hitherto considered to be remote now manifest TBT contamination and effects. Such observations imply further restrictions on the use of organotin-based paints are required. However, the paramount lesson learned from TBT should be that potential replacement compounds must be properly investigated prior to their introduction in order to avoid another global pollution experiment.

QUESTIONS

1. What are the main sources and sinks for dissolved and particulate metals entering the ocean?
2. Define residence time and outline the factors that influence the residence time of an element in the ocean. Provide examples of oceanic residence times for elements that span the time scale.
3. Define salinity. What are the notable variations in sea-salt, in terms of concentration and composition, in the world ocean?
4. Using data available in Table 4, calculate the concentration of H_2CO_3 in seawater at $5\,°C$ in equilibrium with atmospheric CO_2 at 370 ppm.
5. Calculate the speciation of H_2S in seawater at pH 8.3 at $25\,°C$ given that $pK_1 = 7.1$ and $pK_2 = 17.0$.
6. Explain how DMS of marine origin might affect global climate.
7. What are the major nutrients in the ocean? Describe their concentration profiles and account for any differences according to chemical behaviour.
8. Describe the different types of depth profile that various metals exhibit and explain how differences in their profiles originate.
9. Assuming that $\gamma = 1$ for all species, calculate the speciation of lead in seawater at $25\,°C$ given the following values for stability constants for the chlorocomplexes ($K_1 = 10^{7.82}$, $\beta_2 = 10^{10.88}$, $\beta_3 = 10^{13.94}$, $\beta_4 = 10^{16.30}$). Note that $[Cl^-]$ is 0.559 mmol l^{-1}.
10. With specific reference to question 9, outline the potential limitations of using equilibrium models to explain chemical behaviour of trace metals in seawater.
11. Describe the classification of marine sediments and give examples of each sediment type.

12. Why do suspended particles exhibit a surface charge and how could this characteristic moderate the composition of an ocean's waters?

13. With respect to estuarine chemistry, describe conservative and non-conservative behaviour and provide examples of cationic and anionic species for each category.

14. What natural processes are responsible for weathering an oil slick?

15. Why is tributyltin considered such a potent pollutant in the marine environment?

CHAPTER 5

Land Contamination and Reclamation

B. J. ALLOWAY

1 INTRODUCTION

In many parts of the world, especially in the more technologically developed countries, soil, which forms the surface layer of the land, has been exposed to varying degrees of contamination from a wide variety of chemicals for many years, especially since the Second World War. Soil constitutes one of the three key environmental media, the other two being air and water, but the soil differs significantly from these two media because it is predominantly solid and has the capacity to retain many types of pollutants. These retention (sorption) mechanisms cause soil to function as sinks for contaminants. In this way soil acts as a filter, which reduces or prevents contaminants reaching the groundwater. Furthermore, the diverse populations of micro-organisms make soils powerful bioreactors which can degrade many hazardous organic chemical contaminants in addition to the normal biogeochemical cycles which are a vital component of terrestrial ecosystems.

Some contaminants which are sorbed and not degraded will gradually accumulate and could reach concentrations that are potentially harmful to the functioning of the soil–plant system or to consumers of crops which take up these chemicals from contaminated soils. In contrast, pollutant concentrations in water and air become diluted through dispersion in these fluid media and, although larger volumes of these media can be affected by pollution, the effect is usually of limited duration.

The range of chemicals found contaminating soils is vast and can include, for example: heavy metals, such as lead from paints and motor vehicle exhausts, polycyclic aromatic hydrocarbons (PAHs) formed during the incomplete combustion of organic materials such as wood and coal, synthesized organic chemicals such as chlorinated solvents or pesticides from leakages or direct application, radionuclides in fallout from the testing of nuclear weapons or accidents at nuclear power stations, and contaminants (such as Cd) reaching agricultural land as impurities in fertilizers and manures. Some of the most contaminated land is found at the sites of former or current industrial operations. These can include metalliferous mines and smelters, chemical works, gas manufacturing plants, petroleum refineries, and scrap yards. At these types of sites, the contaminants may be raw materials, process-related chemicals such as catalysts and electrodes, manufactured chemical products, and a wide range of wastes.

The distinction between 'pollution' and 'contamination' is used by some authors to indicate the relative severity of adverse effects. 'Contamination' is sometimes used for a situation where a substance resulting from human activity is present in the environment but not causing any obvious harmful effects. In contrast, the more pejorative term 'pollution' is often used when a substance is present and having a harmful effect. Unfortunately, although this distinction is very convenient, it often does not hold true when more detailed studies are made of contamination in soils or other components of the environment. Although not so obvious, toxic or other harmful effects can often be found in situations which have been previously classed as contamination. However, it is the convention to refer to all cases of pollution or contamination of soils/land as contamination. Therefore 'contaminated land' covers the whole range from the presence of low concentrations of a chemical from an extraneous source right through to a case of severe toxicity in plants and/or hazards to animals and humans.

In the UK, contaminated land is defined as 'land which represents an actual or potential hazard to health or the environment as a result of current or previous use'. A more severe case is presented by 'derelict land' which is defined as 'land so damaged by industrial or other development that it is incapable of beneficial use without treatment'. Although it has been customary to distinguish between contamination and pollution, the former being where a substance is recognized as being present but not causing a recognizable problem and the latter where a problem (usually toxicity) is recognized, it is the convention in reference to land that the term 'contaminated' is used for all situations. It has been estimated that around 40,000 ha in England and Wales constitute 'contaminated land' and that almost all the land surface in the UK is contaminated to at least a slight extent. Atmospheric deposition of contaminants is ubiquitous although it will be of higher magnitude close to, or downwind of, major sources, but soils become contaminated by other means including direct application of materials containing contaminants and by flooding.

In view of the widespread occurrence of land contamination, it is important that soils at sites suspected of being contaminated are sampled appropriately and analysed to determine the nature of the contaminants and the concentrations involved in order that a risk assessment can be made. In addition to the risk of toxicity to ecosystems, crops, livestock, and humans, damage to materials and services are also important considerations. These include the sulfate attack of concrete, degradation of plastics used in piping and insulation of electric cables by hydro-carbons of various types, and the risk of fires or explosions from gases and other combustible materials. The risks to humans include respira-tory and other disorders from the inhalation of toxic fumes and dusts, direct ingestion of toxic compounds in contaminated drinking water or on particles of soil, absorption of chemicals through the skin, and the consumption of food plants which have accumulated excessive concen-trations of potentially harmful substances. In general, severe contamina-tion, where acutely toxic conditions prevent the growth of most types of food crops and other vegetation, presents less of a health risk (with regard to ingestion) than conditions where a significant accumulation of potentially toxic substances such as cadmium by food crops can occur without causing obvious toxic symptoms. In the latter situation, crops are likely to be consumed without the chronic toxicity risk being appreciated, whereas crop plants which die or display symptoms of stress are less likely to be consumed.

In addition to contamination, which occurs as a result of human activity (anthropogenic), anomalously high concentrations of potentially harmful inorganic substances can occur in soils as a result of geochemical processes without any human involvement. Although not strictly 'con-tamination' or pollution, these geochemical anomalies can cause toxicity and material degradation problems. Examples include: marine black shale rocks which can contain relatively high concentrations of harmful elements such as cadmium, lead, uranium, and zinc and sulfides.

This chapter aims to show the breadth of the subject of soil contam-ination and its reclamation and to introduce some of the fundamental concepts relating to the behaviour of contaminants in soils. However, the space available does not permit the presentation of much detail, and the reader is recommended to consult other more specialist texts for further information.

2   SOIL: ITS FORMATION, CONSTITUENTS, AND PROPERTIES

Soil is the complex biogeochemical material which forms at the interface between the earth's crust and the atmosphere, and differs markedly in physical, chemical, and biological properties from the underlying weathered rock from which it has developed. Soil comprises a matrix of mineral particles and organic material, often bound together as aggregates, and populated by a wide range of micro-organisms, soil animals, and plant roots. The spaces between particles and aggregates form a system of pores, which are filled with aqueous soil solution and gases. The physical, chemical, and biological properties of soils are highly relevant to considerations of contaminated land and its reclamation.

2.1 Soil Formation

The soil profile (a vertical section from the surface to the underlying weathered rock) is the unit of study in pedology (which is the study of the origin, occurrence and classification of soils). Natural pedogenic processes bring about the differentiation of the soil material into distinct horizontal layers, called horizons, and the characteristics and assemblages of these horizons forms the basis of soil morphological classification. A typical soil profile with a summary of the distribution of contaminants is shown in Figure 1.

The type of soil that forms at any site is determined by climate, the nature of the parent rock material on which it forms, the landscape

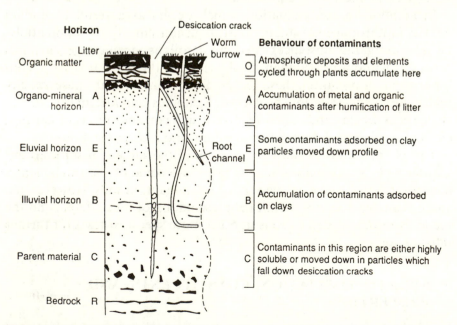

Figure 1   *Diagram of a soil profile showing horizon nomenclature and the general distribution of contaminants*

(especially whether it is a free-draining or water collecting site), the vegetation, and the time period over which the soil has been forming. These factors control the intensity and type of pedogenic processes operating at any site and determine the nature of the soil profile that develops. The major pedogenic processes include: leaching, elution, podzolization, calcification, ferralitization, laterization, salination, solodization, gleying, and peat formation and several of these usually interact in any soil although one is visually predominant.[1] With the exception of the last two, most pedogenic processes involve the vertical movement of solutes and material (usually the products of weathering) either down or up the soil profile, depending on the overall balance of precipitation and evaporation. Movement is predominantly downward in humid regions and upward in arid and semi-arid areas where evaporation exceeds precipitation. Gleying is the creation of reducing conditions in soils, due to intermittent or permanent waterlogging, and leads to the reduction and dissolution of oxides of Fe and Mn.

The pedogenic processes occurring within a soil will have a major influence on the behaviour of contaminants. For example, processes involving leaching down the profile will tend to transport soluble contaminants toward the water table. However, this behaviour is modified by reactions of the contaminant chemicals with the solid matrix of the soil. Soil organic matter, especially colloidal humus, has a strong capacity to sorb a wide range of substances, including heavy metals and many different types of organic contaminants.

2.2 Soil Constituents

It is important to note that true (pedological) soils have many features in common, and several predictions can be made about the behaviour of contaminants within them. However, in some types of contaminated land, especially that at long-established industrial sites (either active or derelict), the material forming the ground under and around buildings may differ markedly from a pedological soil. This is because many industrial sites have undergone continual development where manufacturing processes, products, and wastes have changed over many years. Frequently this involved demolishing old buildings and using old construction material and wastes as foundations for new operations. Thus, much of the material in the top 2 or 3 metres may comprise building rubble, underground pipes and tanks as well as contaminant chemicals. It is essentially 'made land' and bears little similarity to undisturbed (pedological) soils. However, although some of these con-

---

[1] E. M. Bridges, 'World Soils', Cambridge University Press, Cambridge, 1978.

stituents may differ from the normal range of soil minerals and humus, the principles affecting the physical and chemical behaviour of contaminants are still the same although it will be necessary to measure the properties of these 'non-soil' constituents at each site. Soil containing large amounts of concrete or cement rubble may be highly permeable and so prone to leaching, and it is also likely to have a high pH from the alkaline constituents of the cement. The alkaline cement is itself a soil contaminant.

2.2.1 The Mineral Fraction. The soil clay fraction comprises platelike particles <2 μm in diameter and is formed mainly of clay minerals with some small particles of other minerals such as Fe, Al, and Mn oxides. Clay minerals are secondary minerals which have been synthesized from the weathering products of primary minerals. Clay minerals are sheet silicates formed from two basic components: a silica sheet formed of Si—O tetrahedra, and a gibbsite sheet comprising Al—OH octahedra. Isomorphous substitution of either Al^{3+} for Si^{4+}, or Mg^{2+} for Al^{3+} in the structural sheets of some clay minerals gives rise to excess negative charges on their surface. This permanent excess negative charge is independent of the soil pH and provides clay minerals with their ability to adsorb cations. This adsorption on clay minerals and organic matter is referred to as cation exchange and a soil's capacity for this is referred to as its cation exchange capacity (CEC) which is measured in centimoles charge per kg ($cmol_c$ kg^{-1} = 10^{-2} mol cation exchange capacity per kg of soil). The most commonly occurring types of clay minerals are:

Kaolinite —a highly stable 1:1 (silica:gibbsite) mineral, which is non-swelling with a relatively small surface area (5–100 m^2 g^{-1}) and a low adsorptive capacity (cation exchange capacity [CEC] 2–20 $cmol_c$ kg^{-1}).

Smectites —are 2:1 (silica:gibbsite) minerals which swell on wetting and shrink on drying, they have a large surface area (700–800 m^2 g^{-1}) due to the access of soil solution to all lamellae surfaces which, together with isomorphous substitution, contributes to their relatively high adsorptive capacity (CEC 80–120 $cmol_c$ kg^{-1}). Some of the better known members of this group of minerals include montmorillonite and bentonite.

Illite —a 2:1 clay like the smectites with a lower adsorptive and swelling/shrinking capacity and properties intermediate between kaolinite and smectites (surface area 100–200 m^2 g^{-1}, a CEC 10–40 $cmol_c$ kg^{-1}).

Vermiculite—a 2:1 mineral with a very high degree of isomorphous substitution of Mg^{2+} for Al^{3+} in the gibbsite sheet giving it a high CEC (100–150 $cmol_c$ kg^{-1}).

Other ubiquitous secondary minerals in soils include oxides of Fe and Al which are the ultimate residual weathering products of soils. They are high stable under the oxidizing conditions found in freely draining soil environments and the predominant brown colour of soils is due to the presence of Fe oxides. The amount of these oxides present in a soil depends on the mineralogy of the parent rock, the degree of weathering, and the oxidation–reduction (redox) conditions. The ferromagnesian minerals such as olivine, augite, and biotite mica found in basic igneous rocks tend to be the richest primary sources of Fe. Poorly drained and waterlogged soils have reducing conditions which cause the dissolution of oxides and tend to be greyish in colour. The main forms of Fe oxide are ferrihydrite ($Fe_2O_3.2FeOOH.nH_2O$) the freshly deposited form, and goethite (α-FeOOH). The Al oxides include amorphous $Al(OH)_3$ which slowly crystallizes to gibbsite (γ-$Al(OH)_3$). These oxides tend to occur as precipitates within the clay-sized fraction of soil minerals ($<2\ \mu m$). The hydrous oxides have pH-dependent surface charges; in general they are positively charged under acid conditions and negatively charged under alkaline conditions. Manganese oxides have important adsorptive properties but are much less abundant than those of Fe and Al.

Many other minerals may be found in soils, the most ubiquitous being resistant quartz grains which form much of the sand-sized fraction of soils and are relatively unreactive. Calcite ($CaCO_3$) is a major soil mineral constituent in soils developed on limestones in humid areas and in many soils in semi-arid regions. It buffers the soil to a pH usually between 7 and 8.5 and also contributes to the sorptive properties of soils by chemisorption and precipitation of metal carbonates (*e.g.* $CdCO_3$). Traces of incompletely weathered primary minerals can also be found in soils.

2.2.2 Soil Organic Matter. The presence of organic matter and living organisms is a characteristic of soil and helps distinguish it from regolith (weathered rock). All soils, in the pedological sense of the word, contain organic matter in their surface horizons but the amount and type may vary considerably. Soils in arid environments tend to contain only small amounts of organic matter ($<1\%$), whereas those in humid temperate regions have higher contents (1–10%). The organic matter present in soils can be classified as either humic or non-humic. Humic compounds are highly polymerized, colloidal, products of microbial decomposition of plant material, especially lignins. Non-humic material includes undecomposed or partially decomposed fragments of plant tissues and soil organisms. These latter include a wide range of species of bacteria, fungi, protozoa, actinomycetes, and algae, and many species of mesofauna, such as earthworms, which fill an important ecological niche in the comminution of plant litter and its incorporation into the soil.

Although organic matter only tends to form a small percentage of the mass of the soil (usually less than 10%), it has a very great influence on soil chemical and physical properties especially with regard to the behaviour of contaminants. Humus has a high, but pH-dependent, CEC (up to 200 $cmol_c\ kg^{-1}$) and a strong sorptive capacity for many different types of contaminants, including non-polar organic molecules and heavy metals.[2]

2.3   Soil Properties

Space does not permit a detailed discussion of the full range of soil properties here and, therefore, only those properties which have a significant effect on the behaviour of soil contaminants will be discussed. Readers are recommended to consult other specialized text books (*e.g.* References 3–5) for more details of the chemical, physical, and biological properties of soils.

2.3.1   Soil Permeability.   The voids between soil particles and aggregates form a continuous system of pores within the soil profile. Pores with a diameter greater than 30 μm tend to drain under gravity and are normally filled with air in dry weather. Pores smaller than 30 μm diameter tend to retain water against gravity and much of this may be available to plant roots. Soluble contaminants infiltrating the soil profile will be subject to internal drainage within the profile, diffusion between regions of different solute concentration and, in most cases, adsorption onto the organo-mineral colloidal complex. In addition to the inter-particle pores, larger (macropores) also exist; these are created by worm burrows, root channels, desiccation cracks, or deep soil disturbance and lead to more rapid movement of contaminants down the profile, either in solution, or adsorbed on suspended soil particles.

Dense non-aqueous phase liquids (DNAPLs) such as solvents tend to infiltrate relatively rapidly through soil profiles and reach the groundwater. The permeability of soil can vary widely as a result of the nature of

[2] S. J. Ross, 'Soil Processes: A Systematic Approach', Routledge, London, 1989, p. 444.
[3] N. Brady, 'The Nature and Property of Soils', Macmillan, New York, 10th Edn., 1990.
[4] D. L. Rowell, 'Soil Science: Methods and Applications', Longman Scientific and Technical, Harlow, 1994, p. 350.
[5] R. E. White, 'Introduction to the Principles and Practice of Soil Science', Blackwell, Oxford, 3rd Edn., 1997.

the soil constituents, the effective diameter of the system of pores, and soil management. Clay-rich soils tend to have relatively low permeabilities whereas sandy or fissured soils are more permeable. Compaction of soils by machinery, cultivations, or excavations when the soil is too wet, can reduce its permeability and may lead to waterlogging and the onset of anoxic conditions.

2.3.2 Soil Chemical Properties

pH. The pH of the soil is the most important physico-chemical parameter affecting plant growth and the behaviour of ionic contaminants in soils. Soil pH is a measurement of the concentration of H^+ ions in the soil solution present in the soil pores which is in equilibrium with the negatively charged surfaces of the soil particles. Unless stated otherwise, soil pH values are usually measured in distilled water but the use of dilute electrolytes (*e.g.* $CaCl_2$) more closely reflects the field situation.

Soil pHs are normally within the range 4–8.5 although the extreme range found over the world is 2–10.5. In general, soils in humid regions tend to have pH values between 5 and 7 and those in arid regions between 7 and 9. The median pH for more than 3000 soils from all over the USA was reported to be 6.1 (range 3.9–8.9)[6] and for nearly 6000 samples in the UK the median pH was 6.0 (3.1–9.2).[7] In temperate regions, such as the UK, the optimum pH for arable soils is 6.5 and for grassland it is 6.0. Soil pH conditions can be raised by liming with $CaCO_3$ and can be reduced by applying elemental sulfur although the latter is rarely practised.

In general, bacteria do not tolerate very acid conditions, so if it is intended to promote the microbial degradation of organic contaminants the soil should be maintained at a pH between 6 and 8. The mobility and bioavailability of most divalent heavy metals are greater under acid conditions and therefore liming is a way of reducing their bioavailability (except for Mo because the MoO_4^{2-} anion is more available at high pH).

Redox conditions. The balance of oxidation–reduction conditions in soils mainly affects the speciation of elements such as C, N, O, S, Fe, and Mn although Ag, As, Cr, Cu, Hg, and Pb are also affected. The redox conditions in a soil also reflect the oxygen supply for plant roots and soil micro-organisms. Respiration by plant roots, soil fauna and micro-organisms consumes a relatively large amount of oxygen. In situations

[6] C. G. S. Holmgren, M. W. Meyer, R. L. Chaney, and R. B. Daniels, *J. Environ. Qual.*, 1993, **22**, 335–348.
[7] S. P. McGrath and P. Loveland, 'Soil Geochemical Atlas of England and Wales', Blackie Academic and Professional, Glasgow, 1992.

of waterlogging, or exclusion of air by over compaction, micro-organisms with anaerobic respiration predominate, causing a change in the products of decomposition of organic matter (volatile fatty acids and ethylene, *etc*.) and in the speciation of susceptible metals. Microbially mediated methylation of some metallic and metalloid contaminants including As, Hg, Sb, Se, and Tl, tends to occur under anoxic conditions and can be an important mechanism for both the loss of the element from the soil and its conversion to more toxic and bioavailable forms.

Redox equilibria are controlled by the aqueous free electron activity and can be expressed either as pε (negative log of electron activity) or E_H (millivolt difference in potential between a Pt electrode and a H electrode). The conversion factor for the two expressions is: E_H (mV) = 5.2 pε.[8]

The combined effect of redox and pH on the behaviour of hydrous oxides in soils can be summarized by stating that either low pH or negative redox values generally result in the dissolution of oxides of Fe and Mn and, conversely, small increases in either pH or redox values can lead to the precipitation of ferrihydrite or Mn oxides. Sulfate anions are reduced to sulfide below pε − 2.0 and this can lead to the formation of sulfides of a wide range of metals. These sulfides tend to be insoluble and act as a temporary sink of the metal until redox conditions change and the sulfides are oxidized. The oxidation of sulfides, such as iron pyrites (FeS_2) results in a marked increase in the acidity of the soil.

The adsorption of contaminants in soils.   The concentration of solutes in the soil solution is determined by the interaction of adsorption and desorption processes. When contaminants are deposited on the soil surface they either react with the colloids in the soil aggregates at the surface or are washed into the soil profile in rain, irrigation water, or snow melt. Soluble contaminants will infiltrate the topsoil and enter the systems of pores, whereas insoluble and hydrophobic organic molecules may either bind to sites on the soil surface and become incorporated into the topsoil by movement of soil particles during cultivation, or, in the case of dense non-aqueous phase liquids (DNAPLs) move rapidly down the profile through large pores and fissures.

Several different adsorption reactions can occur between the surfaces of organic and mineral colloids and the contaminants. The extent to which the reactions occur will be determined by the composition of the soil (especially the clay mineral, Fe and Mn oxide, and organic matter contents), the soil pH, and the nature of the contaminants. The more strongly adsorbed contaminants are less likely to be leached down the soil profile and will tend to have lower bioavailabilities. Ionic contami-

---

[8] W. L. Lindsay, 'Chemical Equilibria in Soils', John Wiley & Sons, Chichester, 1979, p. 449.

nants such as metals, inorganic anions, and certain organic molecules including the bipyridyl herbicides Paraquat and Diquat are adsorbed onto surface charges on soil colloids. Non-polar organic molecules, which includes most of the persistent organic pollutants and pesticides are adsorbed onto humic polymers by both chemical and physical adsorption mechanisms.

Adsorption of cations and anions in soils.   Ion exchange (or non-specific adsorption) refers to the exchange between the counter-ions balancing the surface charge on the soil colloids and the ions in the soil solution. In the case of cations, it is the negative charges on soil colloids that are responsible for adsorption.

Soil organic matter has a higher CEC at pH 7 than other soil colloids and therefore plays a very important part in all adsorption reactions in most soils, even though it is normally present in much smaller amounts than clays. However, sandy soils with low contents of both organic matter and clay minerals tend to have low adsorptive capacities and are a greater danger for contaminants infiltrating through the soil profile to the water table.

The negative charges on the surfaces of soil colloids are of two types: (a) permanent charges resulting from the isomorphous substitution of a clay mineral constituent by an ion with a lower valency, and (b) the pH dependent charges on oxides of Fe, Al, Mn, and Si and organic colloids which are positive at pH values below their isoelectric points and negative above their isoelectric points. Hydrous iron and aluminium oxides have relatively high isoelectric points (pH 8) *in vitro* but lower values within the soil system (*ca.* pH 7) and so tend to be positively charged under most conditions, whereas clay and organic colloids are predominantly negatively charged under alkaline conditions. With most colloids, increasing the soil pH, at least up to neutrality, tends to increase their CEC. Humic polymers in the soil organic matter fraction become negatively charged due to the dissociation of protons from carboxyl and phenolic groups.

The concept of cation exchange implies that ions will be exchanged between the soil solution and the zone affected by the charged colloid surfaces (double diffuse layer). The relative replacing power of any ion on the cation exchange complex will depend on its valency, its diameter in hydrated form, and the type and concentration of other ions present in the soil solution. With the exception of H^+, which behaves like a trivalent ion, the higher the valency, the greater the degree of adsorption. Ions with a large hydrated radius have a lower replacing power than ions with smaller radii. For example, K^+ and Na^+ have the same valency but K^+ will replace Na^+ owing to the greater hydrated size of the Na^+ ion.

The commonly quoted relative order of replaceability of metal cations on the cation exchange complex is:

$$Li^+ = Na^+ > K^+ = NH_4^+ > Rb^+ > Cs^+ > Mg^{2+} > Sr^{2+} =$$
$$Ba^{2+} > La^{3+} = H^+(Al^{3+}) > Th^{4+}$$

Anion adsorption occurs when anions are attracted to positive charges on soil colloids. As stated above, hydrous oxides of Fe and Al are usually positively charged below pH 7 or 8 and so tend to be the main sites for anion exchange in soils. In general, most soils tend to have far smaller capacities for anion exchange than for cation exchange. Some anions, such as nitrates and chlorides are not adsorbed to any marked extent but others, such as orthophosphates tend to be strongly adsorbed.

Specific adsorption is a stronger form of adsorption, involving several heavy metal cations and most anions with surface ligands in forming partly covalent bonds with lattice ligands on adsorbents, especially oxides of Fe, Mn and Al. This adsorption is strongly pH-specific and the metals and anions which are most able to form hydroxy complexes are adsorbed to the greatest extent. The order for the increasing strength of specific adsorption of selected heavy metals is:

$$Cd > Ni > Co > Zn \gg Cu > Pb > Hg$$

In addition to cation exchange reactions, soil organic matter can sequester metals by complexation mechanisms, especially chelation. The general order of complexation of metals is:

$$Pb > Cu > Fe = Al > Mn = Co > Zn$$

2.3.5 Adsorption and Decomposition of Organic Contaminants. The adsorption of non-ionic and non-polar organic contaminants (including many pesticides) occurs mostly on soil humic material. Since the highest content of organic matter occurs in the surface horizons of soils, there is a tendency for most organic contaminants to be concentrated in the topsoil. Migration of organic contaminants down the profiles of soils occurs to the greatest extent in highly permeable sandy, gravelly, or fissured soils with low organic matter contents. High concentrations of dissolved organic compounds (DOC) of humic origin can cause enhanced mobility and leaching of organic (and metallic) contaminants in soils due to the binding of the contaminant to the soluble ligand. Soils vary in their contents of DOC but applications of sewage sludge, animal manure, and composts result in increased concentrations.

In the case of organic pesticides, most are relatively insoluble and do not move down the soil profile, but exceptions to this are the organo-phosphate insecticides, phenoxyacetic acid herbicides, and bipyridylium herbicides, which can be leached. The phenoxyalkanoic acid herbicides, such as 2,4-D, 2,4,5-T, and MCPA, exist as anions at normal values of soil pH and are adsorbed to a limited extent by hydrous oxides and by hydrogen bonding to humic polymers. However, this sorption is less marked than that which occurs with the cationic bipyridyl herbicides (paraquat and diquat) which are inactivated on contact with soil colloids.[2]

Many pesticide and other organic contaminants (see Section 4.2) that reach water courses from soils have often been washed into the water adsorbed onto soil particles in runoff and not leached down through the profile.[2] In some cases where organic wastes containing organic contaminants, such as sewage sludges and some composts, are applied to soils, they act as both a source and a sink of contaminants (*e.g.* PAHs, PCBs, and heavy metals). Positive correlations are frequently found between the concentrations of pesticide residues and the soil organic matter content.

The propensity for a non-polar organic contaminant molecule to be sorbed by soil organic matter can be predicted by determining its octanol–water distribution coefficient (K_{ow}).[9] Substances with low K_{ow} values (< 10) tend to be hydrophilic (water soluble) whereas those with higher values are more hydrophobic and more strongly bound to organic matter (and lipids in living organisms). For example, the herbicide 2,4-D (2,4-dichloro-phenoxyacetic acid) is relatively soluble in water (900 mg l^{-1}) and has a low octanol–water distribution coefficient (log $K_{ow} = 1.57$), whereas the insecticide DDT has a low water solubility (0.002 mg l^{-1}), a high log K_{ow} value (5.98), and accumulates in soil organic matter and lipids in living organisms.

Organic contaminants may be lost from the soil by physical processes including volatilization, photolytic degradation, and leaching, and by microbial degradation. Soil micro-organisms can become adapted to degrade many different types of persistent contaminant molecules by the secretion of extracellular enzymes. However, the C—Cl bond does not occur in naturally occurring compounds and hence the chlorinated organic molecules are more difficult to degrade. Nevertheless, several species of micro-organism have been found to have developed the ability to degrade chlorinated compounds, but there is normally a time lag while they adapt to the new molecules. During this period there is localized selection and multiplication of strains that have a mutation enabling them to break the bonds of the contaminant molecule and also to

[9] M. D. LaGrega, P. L. Buckingham, and J. C. Evans, 'Hazardous Waste Management', McGraw-Hill, New York, 1994, p. 1146.

tolerate the toxicity of either the contaminant or its intermediate decomposition products.

Adsorption of non-polar organic contaminants onto soil organic matter will not occur in the presence of oils. Microbially synthesized surfactants can help to accelerate the rate of degradation of hydrocarbon oils in contaminated soils. For many organic contaminants, adsorption onto soil colloids and the presence of water are important factors promoting decomposition by micro-organisms. Ross lists the types of degradation of organic molecules as: (a) non-biological degradation, including hydrolysis, oxidation and reduction, and photodecomposition; and (b) microbial decomposition, often involving specially adapted micro-organisms.[2] This type of decomposition normally follows a first order (exponential) type of reaction after an initial lag period while the micro-organisms become adapted to the substrate. In most cases, non-biological degradation processes, such as photodecomposition and volatilization, can occur at the same time as microbially catalysed reactions.

The range of factors affecting the degradation of organic contaminants by micro-organisms include: soil pH, temperature, supply of oxygen and nutrients, the structure of the contaminant molecules, their toxicity and that of their intermediate decomposition products, the water solubility of the contaminant, and its adsorption to the soil matrix (and therefore the organic matter content of the soil).[2] Adsorption of organic molecules tends to decrease with increasing temperature because most adsorption reactions are exothermic. Volatilization losses tend to be greatest at high temperatures.

The persistence of organic contaminants in soils is determined by the balance between adsorption onto soil colloids, uptake by plants, and transformation or degradation processes. Organochlorine molecules are regarded as being highly persistent with a duration in the soil of 2–10 or more years. PAHs have been shown to persist for more than 20 years in soils treated with sewage sludge, but different PAH molecules vary in the rate at which they are decomposed.[10]

3   SOURCES OF LAND CONTAMINANTS

Soils receive contaminants from a wide range of sources, including:

(1) Atmospheric fallout from:
    —fossil fuel combustion (oxides and acid anions of S and N; heavy metals; PAHs)

[10] S. R. Wild, K. S. Waterhouse, S. P. McGrath, and K. C. Jones, *Environ. Sci. Technol.*, 1990, **24**, 17-6–1711.

—Pb, PAHs, *etc.* from automobile exhausts

—metal smelting operations (As, Cd, Cu, Cr, Hg, Ni, Pb, Sb, Tl, Zn)

—metal-using industries, including foundries (Cd, Cu, Pb, Zn)

—chemical industries (organic micropollutants, Hg)

—waste disposal by incineration (Cd, TCDDs, TCDFs)

—radionuclides from reactor accidents and the atmospheric testing of nuclear weapons

—large fires (PAHs; Pb, Cr, *etc.* from paint; TCDDs; TCDFs)

(2) Agricultural chemicals:

—herbicides (organic molecules—some containing TCDD; B, and As compounds)

—insecticides (chlorinated hydrocarbons, *e.g.* DDT, BHC)

—fungicides (Cu, Zn, Hg, and organic molecules)

—acaricides (*e.g.* 'Tar Oil')

—fertilizers (*e.g.* Cd and U impurities in phosphates)

(3) Waste Disposal (intentional/unintentional input to soil):

—farm manures (As, Cu, and Zn in pig and poultry manures)

—sewage sludges (heavy metals; organic pollutants; PAHs; pathogens)

—composts from domestic wastes (metals; organics)

—mine wastes (coal mines—SO_4^{2-}; metalliferous mines As, Cd, Cu, Pb, Zn, Ba, U, *etc.*)

—seepage of leachate from landfills (metals; Cl^-; PCBs; *etc.*)

—ash from fossil fuel combustion, incinerators, bonfires and accidental fires (PAHs; TCDDs; metals; *etc.*)

—burial of diseased livestock on farmland

(4) Incidental Accumulation of contaminants:

—corrosion of metal in contact with soil (*e.g.* Zn from galavanized metal; Cu and Pb from roofing, scrapyards, *etc.*)

—wood preservatives from fencing (PCP; PAHs in creosote; As; Cr; Cu)

—leakage from underground storage tanks (petroleum; aviation fuel; chlorinated solvents)

—warfare and military training (hydrocarbons; PAHs from fires, explosives, and their degradation products; metals from munitions and vehicles)

—sports and leisure activities (Pb, Sb, and As from shotgun pellets; Pb from fishing weights; Pb, Cd, Ni, and Hg from discarded batteries; hydrocarbons from spilt petrol and lubricating oil)

(5) Derelict industrial sites—wide range of contaminants from production, waste disposal, and building demolition, *e.g.*:

  —Gas works—phenols; tars (PAHs); cyanides; and As
  —Electrical industries—Cu, Pb, Zn; PCBs; solvents
  —Tanneries—Cr
  —Scrapyards—metals; PCBs; hydrocarbons

A more comprehensive list of the most common contaminating uses of land includes: waste disposal sites, gas works, oil and petroleum refineries, petrol stations, electricity generating stations, iron and steel works, non-ferrous metals processing, metal products fabrication and metal finishing, chemical works, glass making and ceramics, textile plants, leather tanning works, timber and timber products treatment works, manufacture of integrated circuits and semi-conductors, food processing, sewage works, asbestos works, docks and railway land, paper and printing works, heavy engineering installations, installations processing radioactive materials, and burial of diseased farm livestock.[11]

4 CHARACTERISTICS OF SOME MAJOR GROUPS OF LAND CONTAMINANTS

4.1 Heavy Metals

Metals, such as Ag, Cd, Cr, Cu, Hg, Mn, Mo, Ni, Pb, Sb, Tl, U, V, and Zn tend to be relatively strongly adsorbed by soil constituents. Their mobility and bioavailability depend on the soil conditions but some metals, such as Cd and Zn which tend to be less strongly sorbed than Pb and Cu, can be leached down soil profiles, especially under acid conditions. Industrial contaminated land often contains a suite of heavy metal contaminants present in high concentrations. Examples include chlor-alkali industries where Hg electrodes are used in the electrolysis of sea water to produce Cl_2 gas and NaOH. Both air and soil contamination will have occurred at these sites from emissions of volatile Hg and spillages of liquid Hg. Mines, smelters, foundries, paint factories, and scrap yards are likely to have been heavily contaminated but examples of less obvious non-point sources of heavy metals are: sewage sludge disposal to land (range of metals), Pb in shotgun pellets used for game bird and clay pigeon shooting, and Cd in phosphatic fertilizers.

A very wide range of heavy metal concentrations is found in sewage sludges but the general trend in technologically advanced countries is that concentrations have decreased over the last 30 years with the introduction of cleaner technologies and pollution prevention procedures. In the UK the median (50 percentile) Cd content of sludges

---

[11] House of Commons Environment Committee, '1st Report: Contaminated Land', HMSO, London, 1990, Vol. 1.

decreased by 64% (from 9 μg Cd g^{-1} to 3.2 μg Cd g^{-1}) over the period 1983 to 1993. The *maximum* concentrations of metals in sludges reported in the literature are as follows, but most sludges will have very much lower values (μg g^{-1} in dry matter): Ag ($<$960), As ($<$30), Cd ($<$3410), Cr ($<$40 600), Cu (50–8000), Hg ($<$55), Mn (60–3900), Mo ($<$40), Ni ($<$5300), Pb ($<$3600), Sb ($<$34), Se ($<$10), Sn ($<$700), V ($<$400), Zn ($<$49 000).[12] Shotgun pellets are composed predominantly of Pb (90–98%) but can also contain Sb (up to 8%), As, and Ni. More than 1000 t of Pb may be dispersed into soils annually in some countries. Phosphatic fertilizers are used on agricultural soils in all advanced agricultural systems but these can contain considerable amounts of Cd owing to concentrations of up to 640 mg Cd kg^{-1} P occurring in the phosphate rock used for their manufacture.[13]

4.2 Organic Contaminants

There are more than 20 000 organic contaminants known already and this number will increase as analytical methods are further refined and more studies are made of materials containing complex mixtures of organic pollutants, such as industrial wastes, sewage sludges and landfill leachates.

Petroleum hydrocarbons tend to be very ubiquitous contaminants of soils, arising from spillages, leaking tanks, and intentional mixing of petroleum refinery sludges with soil in the process of landfarming, which exploits the degradation brought about by soil micro-organisms. Many of the lower molecular weight hydrocarbon compounds are relatively easily degraded by soil organisms but it takes time and the contaminated soil could pose a fire risk during the period it is contaminated.

Pesticides can be soil contaminants as a result of persistence after use on crops, runoff from treated land, accidental spillages, or pesticide manufacture. Although there are over ten thousand commercial pesticide formulations of around 450 compounds in use, they can be classified under a relativity narrow range of molecular groups. These compounds are used because of their toxicity towards specific pests but there is a risk of harm to soil organisms, other beneficial plants, animals, and humans. Usually, less than 10% (often around 1%) of the pesticide applied reaches its target, the rest is dispersed in the environment. Humans can be affected through the food chain or in contaminated drinking water

[12] 'Heavy Metals in Soils', ed. B. J. Alloway, Blackie Academic and Professional, Glasgow, 2nd Edn., 1995.
[13] M. E. Sumner and M. J. McLaughlan, 'Contaminants and the Soil Environment in the Australasia-Pacific Region', ed. R. Naidu, R. S. Kookana, D. P. Oliver, S. Rogers, and M. J. McLaughlan, Kluwer Academic Publishers, Dordrecht, 1996, p. 125.

which the pesticide, or its decomposition products, reach either by leaching to groundwater or in runoff into surface waters.

Three main groups of insecticides are currently used in agriculture: organochlorines, organophosphates, and carbamates. The organochlorines have been widely used for up to 50 years and are the most persistent of all groups of pesticides. Their persistence decreases in the order:

DDT > dieldrin > lindane (BHC) > heptachlor > aldrin

with half lives of eleven years for DDT down to four years for aldrin.[2] Organophosphates are highly toxic to humans and other mammals but are less persistent in the soil than organochlorines (six month half lives for parathion, diazinon, and demeton). Carbamates are used to control a wide range of pests including molluscs, fungi, and insects but have a similar persistence to organophosphates. Aldicarb (or Temik) is a highly toxic carbamate that is used as both an insecticide and a nematicide on potato and sugar beet crops. It is readily oxidized in the soil but its oxidation products are also highly toxic and readily leached and can cause ecological and human health problems.[14] Some examples of the molecular structures of pesticides and other organic contaminations are shown in Figure 2.

There are six major groups of compounds used as herbicides: phenoxyacetic acids, toluidines, triazines, phenylureas, bipyridyls, and glycines. The most important phenoxyacetic acids are 2,4-D and 2,4,5-T which have a persistence of up to eight months but are of particular environmental significance because they can be contaminated with dioxins (TCDDs). 'Agent Orange', the defoliant used by the US forces in Vietnam, comprised a mixture of 2,4-D and 2,4,5-T and caused widespread contamination of soils by dioxin. The toluidines and triazines are fairly strongly adsorbed and have a persistence of up to twelve months although atrazine contamination of groundwaters is a serious problem in many intensive arable farming areas. The phenylureas tend to be fairly soluble and are rapidly leached. In contrast, the bipyridyls such as paraquat and diquat are cationic and strongly adsorbed on soil colloids (K_d values of $< 4.2 \times 10^{-4}$ on montmorillonite) and are therefore very persistent. Fungicides comprise a more diverse group of compounds including inorganics, such as copper and mercury compounds, and a wide range of organic compounds.

Apart from pesticides, which are synthesized intentionally, a wide range of chlorinated compounds including dioxins (TCDDs) and dibenzodifurans (TCDFs) have been widely dispersed in the environment as a

[14] F. A. M. de Haan, 'Scientific Basis for Soil Protection in Europe', ed. H. Barth and P. L'Hermite, Elsevier, Amsterdam, 1987, p. 211.

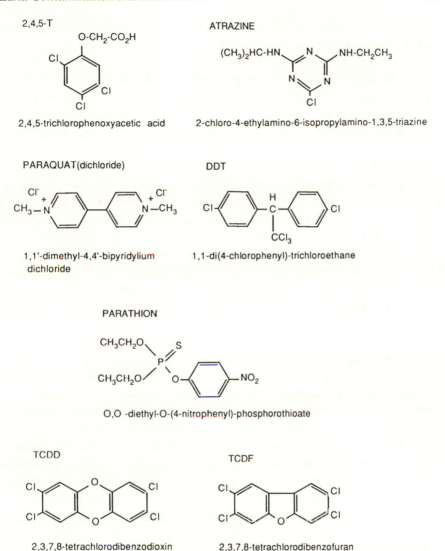

Figure 2   *Examples of the molecular structures of pesticides and organic contaminants*

result of their accidental synthesis at relatively high temperatures, their highly stable structure, and their slow rate of degradation. Typical situations in which they are formed include: synthetic reactions where the temperature conditions become too hot (such as in the manufacture of the herbicide 2,4,5-T) and the incineration of rubbish containing PVC and other sources of chlorides and aromatic compounds. Polychlorinated biphenyls (PCBs) are very stable and were originally manufactured for use in electrical transformers and capacitors and as plasticizers. Although no longer in use, they are found in soils around industrial and domestic waste tips and electronic component factories as well as in remote sites which they have reached by transport in air. Incineration temperatures need to be $> 1200\,°C$ with adequate oxygen in order to ensure the destruction of stable molecules like PCBs. Solvents used for degreasing electrical components such as trichloroethane are important atmospheric and groundwater contaminants (DNAPLs) and may also occur in soils around factories (where they could pose a fire or explosion risk).[15]

Polycyclic aromatic hydrocarbons (PAHs) are very important persistent organic pollutants in the environment. Some of them, such as benzo(*a*)pyrene (BaP) are carcinogenic. They can be found widely distributed as localized areas of contamination in the environment and, unlike many other organic pollutants, are not necessarily associated with industrial and technologically advanced sources. PAHs are produced by fires and incomplete combustion of organic materials. Fire smoke, soot, tar, ashes, and sewage sludges can introduce significant amounts of these compounds into soils and they may persist for more than 20 or 30 years.[15]

4.3   Sewage Sludge

Sewage sludges contain a wide range of environmental contaminants owing to the diverse sources of effluents discharged into sewers. This includes human excretion products, household chemicals, automobile fuels, lubricants and cleaning compounds, stormwater runoff from highways containing PAHs and other fuel combustion products, and effluents from many different industries. PAH concentrations in sewage sludges tend to range between 0.5 and 10 $\mu g\,g^{-1}$ in sludge dry matter with similar levels of PCBs, although $> 1000\ \mu g\,g^{-1}$ are sometimes found.[16]

The organic contaminants frequently found in sewage sludges include:

—halogenated aromatics (PCBs—polychlorinated biphenyls, poly-chlorinated terphenyls, PCNs—polychlorinated naphthalenes, and polychlorobenzenes)
—aromatic amines and nitrosamines
—phenols and halogenated aromatics containing oxygen
—polyaromatic and heteroaromatic hydrocarbons (PAHs) and halogenated aliphatics

[15] B. J. Alloway and D. C. Ayres, 'Chemical Principles of Environmental Pollution', Blackie Academic and Professional, London, 2nd Edn., 1997.
[16] D. Sauerbeck, 'Scientific Basis for Soil Protection', ed. H. Barth and P. L'Hermite, Elsevier, Amsterdam, 1987, p. 181.

—aliphatic and aromatic hydrocarbons
—phthalate esters
—pesticides

Of all the organic contaminants listed here, the PAHs and PCBs are currently considered to constitute the greatest hazard to human health.[16] The range of heavy metal concentrations found in sewage sludges is given above in Section 4.1.

In addition to contaminant chemicals, sewage sludges may contain some pathogens which were not destroyed during the sewage treatment process. Many of these will perish when exposed to extremes of temperature and UV light but care should be taken with this serious potential risk.

A relatively high proportion of the sludge produced in many countries is applied to agricultural land as a means of disposal (67% in the UK). This sludge has many useful properties for agriculture (source of N and P and physical soil conditioner) but its use is limited by its concentrations of persistent contaminants. Sewage sludges are frequently used in the landscaping of derelict land where they act as a growth medium for plants grown for amenity purposes rather than food. The nutrients they contain may also have a beneficial effect on soil micro-organism populations which could help to stimulate the biodegradation of persistent organic pollutants.

5 POSSIBLE HAZARDS FROM CONTAMINATED LAND

Soil contamination can restrict the options available for the reuse of land because of the potential hazards posed by the contaminants. These can include:

Hazard	Examples of Contaminants
(1) Direct ingestion of contaminated soil (mainly children or animals)	As, Cd, Hg, Pb, CN^-, coal tars, phenols
(2) Inhalation of dusts and vapours	toluene, benzene, xylene, various solvents, radon, Hg, asbestos, metal-rich particles
(3) Uptake by plants of contaminants hazardous to animals and people through the food chain	As, Cd, Pb, Tl, PAHs

(4) Phytotoxicity SO_4^{2-}, Cu, Ni, Zn, CH_4

(5) Deterioration of building SO_4^{2-}, SO_3^{2-}, Cl^-, coal tar,
    materials and services   mineral oils, organic solvents

(6) Fires and explosions CH_4, S, coal dust, oils, tar,
                                      petroleum, rubber, high calorific
                                        value wastes (*e.g.* old landfills)

(7) Dermal contact (*e.g.* during   coal tar, phenols, radionuclides,
    demolition, children at play)   PAHs, TCDDs, asbestos, *etc.*

(8) Contamination of water CN^-, SO_4^{2-}, DNAPLs (*e.g.*
                                        solvents), soluble metals,
                                        pesticides, fuels

(Adapted from References 17 and 18).

6 METHODS OF SITE INVESTIGATION

In investigations of land suspected of being significantly contaminated, the type of contaminants suspected, their source, the transport mechanism (*e.g.* wind or direct placement), and the nature of the site are all important factors determining the choice of sampling procedure. For example, agricultural land suspected of being contaminated by fallout from a point source of contamination several kilometres away would be investigated differently from a former industrial site where the remaining buildings and other structures indicate the areas most likely to be heavily contaminated. Site investigations involve the collection (by boring or digging) of samples. These are subsequently sent to an appropriately accredited laboratory for determination of a range of contaminants and selected soil parameters (*e.g.* pH) and the process can be very expensive. Therefore, as with most matters, the sampling regime is a compromise between efficiency and cost. It is convenient to divide the investigation of contaminated sites into several categories:

(a) Large areas, including agricultural land, contaminated by atmospheric fallout from distant, or non-point sources: investigations on land of this type will usually need to use a grid pattern of sampling with most emphasis on topsoil (0–15 cm, especially 0–5 cm) but samples will need to be collected from greater depths

[17] M. J. Beckett and D. L. Sims, 'Contaminated Soil', ed. J. W. Assink and W. J. van der Brink, Martinus Nijhoff, Dordrecht, 1986, p. 285.
[18] Interdepartmental Committee on the Redevelopment of Contaminated Land, 'Guidance on the assessment and redevelopment of contaminated land', Guidance Note 59/83, Department of the Environment, London, 1987.

(15–30, 30–45 cm) to determine whether contaminants have been translocated down the profile. The normal practice for agricultural land in the UK is to take samples with a screw auger in a W pattern with 25 cores per 5 ha block. The cores for each depth sample are bulked together to give one sample. If the soil appears to differ in colour or texture in part of the area being considered, each distinct area should be (bulk) sampled separately.

Preliminary sampling surveys of areas affected by fallout from discrete sources are often carried out using transects where samples are collected at regular intervals in straight lines both up and downwind from the source. The siting of the transects is often made along compass bearings but these may have to be modified to take account of the prevailing wind direction, accessibility of the area, and the constraints of coastlines and topography.

(b) Discrete areas of industrial or other obvious contaminating activity and land adjacent to them: in this situation it is vitally important to know the purpose for which the site was used, the processes which were employed, the raw materials and the location of their storage and handling facilities, the products manufactured, any possible byproducts formed, the wastes from enterprises on the site, and the waste disposal practices adopted. This information should be collected first in a preliminary desk study of old plans and other available records, before any attempt is made to take samples on the site. This is because the information on the location of potential areas of major contamination ('hot spots') within the site, such as storage and effluent tanks, bulk raw materials stores, and waste heaps is required in order to design the sampling programme. These potential 'hot spots' should be carefully examined and sampled separately from any systematic survey of the whole site. Any pre-existing pits or other subterranean structures that are found should be sampled, but care should be taken with regard to safety procedures.

The objective of carrying out a site investigation at a derelict industrial site is to locate the most severely contaminated areas ('hot spots'). These areas could be as small as 50 m^2 and are therefore difficult to locate in any systematic sampling procedure. Comparisons of stratified random grids, regular square grids, and herringbone patterns have shown that the latter is generally the most efficient for locating target areas which are circular, pear-shaped plumes, or ellipses but with the exception of linear/elliptical targets the regular square grid

Table 1 *Example of numbers of sampling points for contaminated sites of different areas using the recommendation of BSI DD175*

Area of site (ha)	Recommended number of sampling points	Minimum contaminated area to provide one sample (at $P < 0.05$) (m²)
0.5	15	905
1.0	25	1129
1.5	85	1732

After Hobson.[21]

is adequate for most purposes.[19] Sampling on a grid is the most widely used approach but the number of samples and size of grid appear to be mainly a matter of professional judgement. The British Standards Draft Code of Practice DD175[20] recommends minimum numbers of sampling points for sites of different areas; an example is given in Table 1.

Another approach used by the Interdepartmental Committee for the Redevelopment of Contaminated Land is that the grid spacing should be kept to a size which could be effectively handled if it happened to be missed during sampling (but discovered later).[22] This generally implies that grids of 10–25 m will be used for small sites and 25–50 m for larger sites. A 10 m grid on a 1 ha site would give a 64% probability of finding a contaminated patch of 100 m² (which is the size of a 10 m × 10 m grid). However, studies on a gasworks site have shown that a 25 m grid gave results which were not significantly different to much smaller grids of down to 6.25 m. This obviously has major financial implications, as of course does the possibility of litigation and very high costs of cleaning up a 'hot spot' area which may have been missed in the site investigation.

Ideally, soil samples should be taken from trial pits (with appropriate safety precautions) rather than boreholes. This allows observations to be made about the stratification of materials and other relevant points. The excavation may release gases, vapours, or liquids and so vigilance is necessary and gas sampling may need to be carried out as a precaution.[23] Boreholes can be drilled more rapidly than pits can be excavated thus allowing the site to be surveyed in a shorter time, and they are the only

[19] R. Bosman, L. Voortman, J. Harmsen, and C. Coggan, 'Guidance on the Procedure for the Investigation of Urban and Industrial Sites with Regard to Soil Contamination—ISO 10381 Part 5' (Working Draft), International Organisation for Standardisation, Geneva, 1996, p. 64.
[20] British Standards Institution, 'Draft for Development, DD175: 1988 Code of Practice for the Identification of Potentially Contaminated Land and Its Investigation', BSI, London, 1988.
[21] D. M. Hobson, 'Contaminated Land: Problems and Solutions', ed. T. Cairney, Blackie Academic and Professional, London, 1993, p. 29.
[22] Interdepartmental Committee on the Redevelopment of Contaminated Land, Guidance Note 18/79, Department of the Environment, London, 1983..
[23] E. E. Finnecy and K. W. Pearce, 'Understanding Our Environment', ed. R. E. Hester, Royal Society of Chemistry, London, 1st Edn., 1986, p. 172.

feasible way of sampling below 3 or 4 m at large numbers of sites. In practice, a combination of pits and boreholes is often used. Careful investigation of the ground before the sampling is essential; areas where the soil material differs in colour, texture, moisture content, apparent organic content, and even smell should be noted and included in the sampling.

Samples collected by digger, power augering, and boreholes should be placed in sealed containers to avoid loss of volatiles, although samples collected for analysis of volatile compounds are usually placed in purpose made containers with a space above the sample to allow accumulation of volatiles. This is referred to as head-space analysis and the volatile constituents are usually determined by gas chromatography either on its own (GC) or linked to a mass spectrometer (GC-MS). Care should be taken to avoid contamination of samples during collection, packaging, transportation, and processing for analysis. Samples of soil are normally air dried or oven dried at 30 °C but care should be taken over the possibility of harmful vapours being released from the samples.

The chemical and microbiological analysis of samples collected from sites suspected of being contaminated is beyond the scope of this chapter. However, modern techniques, including inductively coupled optical emission spectrometry (ICP-OES) and atomic absorption spectrometry (AAS) for inorganic contaminants such as heavy metals, and gas chromatography (GC) and gas chromatography linked to a mass spectrometer (GC-MS) for organic contaminants, provide relatively rapid and efficient analytical procedures which can cope with large numbers of samples. In addition, a new generation of rapid, on-site instruments are being developed which avoid the delay that occurs when samples are collected and sent off to a laboratory for analysis. Some of these methods can be used for on-line, real time monitoring, such as sensors installed for monitoring wells or those used for monitoring volatile emissions from a site. These developments in analytical chemistry enable site investigations to be carried out more rapidly but the sampling procedure and location of the sampling points and the choice of analytical determinants is still vitally important.

7 INTERPRETATION OF SITE INVESTIGATION DATA

Once a site has been investigated and shown to be contaminated, a risk assessment needs to be undertaken in order to decide what course of action to take. For example, in a case where agricultural land has been found to be slightly contaminated, it may be decided that regular monitoring is all that is required and no further remedial action, except possibly to investigate the source of the contamination. In the case of sewage sludge application to land, there will be a gradual build-up of contaminants, such as heavy metals. Nevertheless, so long as the national maximum permissible limits are not exceeded, normal agricultural use of the land can continue. In the case of a currently used industrial site that has been shown to have been contaminated, if the contaminants are not being transferred off the site and are not currently causing a hazard either to human health or to ecosystems, it may be acceptable for that site to be left untouched so long as further contamination is prevented and regular monitoring is carried out. One possibility is to install a containment system which prevents leaching and other

Table 2 *UK Department of the Environment Trigger Concentrations for Environmental Contaminants[18,23] (total concentrations except where indicated)*

Contaminant	Proposed uses	Threshold (Trigger concentrations $\mu g\ g^{-1}$)	Action
Contaminants which may pose hazards to health			
As	Gardens, allotments	10	a
	parks, playing fields, open space	40	
Cd	Gardens, allotments	3	
	parks, playing fields, open space	15	
Cr (hexavalent[b])	Gardens, allotments	25	
	parks, playing fields, open space	1000	
Cr	Gardens, allotments	600	
	parks, playing fields, open space	1000	
Pb	Gardens, allotments	500	
	parks, playing fields, open space	2000	
Hg	Gardens, allotments	1	
	parks, playing fields, open space	20	
Se	Gardens, allotments	3	
	parks, playing fields, open space	6	
Contaminants which are phytotoxic but not normally hazardous to health			
B (water soluble)	Any uses where plants grown	3	
Cu (total)	Any uses where plants grown	130	
(extractable[c])		50	
Ni (total)	Any uses where plants grown	70	
(extractable)		20	
Zn (total)	Any uses where plants grown	300	
(extractable)		130	

[a] Action concentration yet to be specified.
[b] Hexavalent Cr extracted by 0.1 M HCl adjusted to pH 1 at 37.5 °C.
[c] Extracted in 0.05 M EDTA

Table 3 *UK Department of the Environment Trigger Concentrations for Contaminants associated with former coal carbonization sites*[23]

		Threshold	Action
Contaminant	Proposed use	(Trigger concentrations μg g^{-1})	
PAHs	Gardens, allotments	50	500
	Landscaped areas	1000	10 000
Coal tar	Gardens, allotments	200	–
	Landscaped areas, open space	500	–
	Buildings, hard cover	5000	–
Phenols	Gardens, allotments	5	200
	Landscaped areas	5	1000
Free cyanide	Gardens, allotments, landscaped areas	25	500
	Buildings, hard cover	100	500
Complex cyanides	Gardens, allotments	250	1000
	Landscaped areas	250	5000
	Buildings, hard cover	250	Nil
Thiocyanate	All uses	50	Nil
Sulfate	Gardens, allotments, landscaped areas	2000	10 000
	Buildings	2000	50 000
	Hard cover	2000	Nil
Sulfide	All uses	250	1000
Sulfur	All uses	5000	5000
Acidity	Gardens, *etc.*	pH < 5	pH < 3

movement of contaminants out of the contained volume of soil, but the engineering task of installing this can be expensive. However, where a derelict site is to be redeveloped, it will be necessary to remediate the site to an approved quality standard. This may be a universal standard, such as those used in the Netherlands where, at least until recently, all land should be cleaned to a standard appropriate for any future use, including the most demanding, such as domestic gardens where vegetables may be grown and children may eat some soil. Elsewhere, a 'fitness for purpose' approach is used where more stringent quality standards are used for domestic gardens than for industrial sites where food crops will not be grown.

Examples of the critical concentrations of contaminants at sites which are to be redeveloped are given in Tables 2 and 3. For comparison, some indicative values from the soil quality standards used in the Netherlands include: Target Values (μg g^{-1}) for As 29, Cd 0.8, Cr 100, Cu 36, Hg 0.3, Ni 35, Pb 85, and Zn 140. The intervention concentrations (at which action must be taken) for these elements are (μg g^{-1}): As 50, Cd 20, Cr 800, Cu 500, Hg 10, Ni 500, Pb 600, and Zn 3000.[24]

[24] Ministry of Housing, Physical Planning and Environment, Directorate General for Environmental Protection (Netherlands), Environmental Standards for Soil and Water, Leidschendam, 1991.

It must be mentioned that in some cases, 'natural attenuation' of contamination may be occurring at some sites and this implies that either the movement of a contaminant has been restricted by natural permeability boundaries and sorption systems or that certain organic contaminants are being degraded without any further human intervention. This situation is obviously important to recognize but may not be of wide practical significance, except perhaps if it is occurring in part of an actively used site where a change of land use is not likely and therefore sufficient time may be available for the processes to work. In all cases of land contamination, it is very important that, as far as possible, further active contamination should be prevented. This may mean emptying leaking tanks, or changing manufacturing or material handling processes. Unfortunately, with some industrial processes, such as metalliferous smelting, it is impossible to completely eradicate atmospheric pollution although it can be kept to a minimum.

8   RECLAMATION OF CONTAMINATED LAND

The treatment of contaminated land is a very large and rapidly developing subject area with many new technologies being developed. The methods used can be classed as *ex situ* and *in situ* and examples are as follows.

8.1   *Ex situ* Methods

8.1.1   'Dig and Dump'.   This is the colloquial name for the process of excavation of the contaminated soil and its disposal at a licensed landfill. A variation on this is to excavate and incinerate the severely contaminated soil but this would be much more expensive than landfilling. Incineration to temperatures in excess of 1000 °C (up to 2500 °C) with adequate oxygen is the most effective way of destroying highly persistent pollutants such as PCBs and TCDDs (dioxins). The mineral residue of the soil would be converted to a fused silica-rich ash unsuitable for landscape purposes, which would probably need to be disposed of in a landfill. Another variation is to solidify the contaminated soil, usually with cement, or another suitable binder. This is convenient for heavy metal contamination since it renders the metals immobile and the soil can possibly be used for road construction.

8.1.2   Soil Cleaning.   In this method contaminated soil is excavated (as above) and transported to a cleaning facility which could be a soil-washing plant or a bioreactor. One of the following treatments is then carried out.

(a) The soil is washed with selected extractants such as acids, chelating agents or surfactants to remove certain inorganic or organic contaminants.

(b) Bioreactors usually comprise a vessel in which a suspension of soil and water can be stirred and conditions optimized for the degradation of organic contaminants. Augmentation of the microbial population present in the soil is possible but it is usually found that the indigenous organisms can carry out the degradation.

(c) Biodegradation of stripped contaminated soil in beds or wind-rows, called 'biopiles' is also carried out. This has the same objectives as other types of bioremediation but is *ex situ* in that the soil is moved to a site where it can be more easily turned and kept at the optimal temperature.

The cleaned soil usually comprises course mineral particles which cannot be used for supporting plant growth unless mixed with a source of organic matter (*e.g.* compost) and a supply of plant nutrients.

8.2 *In situ* Methods

The development of *in situ* methods over the last 20 years has reflected a great deal of exciting innovative science and technology. This has been encouraged by policies of government organizations in countries such as the USA and the Netherlands and several successful techniques have been developed with others still undergoing investigation. There is a strong financial incentive to develop techniques which do not necessitate the expense of digging out contaminated soil and transporting it to a safe disposal site, such as a licensed landfill. However, in some cases this latter course of action is sometimes still required where the hazard is too great to leave the site for a long period while less intensive remediation takes place. Some of the *in situ* methods are described below.

8.2.1 *Physico-chemical Methods*
Covering. Covering of the contaminated soil by a layer of clean soil so that plants can be grown in the uncontaminated 'cover loam'. The depth of covering is usually at least a metre but, as with all aspects related to contaminated land, cost is once again a major consideration. A plastic or geotextile membrane is often used to separate the contaminated material from the overlying cover soil. It is important not to disturb the under-lying contaminated soil (by excavations or planting deep rooting trees); however, there is a possibility that some contaminants may migrate

upwards through the soil profile during periods of prolonged dry weather when evapotranspiration is greater than precipitation.

Dilution. In cases of mild to moderate contamination, it may be possible to dilute the concentration of contaminants by deep ploughing and mixing the underlying uncontaminated soil with the contaminated topsoil. This can be done in the case of heavy metals such as Pb which are sorbed in a relatively immobile and unavailable form, but the method is unsuitable for mobile or highly toxic chemicals.

Reduction of availability. The plant availability of the contaminants, such as heavy metals can be minimized by liming or adding adsorptive minerals to the soil. However, this does not reduce the risk to children who might intentionally or accidentally eat soil.

By raising the pH of most soils to 7, most cationic heavy metals, including: Cd, Cu, Cr, Ni, Pb and Zn can be rendered more strongly adsorbed and less available for uptake by plants. Manipulation of the redox conditions is possible, for example by flooding or drainage, but is not usually a practicable option, except for the creation of wetlands to render contaminants such as heavy metals unavailable due to the formation of insoluble sulfides. However, the waterlogged conditions must be permanent because the sulfides would oxidize and become soluble and plant available, as in the case of Cd in contaminated paddy soils in Japan which go through a wetting and drying cycle.

Washing. *In situ* soil washing can be carried out by setting up a sprinkler irrigation system which can be used to deliver various extracting solutions to soil. These could include dilute acids or chelating agents for leaching out heavy metals such as Cd, but the site must have suitable underdrainage to remove the contaminant-carrying leachate away to a safe disposal route. This often necessitates the construction of a system of drainage pipes which deliver the leachate to a sump where it is treated to remove the contaminant metals. This can be done by exchange or chelating resins or by precipitation.

Soil vapour extraction. Soils contaminated with highly volatile compounds, such as solvents can be ameliorated by a technique known as soil vapour extraction. This is only suited to soils with a relatively high permeability comprising large pores and/or fissures and involves inserting a system of perforated pipes in the contaminated layer of soil so that air can be drawn through by suction pump. The extracted air containing the volatile compounds is then passed through a column of activated carbon to sorb and remove the organic contaminants.

Air stripping. This technique is used to remove volatiles contaminating groundwater and may need to be carried out alongside other techniques of soil cleaning. Contaminated groundwater extracted from a well is brought to the surface and has air bubbled through it to enhance the volatilization of the organic contaminant. This is usually carried out in a tower containing a system of baffles which facilitate the aeration. The volatile compounds are removed from the sparging air by activated carbon filters.

8.2.2 Biological Methods

Bioremediation. The most widely used method so far has been the *in situ* bioremediation of soils contaminated with organic chemicals, such as petroleum hydrocarbons, using the indigenous soil micro-organisms. By optimizing the soil conditions for the micro-organisms (mainly bacteria) through adjusting the pH, temperature (plastic tunnels), and nutrient supply and by regular cultivation to promote aeration, it has been shown that many organic contaminants can be effectively degraded. In the Netherlands there is a government backed research programme on the *in situ* bioremediation of contaminated soils and waters. The acronym for this programme is NOBIS and the work carried out so far has demonstrated that *in situ* bioremediation is an effective clean-up method.

Bioventing. Bioventing is a method which combines the principles of *in situ* bioremediation with those of soil vapour extraction. This technique aims to optimize the biodegradation of certain organic contaminants, including volatile and semivolatile compounds, halogenated organics, and PAHs, through the addition of oxygen and nutrients to the soil to stimulate the indigenous micro-organisms but it is also possible to augment these by the addition of cultures of bacteria or fungi which have the ability to degrade certain recalcitrant molecules. The air and nutrients are blown into a system of perforated pipes in the soil and soil vapour is also extracted by suction from a series of boreholes (air wells). The area of contaminated land may be covered by an impermeable cap to prevent air being drawn in from the surface so that all movement of air is through the contaminated zone.[25]

Crop growth. The amelioration of soils contaminated with heavy metals, such as Cd or Zn could possibly be brought about by growing hyper-accumulating crop genotypes that take up relatively high concentrations of these and other metals. Although only in the development stage, this technique is expected to provide an environmentally sound

[25] Royal Commission on Environmental Pollution, 19th Report, 'Sustainable Uses of Soil', HMSO, London, 1996.

clean-up procedure. The harvested plant material containing the accumulated contaminant metals would need to be disposed of safely, possibly by burning and landfilling the metal-rich ash.

8.3   Specific Techniques for Gasworks Sites

As an example of the number of techniques which are either currently available or which are being developed for cleaning up contaminated soils, Brown and co-workers[26] provide a list of methods applicable to the remediation of former gasworks sites, where the predominant organic contaminants are PAHs; these methods include:

(a) Co-burning—Highly contaminated soils and tars are mixed with combustible materials and burnt in a modified boiler as a fuel.
(b) Thermal desorption—Contaminated soils are heated to volatilize certain organic contaminants such as tars.
(c) Coal tar removal—Pulverized coal slurry is used to adsorb tars. The coal slurry is separated from the cleaned soil and used as a fuel.
(d) Surfactant flushing—Soil is washed with mixtures of surfactants to remove PAHs. Surfactant-based foams can also be used for this.
(e) *In situ* ozonation—Ozone gas is used to oxidize PAHs directly.
(f) Enhanced biodegradation—Various methods are used, including: growing the biodegrading bacteria in a unipolar magnetic field; using surfactants to increase the exposure of sorbed PAHs to bacterial degradation; using fungi to degrade the more recalcitrant complex PAHs which are very resistant to bacterial activity; and, finally, adding chemical oxidants to render PAHs more biodegradable.

9   CASE STUDIES

This section gives some brief examples of actual contaminated land cases which illustrate the various principles outlined earlier.

9.1   Gasworks Sites

The manufacture of gas (coal gas) from coal commenced in the early years of the nineteenth century and most industrialized communities had their own gasworks to provide fuel for lighting and heating. It is

[26] R. A. Brown, M. Jackson, and M. Loucy, Special issue: 'International Symposium and Trade Fair on the Clean-up of Manufactured Gas Plants', *Land Contam. Reclam.*, 1995, 3, 4, 2.1–2.2.

estimated that in the UK there may have been up to 5000 sites contaminated from this land use (both major gasworks and small gas manufacturing plants to serve a large factory). The Netherlands has 234 gasworks sites needing remediation, and there are 41 in Canada. For almost 200 years gasworks sites (also known as manufactured gas plants—MGPs) have been closed down and the land used for other purposes. However, it is only in the last few decades that the scale of the contamination and its potential hazard have been recognized. At several sites where former gasworks were developed for housing many years ago, it has been necessary to carry out major retrospective clean-up procedures owing to the possible hazards to the residents. One example of this is the Kralingen works, near Rotterdam in the Netherlands. In addition to the gasworks sites themselves, other sites such as old waste dumps have become contaminated with highly toxic wastes from gas manufacture and there may be a total of many thousand of sites (up to 100 000) in the UK affected in some way by gas manufacture.[27]

The gas manufacturing process involved the dry heating of coal in an air-free environment, which produced crude gas, tar, coke, and cinders. The crude gas was subsequently purified by passing through tar separators, condensers, wet purification to remove NH_3, HCN, phenols, and creosols, and, finally, dry purification with ferric oxide to remove sulfur and cyanide compounds. These purification processes resulted in several highly toxic compounds being accumulated at gasworks sites. The tars contain PAHs, hydrocarbons, phenols, benzene, xylene, and naphthalene and the ferric oxide was converted to $Fe_4(Fe(CN)_6)_3$ (hexacyanoferrate—'prussian blue') and sulfides but residues of oxide also contained Pb, Cu, and As. These materials can be found on old gasworks sites and at former waste dumps near to gasworks. The tars were utilized for various by-products. Being dense liquids they tended to accumulate in voids and pits within the site.

The PAHs are persistent organic pollutants and are very toxic. Some are carcinogens and they can enter the body by ingestion, dermal contact, or inhalation. As a result of the ubiquitous occurrence of gasworks-contaminated land and the importance of PAHs as contaminants, many of the *in situ* bioremediation techniques (outlined earlier) have been developed to deal with these chemicals. Cyanides are potentially very toxic but, fortunately, the characteristic blue colour (colloquially referred to as 'blue billy') is a good indicator of their presence; unfortunately this makes it attractive to children to handle. A green coloured Fe cyanide compound is also found at these sites—$Fe_2(CN)_6$

[27] A. O. Thomas and J. N. Lester, *Sci. Total Environ.*, 1994, **152**, 239–260.

('Berlin green').[28] In addition to solid forms, soluble cyanide compounds are also found in the groundwater at these sites along with PAHs, phenols, and several other organic pollutants. In addition to these characteristic contaminants, other common contaminants such as asbestos are also found at old gas works sites.[27]

A survey of eight former gasworks sites in the UK showed the following maximum values for key contaminants ($\mu g\ g^{-1}$): sulfate $<250\,000$, elemental S $<250\,000$, sulfide <4000, free CN <64, total CN <8000, total phenols <1000, toluene extract (tars/PAHs) $<250\,000$, and total heavy metals: As <250, Cd <64, Pb <4000, and Cr <250.[23]

9.2 Soil Contaminated by Landfilling and Waste Disposal

Love Canal, near Niagara Falls in New York State, USA is the name given to a section of unfinished canal which was used in the 1940s by a chemical company as a disposal site for approximately 20 000 tonnes of chemical wastes including intermediates from the manufacture of chlorinated pesticides. The covered waste tip was subsequently used for the construction of a school and houses. Although there had been various reports of irritating fumes affecting children, the problem came to a head after heavy rainfall in the winter of 1977–78 which caused a considerable rise in the water table. Some drums broke though the soil cap over the waste and the basements of houses were penetrated by harmful chemicals in the groundwater. The United States Environmental Protection Agency (EPA) carried out a series of investigations in 1978 and found 82 different contaminant chemicals in the groundwater, of which 27 were on the EPA's Priority List and 11 of these were known to be carcinogenic. The site was declared a disaster area (the first declared contamination disaster in US history) and 239 families were evacuated initially although many more were evacuated later after further investigations. Interestingly, the EPA later found 24 industrial sites more severely contaminated than this in the USA. By the early 1980s, 170 000 waste disposal sites in the USA had been documented by the EPA, of which 2100 had received industrial wastes.

In 1980, in the Netherlands, a major pollution problem was discovered in the village of Lekkerkirk which had been built in the early 1970s on reclaimed land beside the River Lek. The surface of this land had been raised by up to 3.5 m by layers of household and demolition waste, covered by 0.7 m of sand. However, this elevation was still lower than that of the adjacent river and so the groundwater tended to flow upwards

---

[28] J. Shefchek, I. Murarka, and A. Battaglia, Special issue: 'International Symposium and Trade Fair on the Clean-up of Manufactured Gas Plants', *Land Contam. Reclam.*, 1995, **3**, 4, 2.18–2.20.

through the waste towards the surface. In 1978, the surface soil was found to be contaminated by volatile compounds which caused the deterioration of buried plastic drinking water pipes and contaminated the water supply as well as causing toxicity in garden plants and noxious odours. In 1980, investigations revealed that 75% of the houses in the village were affected and these had to be evacuated and the residents found alternative housing. The evacuated houses were supported on hydraulic jacks while $87\,000\ m^3$ of fill mainly contaminated with paint solvents (toluene and lower boiling point solvents), resins, and heavy metals (Cd, Hg, Pb, Sb and Zn) were removed and subsequently disposed of by incineration. The maximum concentrations found were ($\mu g\ g^{-1}$): toluene <1000, lower boiling point solvents <3000, and metals: Cd <97, Hg <8.2, Pb <740, and Zn <1670.

9.3 Heavy Metal Contamination from Metalliferous Mining and Smelting

Zinc was mined around the village of Shipham in Somerset, England during the eighteenth and nineteenth centuries and very high concentrations of Pb, Zn, and Cd were left in the soils and waste heaps around the village.[29] In the late 1970s it was realized that houses built on former mining land between 1951 and 1960 had high concentrations of potentially toxic elements such as Cd in the garden soil and that there was a risk to the health of the householders, who regularly consumed vegetables and fruit grown in these gardens. Some of the highest Cd concentrations reported in agricultural and horticultural soils (up to $470\ \mu g\ Cd\ g^{-1}$) were found, together with $108–6540\ \mu g\ Pb\ g^{-1}$ and $250–37\,200\ \mu g\ Zn\ g^{-1}$. Although the mean Cd concentration in almost 1000 samples of vegetables and fruit was nearly 17 times higher than the national average concentration of $0.015\ \mu g\ Cd\ g^{-1}$, no obvious adverse health effects were found in the group of 500 volunteers (about half of the village population) who underwent detailed health checks. Nevertheless, these soils will remain indefinitely contaminated and the residents of houses with contaminated gardens have been advised not to consume home-grown produce. However, the uptake of Cd at this site was proportionally low owing to a relatively high concentration of free $CaCO_3$ in the soil which buffered the pH at values above 7.0 and sorbed much of the potentially bioavailable metals.

In contrast to Shipham, in the paddy rice growing district of the Jinzu Valley in Toyama Province of Japan after the Second World War, elderly women who had borne several children were found to be suffering from a severe skeletal disorder called Itai-itai (which literally translates as 'ouch

---

[29] H. Morgan, The Shipman Report—Special Issue, *Sci. Total Environ.*, 1988, **75**, 1, 43.

ouch'!) because of the pain experienced. This disease has been found to be caused by excess Cd and a relative deficiency of Ca and protein in the diet. The Cd and other metal contaminants reached the soil from a Zn–Pb mine and smelter upriver from the paddy fields. Concentrations of Cd in the soils were much lower than those in Shipham (*ca.* 5 μg Cd g^{-1}) but the availability of the metal was much higher due to the alternating water-logging and drying out of the paddy soils which initially led to precipitation of Cd as CdS in the gleyed soil followed by a rapid increase in Cd mobility and bioavailability when the soils returned to an oxygenated state and the sulfide dissolved releasing Cd, Zn, and SO_4^{2-}. Unlike in Shipham, the subsistence farmers of the Jinzu Valley consumed a diet largely grown on the contaminated paddy land. The main effects were only in women who had given birth to several babies and this is associated with the dynamics of Ca in the skeleton of pregnant women. These cases reveal that many other factors need to be considered, as well as the type and amount of contamination, in order to assess the impacts on human health.

9.4 Heavy Metal Contamination of Domestic Garden Soils in Urban Areas

Urban areas tend to have higher concentrations of heavy metal contaminants such as Pb than rural areas due to the greater amounts of atmospheric deposition. However, garden soils anywhere can be contaminated from bonfire ash when old painted wood has been burnt, when metal containing rubbish (such as old Pb-sheathed electric cable) are buried in the soil, when Pb-containing petrol is spilt, or when composts, manures, and even mineral fertilizers (especially Cd-containing phosphatic fertilizers) are applied to soil. A survey of heavy metals in soils from 3550 domestic gardens in several towns and cities in England and Wales reported the following maximum concentrations (μg g^{-1}):[30] Pb 14 125 (geometric mean 230), Cd 17 (geometric mean 1.2), Cu 16 800 (geometric mean 53) and Zn 14 568 (geometric mean 260). In Greater London, the concentrations tended to be highest of all. In 579 gardens in Greater London, geometric mean concentrations of heavy metals were: Pb 647, Cd 1.3, Cu 73.0, and Zn 424. Heavy metal concentrations were higher in the gardens of older houses (> 35 years), houses within 500 m of commercial garages (Pb in exhausts), and those in close proximity to demolition sites, waste tips, or metallurgical industries and those within 10 m of a road. In addition to metals, garden soils are often contaminated with atmospherically deposited PAHs, pesticides, solvents, and other contaminants.

[30] I. Thornton, E. Culbard, S. Moorcroft, J. Watt, M. Wheatley, and M. Thompson, *Environ. Technol. Lett.*, 1983, **6**, 137.

9.5 Land Contamination by Solvents, PCBs, and Dioxins Following a Fire at an Industrial Plant

A fire in Cheshire, UK, at a works (Chemstar) that recovered solvents from chemical wastes, resulted in a high level of soil contamination by solvents, benzene, and PCBs from the wastes and dioxins formed from constituents in the wastes during the combustion process. Concentrations (μg g^{-1}) of up to 208 benzene, 304 total solvents, 1160 PCBs, and 168 toxic equivalents of dioxin and furans (2,3,7,8-TCDD) were found. These dioxin/furan concentrations are some of the highest reported in the literature.[31] For comparison, in a well documented case of dioxin contamination at Times Beach, Missouri, USA, where contaminated waste oil had been sprayed onto a horse arena to reduce dust problems in the dry summer, concentrations of 33 μg g^{-1} toxic equivalents 2,3,7,8-TCDD caused the death of 48 horses, and many other types of animals also died.[32] At the Chemstar site the main animals to be affected were guinea pigs, which are well known to be highly sensitive to dioxin toxicity.[31]

10 CONCLUSIONS

Soils can be contaminated by a vast range of chemicals, and the behaviour of the contaminant in the soil depends on both its own chemical properties and the physical and chemical properties of the soil. Heavy metals tend to persist in soil for a long time, amounting to centuries in some cases. Organic contaminants have the potential to be degraded by soil micro-organisms but the rate at which this occurs can range from days to years or decades. Some of the highly persistent chemicals such as PCBs, dioxins and PAHs may persist for more than 20 years. However, it is generally found that most of the persistent organic pollutants are not readily taken up by plants. Ingestion of contaminated soil is likely to be a more important exposure route than accumulation into crops.

The predominant factors affecting the sorption of contaminants by a soil are the pH and mineralogy for heavy metals and the organic matter content for organic chemicals. Industrially contaminated land can contain large amounts of non-pedological material, such as lumps of construction material, and also numerous voids and fissures, which could facilitate movement towards the water table. It is important to remember that it is impossible to avoid some form of contamination in

[31] T. Craig and R. Grzonka, *Land Contam. Reclam.*, 1994, **2**, 1, 19–26.
[32] S. E. Manahan, 'Environmental Chemistry', Brooks/Cole (Wadsworth), Monterey, CA, 1984, p. 612.

soils in technologically advanced countries. This is because atmospheric deposition (sometimes following long-distance transboundary transport) and contaminants in agricultural materials will result in at least slight accumulations of materials as a result of human activity.

Questions

1. Explain why soil and sediments act as sinks for organic and inorganic contaminants, unlike the other environmental media, air and water, and consider the importance of the following with regard to the retention of pollutants:
   (i) soil pH
   (ii) soil organic matter
   (iii) clay minerals and oxides of Fe, Al and Mn.
2. Compare the expected behaviour of cadmium and lead contaminants in soils and their bioavailability to food crops.
3. Consider the possible fate of organic pollutants entering the soil and discuss the importance of sorption, volatilization, leaching, and degradation.
4. Compare the factors controlling the mobility and bioavailability of an organic contaminant, such as a PAH (*e.g.* benzo(*a*)pyrene), and a heavy metal such as Cu.
5. What lessons can be learnt from the Love Canal and Lekkerkirk cases with regard to the disposal of chemical wastes and risk assessment.
6. Former gasworks sites exist in many towns and cities in industrially developed countries. Discuss their potential for redevelopment and the main contamination hazards associated with these sites. Consider how a site investigation should be carried out at a former gasworks and the main contaminants which should be analysed for.
7. Discuss the importance of soil quality criteria and guidelines values for safe concentrations of possible contaminants. Compare the arguments for a multifunctionality approach compared with a fitness for purpose approach.
8. 'A knowledge of soil science is an important requirement when planning an investigation and possible clean-up of a site which is suspected of being significantly contaminated.' Discuss.

CHAPTER 6

Environmental Cycling of Pollutants

ROY M. HARRISON

1 INTRODUCTION: BIOGEOCHEMICAL CYCLING

The earlier chapters of this book have followed the traditional sub-division of the environment into compartments (*e.g.* atmosphere, oceans, *etc.*). Whilst these sub-divisions accord with human perceptions and have certain scientific logic, they encourage the idea that each compartment is an entirely separate entity and that no exchanges occur between them. This, of course, is far from the truth. Important exchanges of mass and energy occur at the boundaries of the compartments and many processes of great scientific interest and environmental importance occur at these interfaces. A physical example is that of transfer of heat between the ocean surfaces and the atmosphere, which has a major impact upon climate and a great influence upon the general circulation of the atmo-sphere. A chemically based example is the oceanic release of dimethyl sulfide to the atmosphere, which may, through its decomposition products, act as a climate regulator (see Chapter 4).

Pollutants emitted into one environmental compartment will, unless carefully controlled, enter others. Figure 1 illustrates the processes affecting a pollutant discharged into the atmosphere.[1] As mixing processes dilute it, it may undergo chemical and physical transforma-tions before depositing in rain or snow (wet deposition) or as dry gas or particles (dry deposition). The deposition processes cause pollution of land, freshwater, or the seas, according to where they occur. Similarly, pollutants discharged into a river will, unless degraded, enter the seas. Solid wastes are often disposed into a landfill. Nowadays these are carefully designed to avoid leaching by rain and dissemination of pollutants into groundwaters, which might subsequently be used for potable supply. In the past, however, instances have come to light where

[1] W. H. Schroeder and D. A. Lane, *Environ. Sci. Technol.*, 1988, **22**, 240.

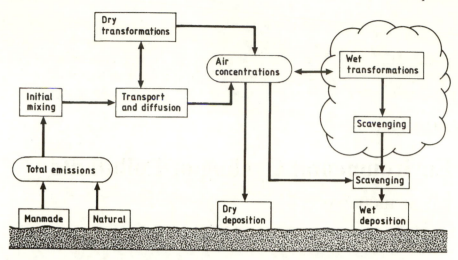

Figure 1 *Schematic diagram of the atmospheric cycle of a pollutant*[1]
(Reprinted from Environmental Science and Technology by permission of the
American Chemical Society)

insufficient attention was paid to the potential for groundwater contamination, and serious pollution has arisen as a result.

Another important consideration regarding pollutant cycling is that of degradability, be it chemical or biological. Chemical elements (other than radioisotopic forms) are, of course, non-degradable and hence once dispersed in the environment will always be there, although they may move between compartments. Thus, lead, for example, after emission from industry or motor vehicles, has a rather short lifetime in the atmosphere, but upon deposition causes pollution of vegetation, soils, and waters.[2] On a very long time-scale, lead in these compartments will leach out from soils and transfer to the oceans, where it will concentrate in bottom sediments.

Some chemical elements undergo chemical changes during environmental cycling which completely alter their properties. For example, nitrate added to soil as fertilizer can be converted to gaseous nitrous oxide by biological denitrification processes. Nitrous oxide is an unreactive gas with a long atmospheric lifetime which is destroyed only by breakdown in the stratosphere. As will be seen later, nitrogen in the environment may be present in a wide range of valence states, each conferring different properties.

Some chemical compounds are degradable in the environment. For example, methane (an important greenhouse gas) is oxidized via carbon

[2] R. M. Harrison and D. P. H. Laxen, 'Lead Pollution: Causes and Control', Chapman & Hall, London, 1981.

monoxide to carbon dioxide and water. Thus, although the chemical elements are conserved, methane itself is destroyed and were it not continuously replenished would disappear from the atmosphere. The breakdown of methane is an important source of water vapour in the stratosphere, illustrating another, perhaps less obvious, connection between the cycles of different compounds.

Degradable chemicals which cease to be used will disappear from the environment. PCBs are no longer used industrially to any significant degree, having been replaced by more environmentally acceptable alternatives. Their concentrations in the environment are decreasing, although because of their slow degradability (*i.e.* persistence), it will take many years before their levels decrease below analytical detection limits.

The transfer of an element between different environmental compartments, involving both chemical and biological processes, is termed biogeochemical cycling. The biogeochemical cycles of the elements lead and nitrogen will be discussed later in this chapter.

1.1 Environmental Reservoirs

To understand pollutant behaviour and biogeochemical cycling on a global scale, it is important to appreciate the size and mixing times of the different reservoirs. These are given in Table 1. The mixing times are a very approximate indication of the time-scale of vertical mixing of the reservoir.[3] Global mixing can take very much longer as this involves some very slow processes. These mixing times should be treated with considerable caution as they oversimplify a complex system. Thus, for example, a pollutant gas emitted at ground level mixes in the boundary layer (*ca.* 1 km) on a time-scale typically of hours. Mixing into the free troposphere (1–10 km) takes days, whilst mixing into the stratosphere (10–50 km) is on the time-scale of several years. Thus, no one time-scale describes atmospheric vertical mixing, and the same applies to other reservoirs. Such concepts are useful, however, when considering the behaviour of trace components. For example, a highly reactive hydrocarbon emitted at ground level will probably be decomposed in the boundary layer. Sulfur dioxide, with an atmospheric lifetime of days, may enter the free troposphere but is unlikely to enter the stratosphere. Methane, with a lifetime of several years, extends through all of the three regions.

---

[3] P. Brimblecombe, 'Air Composition and Chemistry', Cambridge University Press, Cambridge, 2nd Edn., 1996.

Table 1   *Size and vertical mixing of various reservoirs (from Brimblecombe[3])*

	Mass (kg)	Mixing time (years)
Biosphere[a]	4.2×10^{15}	60
Atmosphere	5.2×10^{18}	0.2
Hydrosphere	1.4×10^{21}	1600
Crust	2.4×10^{22}	$>3 \times 10^{7}$
Mantle	4.0×10^{24}	$>10^{8}$
Core	1.9×10^{24}	

[a]Plants, animals, and organic matter are included but coal and sedimentary carbon are not. The mixing time of carbon in living matter is about 50 years.

It should be noted from Table 1 that the atmosphere is a much smaller reservoir in terms of mass than the others. The implication is that a given pollutant mass injected into the atmosphere will represent a much larger proportion of total mass than in other reservoirs. Because of this, and the rather rapid mixing of the atmosphere, *global* pollution problems have become serious in relation to the atmosphere before doing so in other environmental media. The converse also tends to be true, that once emissions into the atmosphere cease, or diminish, the beneficial impact is seen on a relatively short time-scale. This has been seen in relation to lead, for instance, where lead in Antarctic ice (derived from snow) has shown a major decrease resulting from diminishing emissions from industry and use of leaded petrol[4] (Figure 2). Improved air quality in relation to CFCs will take longer to achieve because of the much longer atmospheric lifetimes (> 100 years) of some of these species (see Chapter 1).

1.2   Lifetimes

A very useful concept in the context of pollutant cycling is that of the lifetime of a substance in a given reservoir. We can think in terms of substances having sources, magnitude S, and sinks, magnitude R. At equilibrium:

$$R = S$$

An analogy is with a bath; the inflow from a tap (S) is equal to the outflow (R) when the bath is full. An increase in S is balanced by an increase in R. If the total amount of substance A in the reservoir

[4] J.-P. Candelone, S. Hong, C. Pellone, and C. F. Boutron, *J. Geophys. Res.*, 1995, **100**, 16 605–16 616.

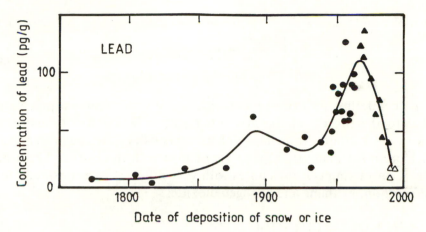

Figure 2   *Changes in lead concentrations in snow/ice deposited in central Greenland from 1773 to 1992*
(Adapted from Candelone et al.[4])

(analogy = mass of water in the bath) is A, then the lifetime, τ is defined by:

$$\tau = \frac{A \text{ (kg)}}{S \text{ (kg s}^{-1})} \tag{1}$$

In practical terms, the lifetime is equal to the time taken for the concentration to fall to $1/e$ (where e is the base of natural logarithms) of its initial concentration, if the source is turned off. If the removal mechanism is a chemical reaction, its rate may be described as follows:

$$R' = \frac{d[A]}{dt} = k[A] \tag{2}$$

(In this case $d[A]/dt$ describes the rate of loss of A if the source is switched off; obviously with the source on, at equilibrium $d[A]/dt = 0$). The latter part of equation (2) assumes first order decay kinetics, *i.e.* the rate of decay is equal to the concentration of A, termed $[A]$, multiplied by a rate constant, k. As discussed later this is often a reasonable approximation.

Taking equation (1) and dividing both numerator and denominator by the volume of the reservoir, allows it to be rewritten in terms of concentration. Thus:

$$\tau = \frac{[A] \text{ (kg m}^{-3})}{S' \text{ (kg m}^{-3} \text{ s}^{-1})} \tag{3}$$

since $S' = R'$

$$\tau = \frac{[A]}{k[A]} = k^{-1} \tag{4}$$

Thus the lifetime of a constituent with a first order removal process is equal to the inverse of the first order rate constant for its removal. Taking an example from atmospheric chemistry, the major removal mechanism for many trace gases is reaction with the hydroxyl radical, $OH^{\bullet}$. Considering two substances with very different rate constants[5] for this reaction, methane and nitrogen dioxide:

$$CH_4 + OH^{\bullet} \rightarrow CH_3^{\bullet} + H_2O \tag{5}$$

$$\frac{-d}{dt}[CH_4] = k_2[CH_4][OH^{\bullet}] \qquad k_2 = 6.2 \times 10^{-15}\,cm^3\,molec\,s^{-1} \tag{6}$$

$$NO_2 + OH^{\bullet} \rightarrow HNO_3 \qquad k_2 = 1.4 \times 10^{-11}\,cm^3\,molec\,s^{-1} \tag{7}$$

Making the crude assumption of a constant concentration of $OH^{\bullet}$ radical[6] (more justifiable for the long-lived methane, for which fluctuations in $OH^{\bullet}$ will average out, than for short-lived nitrogen dioxide),

$$\frac{-d}{dt}[CH_4] = k_2[CH_4][OH^{\bullet}]$$
$$= k_1'[CH_4]$$

where
$$k_1' = k_2[OH^{\bullet}]$$

Worked example

What are the atmospheric lifetimes of CH_4 and NO_2 if the diurnally averaged concentration of $OH^{\bullet}$ radical is $1 \times 10^6\,molec\,cm^{-3}$?

$$k_1' = 6.2 \times 10^{-15} \times 1 \times 10^6$$
$$= 6.2 \times 10^{-9}\,s^{-1}$$

Then from equation (4)

$$\tau = k^{-1}$$
$$= (6.2 \times 10^{-9})^{-1}\,s$$
$$= 5.1 \text{ years for } CH_4$$

[5] B. J. Finlayson-Pitts and J. N. Pitts Jr., 'Atmospheric Chemistry', John Wiley and Sons, Chichester, 1986.
[6] C. N. Hewitt and R. M. Harrison, *Atmos. Environ.*, 1985, **19**, 545.

By analogy, for nitrogen dioxide, the lifetime,

$$\tau = 20 \text{ hours}$$

This general approach to atmospheric chemical cycling has proved useful in many instances. For example, measurements of atmospheric concentration, $[A]$, for a globally mixed component may be used to estimate source strength, since

$$S' = R' = \frac{-d[A]}{dt} = k_2[A][OH^\bullet]$$

and

$$S = S' \times V$$

where V is the volume of atmosphere in which the component is mixed. Source strengths estimated in this way, for example for the compound methyl chloroform, CH_3CCl_3, known to destroy stratospheric ozone, may be compared with known industrial emissions to deduce whether natural sources contribute to the atmospheric burden.

1.2.1 Influence of Lifetime on Environmental Behaviour. Some knowledge of environmental lifetimes of chemicals is very valuable in predicting their environmental behaviour. In relation to the atmosphere, there is an interesting relationship between the spatial variability in the concentrations of an atmospheric trace species and its atmospheric lifetime.[3] Compounds such as methane and carbon dioxide with a long lifetime with respect to removal from the atmosphere by chemical reactions or dry and wet deposition (see Section 2 of this chapter) show little spatial variability around the globe, as their atmospheric lifetime (several years) exceeds the time-scale of mixing of the entire troposphere (of the order of a year). On the other hand, for a short-lived species such as nitrogen dioxide, removal by chemical means or dry or wet deposition occurs much more quickly than atmospheric mixing and hence there is very large spatial variability, with concentrations sometimes exceeding 100 ppb in urban areas, whilst remote atmosphere concentrations can be at the level of a few ppt. By analogy, short-lived species also show a much greater hour-to-hour and day-to-day variation in concentration at a given measuring point than long-lived species for which local sources impact only to a modest degree on the existing background concentration.

This illustration using the atmosphere can be taken somewhat further in relation to other environmental media. Lifetimes of highly soluble species such as sodium and chloride in the oceans are long compared to the mixing times and therefore variations in salinity across the world's oceans are relatively small (see Chapter 4). Where soils are concerned, mixing times will generally far exceed lifetimes and extreme local hot spot concentrations can be found where soils have become polluted.

Lifetime also influences the way in which we study the environmental cycles of pollutants. In the case of reactive atmospheric pollutants, it is the reaction rate, or rate of dry or wet deposition, which determines the lifetime. We are therefore concerned mainly with the rates of these processes in determining the atmospheric cycle. In the case of longer-lived species, such as persistent organic compounds like PCBs and dioxins, chemical reaction rates are rather slow and these compounds can approach equilibrium between different environmental media such as the atmosphere and surface ocean or the atmosphere and surface soil, with evaporation exceeding deposition during warmer periods and wet and dry deposition replacing the contaminant into the soils or oceans in cooler weather conditions. Both the kinetic approach dealing with reaction rates and the thermodynamically based approach considering partition between environmental media will be introduced in this chapter. In general the kinetic or reaction rate approach will be most appropriate to the study of short-lived reactive substances, whilst the equilibrium approach will be more applicable to long-lived substances.

2   RATES OF TRANSFER BETWEEN ENVIRONMENTAL COMPARTMENTS

2.1   Air–Land Exchange

The land surface is an efficient sink for many trace gases. These are absorbed or decomposed on contact with plants or soil surfaces. Plants can be particularly active because of their large surface area and ability to absorb water-soluble gases. The deposition process is crudely described by the deposition velocity, v_d,

$$v_d(\text{cm s}^{-1}) = \frac{\text{Flux } (\mu\text{g m}^{-2}\text{ s}^{-1})}{\text{Atmospheric concentration } (\mu\text{g m}^{-3})}$$

The term *flux* is analogous to a flow of material, in this case expressed as micrograms of substance depositing per square metre of ground surface per unit time. In the case of rough surfaces the square metre of area refers to the area of a hypothetical horizontal flat surface beneath the true

Table 2 *Some typical values of deposition velocity*

Pollutant	Surface	Deposition velocity (cm s^{-1})
SO$_2$	Grass	1.0
SO$_2$	Ocean	0.5
SO$_2$	Soil	0.7
SO$_2$	Forest	2.0
O$_3$	Dry grass	0.5
O$_3$	Wet grass	0.2
O$_3$	Snow	0.1
HNO$_3$	Grass	2.0
CO	Soil	0.05
Aerosol (<2.5 μm)	Grass	0.1

surface rather than the sum of the area of all the rough elements such as plant leaves which make up the true surface.

Since the deposition process itself causes a gradient in atmospheric concentration, v_d is defined in relation to a reference height, usually 1 m, at which the atmospheric concentration is measured. For reasons described later, v_d is not a constant for a given substance, but varies according to atmospheric and surface conditions. However, some typical values are given in Table 2, which exemplify the massive variability.

For some trace gases, for example, nitric acid vapour, dry deposition represents a major sink mechanism. In this case the process may have a major impact upon atmospheric lifetime.

Worked example

Dry deposition is frequently the main sink for ozone in the rural atmospheric boundary layer. What is the lifetime of ozone with respect to this process?

Assuming a typical dry deposition velocity of 1 cm s^{-1} and a boundary layer height of 1000 m, (H),

$$\frac{-d}{dt}[O_3] \,(\mu g \, m^{-3} \, s^{-1}) = \frac{\text{Flux} \,(\mu g \, m^{-2} \, s^{-1})}{\text{Mixing depth (m)}}$$
$$= \frac{v_d \times [O_3]}{H}$$
$$= k[O_3] \quad \text{where } k = v_d/H$$

By analogy with equation (4),

$$\tau = \frac{h}{v_d}$$
$$= 1000/0.01 \text{ s}$$
$$= 28 \text{ hours}$$

Thus, taking the boundary layer as a discrete compartment, the lifetime of ozone with respect to dry deposition is around 1 day. The lifetime in the free troposphere (the section of the atmosphere above the boundary layer) is longer, being controlled by transfer processes in and out, and chemical reactions. The stratospheric lifetime of ozone is controlled by photochemical and chemical reaction processes.

Dry deposition processes are best understood by considering a resistance analogue. In direct analogy with electrical resistance theory, the major resistances to deposition are represented by three resistors in series. Considering the resistances in sequence, starting well above the ground, these are as follows:

(i) r_a, the aerodynamic resistance describes the resistance to transfer downwards towards the surface through normally turbulent air;

(ii) r_b, the boundary layer resistance describes the transfer through a laminar boundary layer (approximately 1 mm thickness) at the surface;

(iii) r_s, the surface (or canopy) resistance is the resistance to uptake by the surface itself. This can vary enormously, from essentially zero for very sticky gases such as HNO_3 vapour, which attaches irreversibly to surfaces, to very high values for gases of low water solubility which are not utilized by plants (*e.g.* CFCs).

Since these resistances operate essentially in series, the total resistance, R, which is the inverse of the deposition velocity, is equal to the sum of the individual resistances.

$$R = \frac{1}{v_d} = r_a + r_b + r_s \tag{8}$$

Some trace gases have a net source at the ground surface and diffuse upwards; an example is nitrous oxide.

Whether the flux is downward or upward, it is driven by a concentration gradient in the vertical, dc/dz. The relationship between flux, F, and concentration gradient is:

$$F = K_z dc/dz$$

where K_z is the eddy diffusivity in the vertical (a measure of the atmospheric conductance). Fluxes, and thus deposition velocities, can be estimated by measurement of a concentration gradient simultaneously with the eddy diffusivity.[7] It is usually assumed that trace gases transfer in the same manner as sensible heat (*i.e.* convective heat transfer, not radiative or latent heat) or momentum. Thus the eddy diffusivity for either of these parameters is measured usually from simple meterological variables (gradients in temperature and wind speed).

A few substances are capable of showing both upward and downward fluxes. An example is ammonia. Ammonium in the soil, NH_4^+, is in equilibrium with ammonia gas, $NH_{3_{(g)}}$

$$NH_4^+ + H_2O \rightleftharpoons NH_{3_{(g)}} + H_3O^+ \qquad (9)$$

when atmospheric concentrations of ammonia exceed equilibrium concentrations at the soil surface (known as the compensation point), the net flux of ammonia is downwards. When atmospheric concentrations are below the equilibrium value, ammonia is released into the air.[8]

2.2 Air–Sea Exchange

The oceans cover some two-thirds of the earth's surface and consequently provide a massive area for exchange of energy (climatologically important) and matter (an important component of geochemical cycles).

The seas are a source of aerosol (*i.e.* small particles), which transfer to the atmosphere. These will subsequently deposit, possibly after chemical modification, either back in the sea (the major part) or on land (the minor part). Marine aerosol comprises largely unfractionated seawater, but may also contain some abnormally enriched components. One example of abnormal enrichment occurs on the eastern coast of the Irish Sea. Liquid effluents from the Sellafield nuclear fuel reprocessing plant in west Cumbria are discharged into the Irish Sea by pipeline. At one time, permitted discharges were appreciable and as a result radio-isotopes such as ^{137}Cs and several isotopes of plutonium have accumulated in the waters and sediments of the Irish Sea. A small fraction of these radioisotopes were carried back inland in marine aerosol and

[7] J. A. Garland, *Proc. R. Soc. London, Ser. A*, 1977, **354**, 245.
[8] S. Yamulki, R. M. Harrison, and K. W. T. Goulding, *Atmos. Environ.*, 1996, **30**, 109–118.

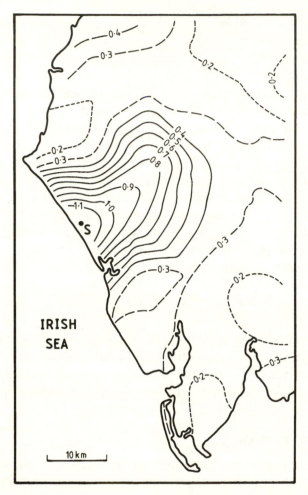

Figure 3   *Concentrations of plutonium in soils of West Cumbria ($^{239+240}$ Pu to 15 cm depth; pCi cm^{-2}). The point marked S indicates the position of the Sellafield reprocessing works*
(From Cawse[10])

deposited predominantly in the coastal zone.[9] Whilst the abundance of ^{137}Cs in marine aerosol was reflective only of its abundance in seawater (an enrichment factor—see Chapter 4—of close to unity), plutonium was abnormally enriched due to selective incorporation of small suspended sediment particles in the aerosol. This has manifested itself in enrichment of plutonium in soils on the west Cumbria coast,[10] shown as contours of $^{239+240}$Pu deposition (pCi cm^{-2}) to soil in Figure 3.

[9] R. S. Cambray and J. D. Eakins, *Nature*, 1982, **300**, 46.
[10] P. A. Cawse, UKAEA Report No. AERE—9851, 1980.

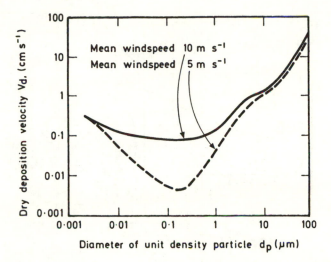

Figure 4 *Calculated values of deposition velocity to water surfaces as a function of particle size and wind speed*

The seas may also act as a receptor for depositing aerosol. Deposition velocities of particles to the sea are a function of particle size, density, and shape, as well as the state of the sea. Experimental determination of aerosol deposition velocities to the sea is almost impossible and we have to rely upon data derived from wind tunnel studies and theoretical models. The results from two such models appear in Figure 4, in which particle size is expressed as aerodynamic diameter, or the diameter of an aerodynamically equivalent sphere of unit specific gravity.[11,12] If the airborne concentration in size fraction of diameter d_i is c_i, then

$$\text{Total flux} = \sum^{i} v_g(d_i)c_i$$

where $v_g(d_i)$ is the mean value of deposition velocity appropriate to the size fraction d_i. Measurements show that whilst most of the lead, for example, is associated with small, sub-micrometre particles, the larger particles compose the major part of the flux.

Airborne concentrations of particulate pollutants are not uniform over the sea. The spatial distribution of zinc over the North Sea[13] averaged over a number of measurement cruises appears in Figure 5.

[11] S. A. Slinn and W. G. N. Slinn, *Atmos. Environ.*, 1980, **14**, 1013.
[12] R. M. Williams, *Atmos. Environ.*, 1982, **16**, 1933.
[13] C. R. Ottley and R. M. Harrison, Eurotrac ASE Annual Report, Garmisch-Partenkirchen, 1990.

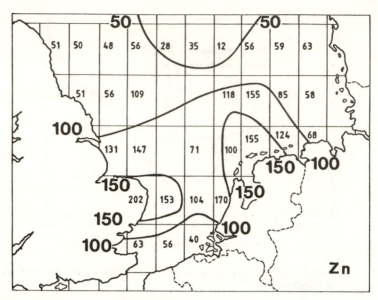

Figure 5 *Spatial distribution of zinc concentrations (in ng m^{-3}) in air over the North Sea during 1989*
(From Ottley and Harrison[13])

Spatial patterns of other metals and many artificial pollutants are similar, reflecting the impact of land-based source regions, with concentrations falling toward the north and centre of the sea.

Because of its position and relatively high pollution loading, the North Sea is a focus of considerable interest. An inventory of inputs of trace metals (*e.g.* Pb, Cd, Zn, Cu, *etc.*) accords similar importance to riverine inputs and atmospheric deposition.[14] Controls have now been applied to many source categories and total inputs of the metals indicated have in general declined appreciably. One particular example is lead, for which most European countries introduced severe controls on use in gasoline (petrol) during the 1980s and atmospheric concentrations have fallen accordingly. Although the data are less clear, it might be anticipated that concentrations in river water will also decline as a result of reduced inputs from direct atmospheric deposition and in runoff waters from highways and land surfaces.

As explained in Chapter 4, the sea may be both a source and a sink of trace gases. The direction of flux is dependent upon the relative concentration in air and seawater.[15] If the concentration in air is C_a, the

[14] R. F. Critchley, *Proc. Int. Conf. Heavy Metals in the Environment*, Heidelberg, Germany; CEP Consultants, Edinburgh, 1983, p. 1109.
[15] P. S. Liss and L. Merlivat, 'The Role of Air–Sea Exchange in Geochemical Cycling', ed. P. Buat-Menard, Reidel, Dordrecht, 1986, p. 113.

equilibrium concentration in seawater, $C_{w(equ)}$ is given by

$$C_{w(equ)} = C_a H^{-1} \qquad (10)$$

where H is the Henry's Law constant. The Henry's Law constant can be expressed as follows:

$$H = \frac{p_s}{S_{aq}}$$

where p_s = saturation vapour pressure and S_{aq} = equilibrium solubility in water.

Worked example

For benzene $\qquad p_s = 12.7 \text{ kPa at } 25\,^\circ\text{C}$

$$S_{aq} = 1.78 \text{ g l}^{-1} = \frac{1.78 \times 1000}{78} \text{ mol m}^{-3}$$

Calculate H for benzene at $25\,^\circ\text{C}$.

For benzene $\qquad\qquad H = \dfrac{p_s}{S_{aq}}$

$$H = \frac{12.7 \times 10^3}{(1.78 \times 10^3)/78}$$

$$= 556 \text{ Pa m}^3 \text{ mol}^{-1}$$

If C_w is the actual concentration of the dissolved gas in the surface seawater and

$$C_w = C_{w(equ)}$$

the system is at equilibrium and no net transfer occurs. If, however, there is a concentration difference, ΔC, where

$$\Delta C = C_a H^{-1} - C_w \qquad (11)$$

there will be a net flux. If

$$C_a H^{-1} > C_w$$

the water is sub-saturated with regard to the trace gas and transfer occurs from air to water. Conversely, gas transfers from supersaturated water to the atmosphere if

$$C_a H^{-1} < C_w$$

The rate at which gas transfer occurs is expressed by

$$F = K_{(T)w} \Delta C \tag{12}$$

where $K_{(T)w}$ is termed the total transfer velocity. This can be broken down into component parts as follows:

$$\frac{1}{K_{(T)w}} = \frac{1}{\alpha k_w} + \frac{1}{H k_a} = r_w + r_a \tag{13}$$

where k_a and k_w are the individual transfer velocities for chemically unreactive gases in air and water phases, respectively and α ($= k_{reactive}/k_{inert}$) is a factor which quantifies any enhancement of gas transfer in the water due to chemical reaction. The terms r_w and r_a are the resistances to transfer in the water and air phases respectively and are directly analogous to the resistance terms in equation (8). For chemically reactive gases, usually $r_a \gg r_w$ and atmospheric transfer limits the overall flux. For less reactive gases the inverse is true and $K_{(T)w} \cong k_w$; the resistance in the water is the dominant term.

Much research has gone into evaluating k_w and $K_{(T)w}$, both in theoretical models, and in wind tunnel and field studies. The results are highly wind speed dependent due to the influence of wind upon the surface state of the sea. The results of some theoretical predictions and experimental studies[16] for CO_2 (a gas for which k_w is dominant) are shown in Figure 6.

In addition to dry deposition, trace gases and particles are also removed from the atmosphere by rainfall and other forms of precipitation (snow, hail, *etc.*), entering land and seas as a consequence. Wet deposition may be simply described in two ways. Firstly,

$$\text{Scavenging ratio} = \frac{\text{Concentration in rain (mg kg}^{-1})}{\text{Concentration in air (mg kg}^{-1})}$$

Typical values of scavenging ratio[17] lie within the range 300–2000. Scavenging ratios are rather variable, dependent upon the chemical nature of the trace substance (particle or gas, soluble or insoluble, *etc.*)

[16] A. J. Watson, R. C. Upstill-Goddard, and P. S. Liss, *Nature*, 1991, **349**, 145.
[17] R. M. Harrison and A. G. Allen, *Atmos. Environ.*, 1991, **25A**, 1719.

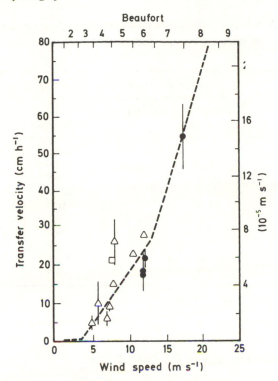

Figure 6 *Air-sea transfer velocities for carbon dioxide at 20 °C as a function of wind speed at 10 metres (m s⁻¹ or Beaufort Scale). The graph combines experimental data (points) and a theoretical line*
(From Watson *et al.*[16]) (Reprinted by permission from *Nature (London)*, **349**, 145; Copyright © 1991 Macmillan Magazines Ltd.)

and the type of atmospheric precipitation. Incorporation of gases and particles into rain can occur both by in-cloud scavenging (also termed rainout) and below-cloud scavenging (termed washout).

Numerical modellers often find it convenient to describe wet deposition by a scavenging coefficient, actually a first order rate constant for removal from the atmosphere. Thus, for trace substance A,

$$\frac{d[A]}{dt} = -\Lambda[A]$$

where Λ is the washout coefficient, with units of s^{-1}. A typical value of Λ for a soluble substance is $10^{-4}\,s^{-1}$ although actual values are difficult to measure and are highly dependent upon factors such as rainfall intensity.

Table 3 *A comparison of the concentration of major elements in 'average' riverine particulate material and surficial rocks*

	Concentrations (g kg^{-1})	
Element	Riverine particulate material	Surficial rocks
Al	94.0	69.3
Ca	21.5	45.0
Fe	48.0	35.9
K	20.0	24.4
Mg	11.8	16.4
Mn	1.1	0.7
Na	7.1	14.2
P	1.2	0.6
Si	285.0	275.0
Ti	5.6	3.8

Adapted from Martin and Meybeck[18]

3 TRANSFERS IN AQUATIC SYSTEMS

When rain falls over land some drains off the surface directly into surface water courses in surface runoff. A further part of the incoming rainwater percolates into the soil and passes more slowly into either surface waters or underground reservoirs. Water held in rock below the surface is termed groundwater, and a rock formation which stores and transmits water in useful quantities is termed an aquifer. Water which passes through soil or rock on its way to a river is chemically modified during transit, generally by addition of soluble and colloidal substances washed out of the ground. Some substances are removed from the water; for example river water often contains less lead than rainwater; one mechanism of removal is uptake by soil.

River waters carry both dissolved and suspended substances to the sea. The concentrations and absolute fluxes vary tremendously. The suspended solids load is largely a function of the flow in the river, which influences the degree of turbulence and thus the extent to which solids are held in suspension and resuspended from the bed, once deposited. Table 3 shows a comparison of 'average' riverine suspended particulate matter and surficial rock composition[18] for the major elements. Elements resistant to chemical weathering or biological activity (*e.g.* aluminium, titanium, iron, phosphorus) show some enrichment in the riverine solids,

[18] J. M. Martin and M. Meybeck, *Mar. Chem.*, 1979, 7, 177–206.

Table 4   *Average concentrations of the major constituents dissolved in rain and river water*

	Concentrations (mg dm^{-3})	
Constituent	Rain water	River water
Na$^+$	1.98	6.3
K$^+$	0.30	2.3
Mg^{2+}	0.27	4.1
Ca^{2+}	0.09	15
Fe		0.67
Al		0.01
Cl$^-$	3.79	7.8
SO$_4^{2-}$	0.58	11.2
HCO$_3^-$	0.12	58.4
SiO$_2$		13.1
pH	5.7	

Adapted from Garrels and Mackenzie[19]

whilst more soluble elements are subject to weathering and are depleted in the solids, being transported largely in solution (sodium, calcium). Some pollutant elements such as the metals lead, cadmium, and zinc tend to be highly enriched in the solids relative to surficial rocks or soils due to artificial inputs.

The dissolved components of river water typically exhibit significantly higher concentrations than in rainwater[19] (Table 4), due to leaching from rocks and soils. Some insight into the processes governing river water composition may be gained from Figure 7. Starting from the point of lowest dissolved salts concentrations, the ratio of Na/(Na + Ca) approaches one. This is similar to rainwater, and is termed the precipitation dominance regime. It is typified by rivers in humid tropical areas of the world with very high rainwater inputs and little evaporation. As the dissolved solids concentration increases the ratio Na/(Na + Ca) declines, indicating an increasing importance for calcium in the rock dominance regime. Here, increased weathering of rock provides the major source of dissolved solids. As dissolved solids increase further, the abundance of calcium decreases relative to sodium as the water becomes saturated with respect to CaCO$_3$, and this compound precipitates. Waters in the evaporation/precipitation regime are typified by rivers in very arid parts of the world (*e.g.* River Jordan) and the major seas and oceans of the world.[20,21]

[19] R. M. Garrels and F. T. MacKenzie, 'Evolution of Sedimentary Rocks', ed. W. W. Norton, New York, 1971.
[20] R. J. Gibbs, *Science*, 1970, **170**, 1088.
[21] R. M. Harrison and S. J. de Mora, 'Introductory Chemistry for the Environmental Sciences', Cambridge University Press, Cambridge, Second Edn., 1996.

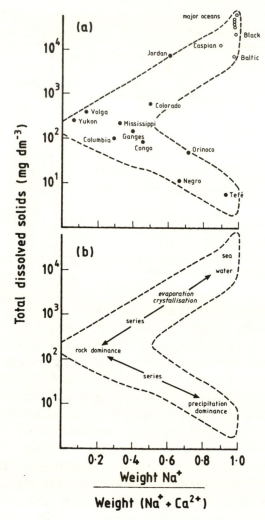

Figure 7 *The chemistry of the Earth's surface waters:* (a) *typical values of the ratio* $Na^+/(Na^+ + Ca^{2+})$ *as a function of dissolved solids concentration for various major rivers and oceans;* (b) *the processes leading to the observed ratios* (From Gibbs[20]) copyright © 1970, American Association for the Advancement of Science

The flux of material in a river to the sea is expressed by:

$$\text{Flux (g s}^{-1}) = \text{Volumetric discharge (m}^3\text{ s}^{-1}) \times \text{Concentration (g m}^{-3})$$

In total, the rivers of the world carry around 4.2×10^{12} kg per year of dissolved solids to the oceans and 18.3×10^{12} kg per year of suspended solids.

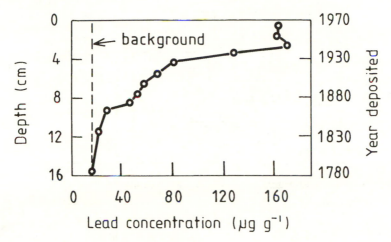

Figure 8 *Lead profile in a lake sediment in relation to depth and the year of incorporation* (From Davies and Galloway[22])

In slow-moving water bodies such as lakes and ocean basins, suspended solids falling to the bottom produce a well stratified layer of bottom sediment. This is stratified in terms of age with the oldest sediment at the bottom (where when suitably pressurized it can form rock) and the newest at the top, in contact with the water. If burrowing organisms do not provide too much disturbance (termed bioturbation), the sediment can preserve a record of depositional inputs to the water body. An example is provided by Figure 8 in which lead is analysed in a sediment core dated from its radioisotope content.[22] The concentration rises from a background in around the year 1800, corresponding to the onset of industrialization. Considerably increased deposition is seen after 1930 due to the introduction of leaded petrol. Whilst some of the lead input is via surface waters, the majority probably arises from atmospheric deposition.

4 BIOGEOCHEMICAL CYCLES

A general model of a biogeochemical cycle appears in Figure 9. Although biota are not explicitly included, their role is a very important one in mediating transfers between the idealized compartments of the model. For example, the role of marine phytoplankton in transferring sulfur from the ocean to the atmosphere in the form of dimethyl sulfide has been highlighted in Chapter 4. Biota play a major role in determining atmospheric composition. Photosynthesis removes carbon dioxide from

---

[22] A. O. Davies and J. N. Galloway, 'Atmospheric Pollutants in Natural Waters', ed. S. J. Eisenreich, Ann Arbor, MI, 1981, p. 401.

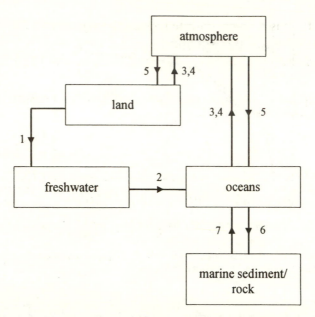

Figure 9 *Schematic diagram of the major fluxes and compartments in a biogeochemical cycle: (1) runoff; (2) streamflow; (3) degassing; (4) particle suspension; (5) wet and dry deposition; (6) sedimentation; (7) remobilization*

the atmosphere and replenishes oxygen. In a world without biota, lightning would progressively convert atmospheric oxygen into nitrogen oxides and thence to nitrate which would reside in the oceans. Biota also exert more subtle influences. In aquatic sediments, micro-organisms often deplete oxygen more quickly than it can be replenished from the overlying water, producing anoxic conditions. This leads to chemical reduction of elements such as iron and manganese, which has implications for their mobility and bioavailability.

Biological reduction processes in sediments may be viewed as the oxidation of carbohydrate (in its simplest form CH_2O) with accompanying reduction of an oxygen carrier. In the first instance, dissolved molecular oxygen is used. The reaction is thermodynamically favoured, as reflected by the strongly negative ΔG.

$$CH_2O + O_2 \rightarrow CO_2 + H_2O \qquad \Delta G = -125.5 \text{ kJ mol}^{-1} \text{ e}^-$$

When all of the dissolved oxygen is consumed, anaerobic organisms take over. Initially, nitrate-reducing bacteria are favoured

$$2CH_2O + NO_3^- + 2H^+ \rightarrow 2CO_2 + H_2O + NH_4^+ \quad \Delta G = -82.2 \text{ kJ mol}^{-1} \text{ e}^-$$

Once the nitrate is utilized, sulfate reduction takes over

$$SO_4^{2-} + H^+ + 2CH_2O \rightarrow HS^- + 2H_2O + 2CO_2 \quad \Delta G = -25.6 \text{ kJ mol}^{-1} \text{ e}^-$$

Finally, methane-producing organisms dominate in a sediment depleted in oxygen, nitrate, and sulfate

$$2CH_2O \rightarrow CH_4 + CO_2 \qquad \Delta G = -23.5 \text{ kJ mol}^{-1} \text{ e}^-$$

Thus highly anoxic waters are commonly sources of hydrogen sulfide, H_2S, from sulfate reduction and of methane (marsh gas). The formation of sulfide in sediments has led to precipitation of metal sulfides over geological time, causing accumulations of sulfide minerals of many elements, *e.g.* PbS, ZnS, HgS, *etc.*

4.1 Case Study 1: The Biogeochemical Cycle of Nitrogen

Nitrogen has many valence states available and can exist in the environment in a number of forms, depending upon the oxidizing ability of the environment. Figure 10 indicates the most important oxidation states and the relative stability (in terms of free energy of formation).[23] The oxides of nitrogen represent the most oxidized and least thermodynamically stable forms. These exist only in the atmosphere. Ammonia can exist in gaseous form in the atmosphere but rather rapidly returns to the soil and waters as ammonium, NH_4^+. Fixation of atmospheric N_2 by leguminous plants leads to ammonia, NH_3. In aerobic soils and aquatic systems, NH_3 and NH_4^+ are progressively oxidized by micro-organisms via nitrite to nitrate. The latter is taken up by some biota and used as a nitrogen source in synthesizing amino acids and proteins, the most thermodynamically stable form of nitrogen. After the death of the organism, microbiological processes will convert organic nitrogen to ammonium (ammonification) which is then available for oxidation or use by plants. Conversion of ammonia to nitrate is termed nitrification, whilst denitrification involves conversion of nitrate to N_2.

Figure 11 shows an idealized nitrogen cycle. The numbers in boxes represent quantities of nitrogen in the various reservoirs, whilst the arrows show fluxes.[23] It is interesting to note that substances involving relatively small fluxes and burdens can have a major impact upon people. Thus nitrogen oxides, NO, NO_2, and N_2O are very minor constituents relative to N_2 but play major roles in photochemical air pollution (NO_2), acid rain (HNO_3 from NO_2), and stratospheric ozone depletion (N_2O).

[23] P. O'Neill, 'Environmental Chemistry', George, Allen, and Unwin, London, 1985.

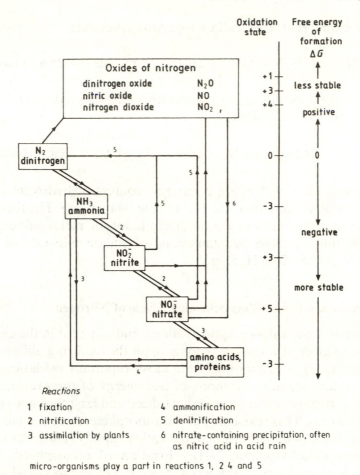

Figure 10   *Chemical forms and cycle of nitrogen*
(From O'Neill[23])

Nitrate from fertilizers represents a very small flux but has major implications in terms of eutrophication of surface waters.

4.2   Case Study 2: Aspects of the Biogeochemical Cycle of Lead

Lead is a simpler case to study than nitrogen due to the small number of available valence states. The major use of lead until recently was as tetraalkyl lead gasoline additives in which lead is present as Pb^{IV}. The predominant compounds used are tetramethyl lead, $Pb(CH_3)_4$, and tetraethyl lead, $Pb(C_2H_5)_4$. These are lost to the atmosphere as vapour from fuel evaporation and exhaust emissions from cold vehicles, but comprise only about 1–4% of lead in polluted air.[2] Leaded gasoline also contains the scavengers 1,2-dibromoethane, CH_2BrCH_2Br and 1,2-

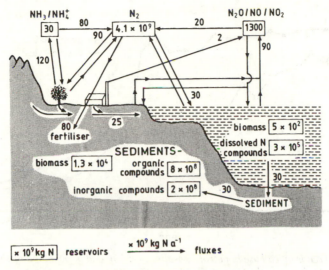

Figure 11 *Schematic representation of the biogeochemical cycle of nitrogen, indicating the approximate magnitude of fluxes and reservoirs* (After O'Neill[23])

dichloroethane, CH_2ClCH_2Cl which convert lead within the engine to lead halides, predominantly lead bromochloride, $PbBrCl$, in which lead is in the Pb^{II} valence state, its usual form in environmental media. About 75% of lead alkyl burned in the engine is emitted as fine particles of inorganic lead halides. Atmospheric emissions of lead arise also from industry; both these and vehicle-emitted lead are declining. Figure 12 shows trends in United Kingdom emissions of lead to atmosphere from leaded petrol.[24] Lead emitted to the atmosphere has a lifetime of around 7–30 days and hence may be subject to long-range transport. Concentrations of trace elements in polar ice provide a historical record of atmospheric deposition. Measurements (Figure 2) have shown a marked enhancement in lead accompanying the increase in leaded gasoline usage, and a major decline in recent years attributable to reduced emissions to atmosphere.[4]

Atmospheric lead is deposited in wet and dry deposition. Lead is relatively immobile in soil, and agricultural surface soils in the UK exhibit concentrations approximately double those of background soil which contain *ca.* 15–20 mg kg^{-1} derived from soil parent materials, other than in areas of lead mineralization where far greater concentrations can be found. Local perturbations to the cycle of lead can be important. For instance, the lead content of garden soils correlates strongly with the age of the house. This is probably due to the

[24] Department of Environment, Transport and the Regions, 'Digest of Environmental Statistics', No. 19, The Stationery Office Ltd., Edinburgh, 1997.

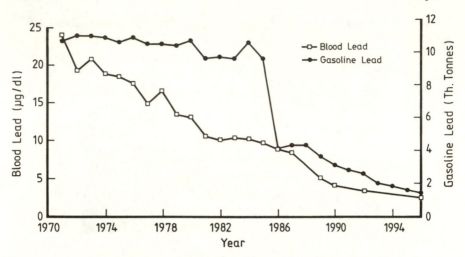

Figure 12   *Trends in lead use in petrol (gasoline) and of lead in the blood of the general population in the United Kingdom, 1970–1995*

deterioration of leaded paintwork on older houses and the former practices of disposing of household refuse and fire ashes in the garden. Lead is also of low mobility in aquatic sediments and hence the sediment may provide a record of historical lead deposition (see Figure 8).

Plants can take up lead from soil, thus providing a route of human exposure. Careful research in recent years has established transfer factors, termed the Concentration Factor, CF, where

$$CF = \frac{\Delta \text{Concentration of lead in plant (mg kg}^{-1} \text{ dry wt.)}}{\Delta \text{Concentration of lead in soil (mg kg}^{-1} \text{ dry wt.)}}$$

The value of CF for lead is lower than for most metals and is typically within the range 10^{-3} to 10^{-2}. Much higher values had been estimated from earlier studies which ignored the importance of direct atmospheric deposition as a pathway for contamination. The direct input from the air to leaves of plants is often as great, or greater than soil uptake.[24,25] This pathway may be described by another transfer factor, termed the Air Accumulation Factor, AAF, where

$$AAF \ (m^3 \ g^{-1}) = \frac{\Delta \text{Concentration of lead in plant (}\mu g \ g^{-1} \text{ dry wt.)}}{\Delta \text{Concentration of lead in air (}\mu g \ m^{-3})}$$

Values of AAF are plant dependent, due to differences in surface characteristics, but values of 5–40 are typical.[25,26] Thus a plant grown

[25] R. M. Harrison and M. B. Chirgawi, *Sci. Total Environ.*, 1989, **83**, 13.
[26] R. M. Harrison and M. B. Chirgawi, *Sci. Total Environ.*, 1989, **83**, 47.

on an agricultural soil with 50 mg kg^{-1} lead will derive 0.25 mg kg^{-1} dry weight lead from the soil (CF $= 5 \times 10^{-3}$), whilst airborne lead of $0.1 \ \mu\text{g m}^{-3}$ will contribute $2.0 \ \mu\text{g g}^{-1}$ ($\equiv \text{mg kg}^{-1}$) of lead (AAF $= 20 \text{ m}^3 \text{ g}^{-1}$). Thus in this instance airborne lead deposition is dominant. The air lead concentration of $0.1 \ \mu\text{g m}^{-3}$ was typical of rural areas of the UK until 1985. Since that time, the drastic reduction of lead in gasoline has led to appreciably reduced lead-in-air concentrations in both urban and rural localities.

Human exposure to lead arises from four main sources:[2,27]

(i) inhalation of airborne particles. The adult human respires approximately 20 m^3 of air per day. Thus for an urban lead concentration of $0.1 \ \mu\text{g m}^{-3}$, *intake* is $2 \ \mu\text{g}$ per day. This is rather efficiently absorbed (*ca.* 70%) and therefore *uptake* is around $1.4 \ \mu\text{g}$ per day in this instance.

(ii) ingestion of lead in foodstuffs. The concentrations of lead in food obviously vary between different foodstuffs and even between different batches of the same food. Typical freshweight concentrations (much of the weight of some foods is water) are from 10 to $50 \ \mu\text{g Pb kg}^{-1}$. Thus a food consumption of 1.5 kg per day represents an *intake* of around $50 \ \mu\text{g}$ per day and an *uptake* (10–15% efficient) of around $6 \ \mu\text{g}$ per day.

(iii) drinking water and beverages. Concentrations of lead in drinking water vary greatly, related particularly to the presence or absence of lead in the household plumbing system. Most households in the UK conform to the EC standard of $50 \ \mu\text{g l}^{-1}$ and a concentration of $5 \ \mu\text{g l}^{-1}$ may be taken as representative. Gastrointestinal absorption of lead from water and other beverages is highly dependent upon food intake. After long fasting, absorptions of 60–70% have been recorded, 14–19% with a short period of fasting before and after the meal, and only 3–6% for drinks taken with a meal. If 15% is taken as typical, for a daily consumption of 1.5 litres, *intake* is $7.5 \ \mu\text{g}$ and *uptake* $1.1 \ \mu\text{g}$.

(iv) cigarette smoking exposes the individual to additional lead.

Whilst both individual exposure to lead and the uptake efficiencies of individuals are very variable, it is evident that exposure arises from a number of sources and control of human lead intake, if deemed to be desirable, requires attention to all of those sources. An additional pathway of exposure, not easily quantified, and not included above is

[27] Royal Commission on Environmental Pollution, 'Ninth Report: Lead in the Environment', HMSO, London, 1983.

ingestion of lead-rich surface dust by hand to mouth activity in young children.

The above calculations estimate that for a typical adult in a developed country, daily uptake of lead from air, diet, and drinking water is respectively 1.4 μg, 6 μg, and 1.1 μg. Exposure to lead from all of these sources has fallen rapidly over the past 20–30 years. Figure 12 contrasts the temporal trends in use of lead in petrol (gasoline) and blood leads in the general population of the UK over the period when much of this decline took place. It is interesting to note that from 1971 to 1985 use of lead in petrol was relatively steady, but blood leads declined by a factor of more than two over this period mainly as a response to reductions in dietary exposure, particularly associated with the cessation of use of leaded solder to seal food cans. A dramatic reduction in gasoline lead usage occurred at the end of 1985 when the maximum permissible lead content of petrol was reduced from 0.4 g l^{-1} to 0.15 g l^{-1}, and there has been a steady reduction in lead use since, with the increased market penetration of unleaded fuel. Despite the ability of a vehicle emitting lead to cause direct lead exposure through the atmosphere, as well as indirect exposure through contamination of food and water, the lack of any obvious step change in blood lead associated with the reduction of lead in petrol shows clearly that at that time leaded petrol was not a major source of exposure for the general population.

5. ENVIRONMENTAL PARTITIONING OF LONG-LIVED SPECIES

For chemical species sufficiently long lived to approach some form of equilibrium between environment media, partition coefficients are an extremely useful means of expressing their likely ultimate distribution. The best known of these is K_{ow}, the octanol–water partition coefficient which is defined as follows:

$$K_{ow} = \frac{\text{concentration in octan-1-ol}}{\text{concentration in water}}$$

Since for many of the compounds such as PCBs, dioxins, and PAH to which this concept is applied, the value of K_{ow} is relatively large, it is often expressed as its logarithm, log K_{ow}. This is taken as a measure of the bioaccumulative tendencies of an organic compound as it approximates to the lipid/water partition coefficient. It is predominantly dependent on water solubility as the variation of solubility for the various organic compounds in octan-1-ol is relatively modest. For classes of compounds such as the dioxins, the value of K_{ow} typically increases with the relative molecular mass of the organic compound. In its simplest use K_{ow} might be used to predict the likely concentration of an organic chemical in fish tissues relative to that in the surrounding water.

A further useful partition coefficient is K_{oc} which expresses the partition of a chemical between water and natural organic carbon and has units of $dm^3 \ kg^{-1}$. The utility of K_{oc} is in describing the likely partitioning of a chemical into soil organic matter or uptake by plants and animals. K_{oc} is closely correlated to K_{ow} but is dependent on the kind of organic carbon considered.

A branch of numerical modelling termed fugacity modelling uses partition coefficients such as K_{oc}, K_{ow} and the Henry's Law constant describing the partition between air and water to predict the distribution of persistent organic chemicals in a model environment. Whilst the approach is not sufficiently sophisticated to give exact predictions of concentrations in environmental media, it is nonetheless very valuable in predicting in general terms the behaviour of chemicals within the environment and comparing the partitioning of related compounds with different physico-chemical properties.

Questions

1. Discuss what is meant by a biogeochemical cycle, describing the major facets and illustrating the processes involved with examples.
2. Discuss the temporal trends in lead emissions and concentrations in the environment and how environmental media can be used to elucidate historical trends in environmental lead.
3. Explain what is meant by an environmental lifetime and derive an expression for environmental lifetime in terms of a chemical rate constant. Compare and contrast the typical atmospheric lifetimes of methane, nitrogen dioxide and the CFCs and explain how this relates to the atmospheric distribution and properties of these compounds.
4. Explain the processes by which trace substances can exchange between the atmosphere and the oceans and show how rates of exchange can be calculated. Give examples of substances whose exchange between these media is important.
5. Explain why the waters in rivers in different parts of the world have differing composition and relate this to the climatology of the region. Explain carefully what is meant by dissolved and suspended solids and explain how both arise.
6. Explain the environmental pathways followed by lead emissions from road traffic after emission to the atmosphere and explain how

this can lead to pollution of a range of environmental media. Indicate the quantitative ways in which such transfer can be expressed.

7. Estimate atmospheric lifetimes for the following:
   (a) methane, if the globally and diurnally averaged concentration of hydroxyl radical is 5×10^5 cm^{-3}.
   (b) nitrogen dioxide in the middle of a summer day when the concentration of hydroxyl radical is 8×10^6 cm^{-3}.
   (c) nitrogen dioxide at nighttime if the sole mechanism of removal is dry deposition with a deposition velocity of 0.1 cm s^{-1}, and the mixing depth is 100 m.

8. If the atmospheric concentration of sulfur dioxide is 10 ppb, calculate the following:
   (a) the atmospheric concentration expressed in μg m^{-3} at one atmosphere pressure and 25 °C.
   (b) the deposition flux to the surface if the deposition velocity is 1.0 cm s^{-1}.
   (c) the atmospheric lifetime with respect to dry deposition for a mixing depth of 800 m.
   (d) the atmospheric lifetime with respect to oxidation by hydroxyl radical if the diurnally averaged OH$^{\bullet}$ radical concentration is 8×10^5 cm^{-3} and the rate constant for the SO$_2$–OH$^{\bullet}$ reaction is 9×10^{-13} cm^3 molec^{-1} s^{-1}.
   (e) the lifetime with respect to wet deposition if the washout coefficient is 10^{-4} s^{-1}.

9. Explain the thermodynamic controls on biological reduction processes in aquatic sediments and explain how these influence the chemical forms of nitrogen in the environment.

CHAPTER 7

Environmental Monitoring Strategies

C. NICHOLAS HEWITT AND ROBERT ALLOTT

1 OBJECTIVES OF MONITORING

The gathering of information on the existence and concentration of substances in the environment, either naturally occurring or from anthropogenic sources, is achieved by **measurement** of the substance or phenomenon of interest. However, single measurements of this type made in isolation are virtually worthless, since temporal and spatial variations cannot be deduced. Rather, it is necessary to *monitor* the parameter of interest by repeated measurements made over time and space, with sufficient sample density, temporally and spatially, that a realistic assessment of variations and trends may be made.

Monitoring of the environment may be undertaken for a number of reasons and it is important that these be defined before sampling takes place. The generalization that 'monitoring is done in order to gain information about the present levels of harmful or potentially harmful pollutants in discharges to the environment, within the environment itself, or in living creatures (including ourselves) that may be affected by these pollutants'[1] may be expanded as follows:

(a) Monitoring may be carried out to assess pollution effects on humans and their environment, and so to identify any possible cause and effect relationships between pollutant concentrations and, for example, health effects, or environmental changes.
(b) Monitoring may be carried out in order to study and evaluate pollutant interactions and patterns. For example source appor-

[1] Department of the Environment, 'The Monitoring of the Environment in the United Kingdom', Report by the Central Unit on Environmental Pollution, HMSO, London, 1974.

tionment[2] and pollutant pathway studies usually rely on environmental monitoring.

(c) Monitoring may be carried out to assess the need for legislative controls on emissions of pollutants and to ensure compliance with emission standards. An assessment of the effectiveness of pollution legislation and control techniques also depends upon subsequent monitoring.

(d) In areas prone to acute pollution episodes, monitoring may be carried out in order to activate emergency procedures.

(e) Monitoring may be carried out in order to obtain a historical record of environmental quality and so provide a database for future use in, for example, epidemiological studies.

(f) Monitoring may also be necessary to ensure the suitability of water supply for a proposed use (industrial or domestic) or to ensure the suitability of land for a proposed use (for example for housing).

A basic problem in the design of a monitoring programme is that each of the above reasons for carrying out monitoring demands different answers to a number of questions. For example, the number and location of sampling sites, the duration of the survey, and the time-resolution of sampling will all vary according to the use to which the collected data are to be put. Decisions on what to monitor, when and where to monitor, and how to monitor are often made much easier once the purpose of monitoring is clearly defined. Therefore it is most important that the first step in the design of a monitoring programme should be to set out the objectives of the study. Once this has been done then the programme may be designed by consideration of a number of steps in a systematic way (see Figure 1) such that the generated data are suitable for the intended use. It is important also that the data produced by a monitoring programme should be continuously appraised in the light of these objectives. In this way, limitations in the design, organization, or execution of the survey may be identified at an early stage.

The aim of this chapter is to present and discuss the most important and relevant considerations that must be taken into account in the design and organization of a monitoring exercise. It is not intended to be a manual or practical guide to monitoring; rather it is hoped that it highlights the types of approaches that may be used and some of the problems likely to be encountered. The inclusion of case studies and references direct the reader to the more specific practical information which is available elsewhere.

[2] P. K. Hopke, 'Receptor Modelling in Environmental Chemistry', John Wiley & Sons, New York, 1985.

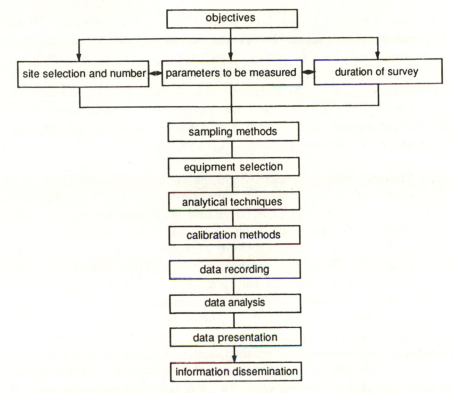

Figure 1   *Steps in the design of a monitoring programme*

2   TYPES OF MONITORING

The Earth's surface is comprised of four distinct media; the atmosphere, the hydrosphere, the biosphere, and the land. Pollutants can occur in any of the solid, liquid, or gaseous phases. However, the environment is not a simple system and consequently each of the four media may contain pollutants in each of the three phases. Monitoring may therefore be necessary for a particular pollutant in a specific phase in a particular environmental compartment (*e.g.* sulfur dioxide in air) or it may encompass two or more phases and/or media (*e.g.* dissolved and particulate phase metals in water). Pollutants in the environment originate from a multitude of different types of sources and the identification of these is a prerequisite to the design of a monitoring programme.

First, pollutant sources may be classified by their spatial distribution as point sources, line sources, or area sources. Point sources include industrial chimneys, liquid waste discharge pipes, and localized toxic waste dumps on land. Line sources may include highways, airline routes, and runoff from agricultural land, while area emissions may arise from extensive urban or industrial complexes. Sources may also be classified as either stationary or mobile, motor vehicles being the obvious example of the latter. Classification may also be made for air pollutant sources on the basis of the height of discharge, *i.e.* at street level, building level, stack level, or above the atmospheric boundary layer level. A further important distinction may be made between 'planned', 'fugitive', and 'accidental' emissions to the environment.

(a) Planned emissions arise when (as is invariably the case) it is economically or technically impossible to completely remove all the contaminants in a discharge and hence the process operation allows pollutants to be discharged to the environment at known and controlled rates. Obvious examples of planned emissions include sulfur dioxide from power generation plants and low-level radioactive effluent during nuclear fuel reprocessing.

(b) Fugitive emissions arise when pollutants are released in an unplanned way, normally without first passing through the entire process. They therefore occur at a point sooner in the process than the stack or duct designed for 'planned' emissions. They generally originate from operations which are uneconomic or impractical to control, have poor physical arrangements for effluent control, or are poorly maintained or managed. An example is the escape of heavy metal contaminated dust from a factory on vehicle tyres, arising from poor dust control and wheel washing arrangements.

(c) Accidental emissions result from plant failure, such as a burst filter bag or faulty valve or from an accident involving either equipment or operator error (*e.g.* the Chernobyl reactor accident). Accidental emissions can give rise to very high mass emission rates and ambient concentrations but they normally occur only infrequently.

Classification of the sources of pollutants in this way allows the distinction of two differing approaches to their monitoring. On the one hand, samples may be taken of the effluent before discharge to, and dispersion in, the environment (source monitoring). Alternatively, samples may be taken of the ambient environment into which discharges occur, for example of the air or receiving waters, without consideration of source strengths and rates. Obviously neither one of these approaches alone can necessarily provide all the data required to resolve a particular problem and often it is desirable to complement one with the other.

2.1 Source Monitoring

2.1.1 General Objectives. Source monitoring may be carried out for a number of reasons:

(a) Determination of the mass emission rates of pollutants from a particular source, and assessment of how these are affected by process variations.
(b) Evaluation of the effectiveness of control devices for pollution abatement.
(c) Evaluation of compliance with statutory limitations on emissions from individual sources.

2.1.2 Stationary Source Sampling for Gaseous Emissions. A common feature of many industrial processes is that effluent output rates exhibit cyclical patterns. These may be related to working shift arrangements or be a function of the operations involved, but both require that source testing or monitoring be planned accordingly. Process operations should be reviewed so that discharges during the period of sampling are representative of the plant output in order to ensure that the samples themselves are representative of the effluent, and that the final pollutant analysis will be a representative measure of the entire output.

Two requirements may be specified for valid source monitoring. First, the sample should accurately reflect the true magnitude of the pollutant emission at a specific point in the stack at a specific instant of time. This requirement is met by adequate sampling instrument design. Secondly, enough measurements should be obtained over time and space so that their combined result will accurately represent the entire source emission. This requires consideration of the emissions both in time and in space, across the entire cross-section of the stack.

In a circular flue, sampling at the centroids of equal area annular segments will ensure that emission variations across the stack cross-section are quantified. In a rectangular flue sample points should be located at the centroids of smaller equal area rectangles. Generally eight or twelve such sampling points are adequate to compensate for any deficiencies in the location of the sampling site with respect to the length of the stack and to non-ideal flow conditions at the site caused by bends, inlets, or outlets. If it is a particulate pollutant which is being sampled within the stack, it is important that an isokinetic sampling regime is maintained.

2.1.3 Mobile Source Sampling for Gaseous Effluents. Vehicle and aircraft emissions are heavily dependent upon the engine operating mode (*i.e.* idling, accelerating, cruising, or decelerating) and the results obtained by sampling must be considered specific to the type of operating cycle used during the test. Emission tests are usually performed with the vehicle on a dynamometer equipped with inertia fly wheels to represent the vehicle weight and brake loading on a level road.

2.1.4   Source Monitoring for Liquid Effluents.   Liquid wastes and effluents often tend, like gaseous effluents, to be inhomogeneous and care is needed in selecting sampling positions. Having considered the location of the site in relation to plant operation (*e.g.* should the site be before or after a particular stage of the process or treatment) it is desirable that a region of high turbulence and/or good mixing be chosen. As for gaseous emissions, several samples may have to be taken across the cross-section of a pipe or channel. Sampling from vertical pipes is less liable to be affected by deposition of solids than sampling from horizontal pipes, and a distance of approximately 25 pipe-diameters downstream from the last inflow should ensure that mixing of the two streams is essentially complete.[3] If suitable homogeneous regions for sampling cannot be found, particularly where suspended materials are present, samples may have to be taken from several positions along the effluent stream.

Where the composition of a liquid effluent is known to vary with time, grab samples may be collected at set intervals, either manually or by use of an automatic sampler. An alternative approach is to sample at intervals varying with the flow rate so that a more representative composite may be obtained.

2.1.5   Source Monitoring for Solid Effluents.   Solid effluents may arise from a number of different processes, including sludge after sewage treatment, ash residue from municipal incinerators, or low-grade gypsum from desulfurization plants attached to coal-fired power stations. In general, solid wastes are even less homogeneous than either liquid or gaseous effluents. Therefore, great effort must be made to ensure that samples are representative of the bulk waste (see Section 3.3). Monitoring of sewage sludge is particularly common due to sludge acting as an efficient sorption material for heavy materials. Typically, 80–100% of the input lead in a sewage treatment plant is incorporated into the sludge, resulting in sludge lead concentrations of 100–3000 mg g^{-1}. Consideration must therefore be given to the concentrations of pollutants in the material before it is used as fertilizer, incinerated, dumped at sea, or used as landfill. The determination of the metal balance of a sewage treatment works may be necessary when considering the fate of the treated effluent and solid waste.

[3] A. L. Wilson, 'Examination of Water for Pollution Control', ed. M. J. Suess, Pergamon, London, 1982, Vol. 1.

In most countries guidelines exist to control the disposal of sewage sludges to land, usually based primarily upon the zinc, copper, and nickel content of the sludge. Hence considerable quantities of other metals, including lead, may be added to land over a normal 30 year disposal period. In the UK the disposal of lead-rich sewage sludges to land is controlled where direct ingestion by animals of contaminated grass or soil can occur.

Until fairly recently most trace metal analysis of environmental samples was designed to give a measure of the total elemental concentration in the sample, as it was felt that this gave an adequate measure of the pollution load for that metal. However, in the past two decades it has become apparent that total metal concentrations are often not sufficient and that information based upon some form of physico-chemical speciation scheme is required. This may include, for example, solubility of the pollutant in acids of different strengths, the size distribution of particles, and the association with organic compounds. This is because the physical, chemical and biological responses to a pollutant will vary according to its physical and chemical speciation. One disadvantage of this type of analysis is that it is complicated and time-consuming compared with total metal determinations. Thus speciation studies are invariably limited to a few samples, where many (tens or even hundreds) would be taken in a total-metal study.

Case Study 1: Organic solvent residues at a landfill site.[4]   Landfill sites are now recognized as sources of toxic and explosive substances, including methane and organic chemicals. The contamination of groundwater by these toxic organic chemicals is of major environmental concern in Europe and North America. At a landfill site studied near Ottawa, Canada, disposal of chlorinated and non-chlorinated solvents, wood preservatives, and small amounts of other wastes occurred between 1969 and 1980. Groundwater samples were collected from monitoring wells within the landfill site using either piezometers or multilevel samplers attached to peristaltic pumps. Analysis was carried out by gas chromatography–mass spectrometry (GC–MS) which enabled the identification and quantification of a wide range of volatile organic compounds, including dioxane (~ 300–$2000 \ \mu g \ l^{-1}$), diethylether (< 2–$658 \ \mu g \ l^{-1}$), trichloroethene (7–$583 \ \mu g \ l^{-1}$), and 1,1,2-trichloro-1,2,2-trifluorethane (Freon F113) (< 5–$2725 \ \mu g \ l^{-1}$). The contaminant of greatest concern was 1,4-dioxane, due to its toxicity and persistence. Freon F113 was the organic chemical found in greatest concentration. Although very persistent in the subsurface, it appeared to have undergone transformation, as one toxic product, F-1113, was identified.

[4] S. Lesage, R. E. Jackson, M. W. Priddle, and P. G. Riemonn, *Environ. Sci. Technol.*, 1990, **24**, 559–566.

2.2 Ambient Environment Monitoring

2.2.1 General Objectives. Monitoring the environment may be carried out for a number of reasons, as outlined in Section 1. However, whatever the purpose of the survey the overriding consideration when designing a programme is to ensure that the samples obtained provide adequate data for the purpose intended. Invariably this means that samples should be representative of conditions prevailing in the environment at the time and place of collection. Thus, not only must the sampling location be carefully chosen but also the sampling position at the chosen location.

The selection of a specific monitoring site requires consideration of four steps: identification of the purpose to be served by monitoring; identify the monitoring site type(s) that will best serve the purpose; identification of the general location where the sites should be placed; and final identification of specific monitoring sites.

2.2.2 Ambient Air Monitoring. Air pollution problems vary widely from area to area and from pollutant to pollutant. Differences in meteorology, topography, source characteristics, pollutant behaviour, and legal and administrative constraints mean that monitoring pro-grammes will vary in scope, content, and duration, and the types of station chosen will also vary. However ambient monitoring sites may be divided into several categories:

- (a) source-orientated sites for monitoring individual or small groups of emitters as part of a local survey (*e.g.* one particular factory).
- (b) sites in a more extensive survey which may be located in areas of highest expected pollutant concentrations or high population density, or in rural areas to give a complete nationwide or regional coverage.
- (c) baseline stations to obtain background concentrations, usually in remote or rural areas with no anticipated changes to land use.

Location of source-orientated monitors. Occasionally the effects and impact of a specific pollutant source are of sufficient interest or importance to warrant a special survey. This will usually include a site at the point of anticipated maximum ground-level concentration, which can be estimated from dispersion calculations (see Section 4.1 below), and also a nearby site to characterize the 'background' conditions in the area. Examination of meteorological records will usually be necessary in order to choose suitable locations for the sites and several computerized models are available for determining the areas of maximum average impact from a point source. Calculation of expected ground-level concentrations using the standard equations discussed in Section 4.1

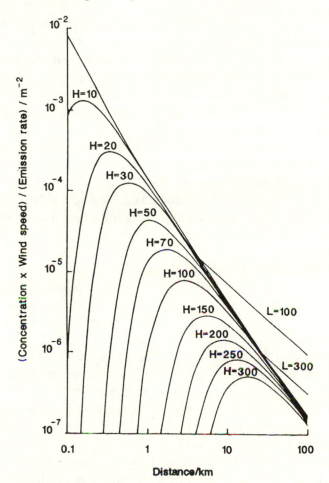

Figure 2 *Normalized ground-level concentrations from an elevated source for neutral stability. The effective stack height (H) is the sum of the release height (e.g. chimney height) and the height gained by the plume due to momentum and buoyancy*

show that the concentration rises rapidly with distance from the source to a maximum and then falls gradually beyond the maximum,[5] as shown in Figure 2.[6] This is for meteorological conditions of neutral stability and different heights of emission (H). The ordinates in this graph represent concentration normalized for emission rate (Q) and wind speed (U) and the various curves are for different source heights (H metres) and different limits to vertical dispersion (L). It is prudent therefore to

[5] 'WMO Operations Manual for Sampling and Analysis Techniques for Chemical Constituents in Air and Precipitation', World Meteorological Organization, No. 299, Geneva, 1971.
[6] D. B. Turner, 'Workbook of Atmospheric Dispersion Estimates', National Air Pollution Control Administration, US Environmental Protection Agency, Research Triangle Park, NC, 1970.

locate the monitoring site somewhat beyond the distance where the maximum concentration is predicted. This allows some margin for error by placing the monitor in a region of relatively small concentration gradients. Obviously it is desirable to have an array of stations at differing distances and directions from the source and typically 4–6 samplers might be considered sufficient for monitoring a single point source.

In some cases pollutants are emitted to the atmosphere from a single source but in a more diffuse manner than from a single stack. Calculations of mass emission rates and distance of maximum ground-level concentration are more difficult to make for such diffuse or fugitive emissions which, in some cases, may have significant impacts on the local air quality.

Location of monitors in larger-scale surveys. Often it is important to know the geographical extent of atmospheric pollution, and to have localized information on source strengths or ground-level concentrations within a plume is not sufficient. For example, the National Survey of Air Pollution (NSAP) monitoring network in the United Kingdom was established in 1961 following recognition of the need for the acquisition of a nationwide, day-to-day, and long term bank of data of sulfur dioxide and smoke concentrations. The original network of 1200 sites was based upon the assumption that it was necessary to monitor in rural areas (150 sites) and different types of urban areas such as high-density residential areas, industrial areas, commercial areas, and smoke-controlled areas. Since the introduction of this simple scheme in 1961 many of the original stations have ceased monitoring, others have been replaced, and additional stations have been added. Reasons for these changes include the need to monitor recently established smoke-control areas, new industrial estates and redeveloped areas, and surveys around new and projected power stations. In 1981 a rationalized long-term network of 150 NSAP stations was established, re-designated as the UK Smoke and Sulphur Dioxide Monitoring Network. However, about 400 existing sites were retained in the short term to provide continuation of monitoring in urban areas where the EC air quality standards for smoke and SO_2 may be approached or exceeded. As concentrations fall to 'acceptable' levels in each urban area so these sites are discontinued.

These sites have, in the last five years, been supplemented by a number of fully automated sites in urban and rural areas of the UK giving on-line, real-time data accessible to the public via the Internet, freephone, and teletext. The pollutants monitored at these sites vary, but the most comprehensively equipped sites determine SO_2, NO, NO_x, O_3, TSP, CO, and a wide range of volatile organic compounds. In contrast the simplest sites only monitor ozone. A full description of this programme is available elsewhere.[7]

Several methods are available for the rationalization of an existing monitoring network, and in the case of the NSAP network spatial correlation analysis was applied to determine which sites might be discontinued without significantly losing overall coverage.

The design of monitoring networks for air pollution has been treated in several different ways. For example monitoring sites may be located in areas of severest public health effects, which involves consideration of pollutant concentration, exposure time, population density, and age distribution. Alternatively the frequency of occurrence of specific meteorological conditions and the strength of sources may be used to maximize monitor coverage of a region with limited sources. An estimate of the ambient dosage (the product of the concentration and the exposure time) computed from source emission data and diffusion models can be used to determine where air monitoring sites should best be located.

Location of regional-scale survey monitors. On the regional or global scale, monitoring is usually concerned with long term changes in background concentrations of pollutants and so the principal siting requirement is that truly representative baseline or background levels may be measured over a long term period without interference from local sources. It has been suggested[5] that baseline stations should be located in areas where no significant changes in land use practices are anticipated for at least 50 years within 100 km of the station, and should be away from population centres, major highways, and air routes. Such locations are hard to find, but Cape Grim in Tasmania is one example.

One example of a network established to monitor distant sources for regional or global effects is the OECD 'Long-Range Transport of Air Pollutants' project which measured chemical components in precipitation and SO_2 and particulate sulfate in air. The long term measurement of carbon dioxide concentrations in air at Mauna Loa in Hawaii is probably the best example of 'baseline' monitoring.

Case study 2: National and regional radon surveys in the UK. Increasing awareness of the importance of human exposure to the naturally occurring radioactive gas radon, has resulted in several national surveys being commissioned for the UK. ^{222}Rn, commonly known as radon, is a colourless odourless gas which results from the decay of ^{238}U and in turn decays to a series of radioactive daughters. Two other isotopes of radon exist, ^{219}Rn and ^{220}Rn (thoron) within the ^{232}Th and ^{235}U radioactive

[7] G. Davison and C. N. Hewitt, 'Air Pollution in the United Kingdom', Royal Society of Chemistry, London, 1997.

decay series. However, it is ^{222}Rn which has the greatest health significance, since it has the largest proportion of alpha emitting, high activity progeny. The inhalation of radon and particularly its daughters is believed to be responsible for a major proportion of the overall annual dose of ionizing radiation received by the UK population.

Several methods exist of the measurement of radon in air and may be divided into active or passive techniques. The active techniques are generally only used for research or special survey purposes and normally consist of a pump which draws air through a filter trapping the radon decay products. The alpha radiation from these daughters is measured by scintillators or semiconductor detectors. The most common method for monitoring radon is the passive alpha track detector and these have been extensively used in the UK.[8] Radon is allowed to diffuse into a small container, but radon daughters present in the ambient air are excluded. Inside the pot is placed a piece of polycarbonate, cellulose nitrate, or allyl diglycol carbonate film. Alpha particles, formed by the decay of radon, damage chemical bonds in the surface of the film. After a period of exposure, the surface of the film is etched with NaOH and tracks appear. The number density of these tracks is proportional to the average radon concentration and may be either counted using a computerized image analysing system under a microscope, or the tracks filled with a scintillant (ZnS–Ag) which fluoresces when exposed to an alpha source in proportion to the number of tracks present. The detectors may be calibrated by exposure to known radon concentrations for known periods of time. Exposure times of weeks to months are required with this method.

To date, the National Radiological Protection Board (NRPB) has surveyed about 200 000 dwellings in the UK. The distribution of observed radon concentrations was log normal, with a population-weighted mean of 20 Bq m^{-3}. On the basis of these measurements, about 0.5% of the housing stock in the country (100 000 houses) is believed to have radon concentrations above the Government's Action Level of 200 Bq m^{-3}, and so requiring remedial work to be taken. This can be done by preventing the gas entering the building (*e.g.* by sealing floors), increasing the ventilation rate (*e.g.* by extractor fan), or removing the radon decay products from the air (*e.g.* by electrostatic precipitator).

The primary influence of bedrock geology on relative radon concentrations throughout the UK is exhibited in Figure 3. High levels of radon in homes (reaching 8000 Bq m^{-3}) were found in areas where the bedrock contained high concentrations of uranium. These included granites and

---

[8] National Radiological Protection Board, 'Exposure to Radon in UK Dwellings', NRPB-R272, HMSO, London, 1994.

Figure 3 *Relative radon concentrations in homes throughout the UK*
(Reproduced with permission from the National Radiological Protection
Board)

areas of mineralization in south-west England and Scotland (up to
2000 ppm U) and some shales and limestones in north and central
England (about 800 ppm U).

Other factors which influence radon levels were also identified, including season, time of day, meteorological conditions, ventilation (*e.g.*
existence of double glazing), usage, and other occupancy habits. These
can result in a difference factor of two between summer and winter
concentrations. Furthermore, radon concentrations may vary between
rooms, with the highest radon levels in basements and, on average, first
floor bedrooms have concentrations two-thirds those of living rooms.

2.2.3 Environmental Water Monitoring. Pollutants enter the aquatic
environment from the air (by dry deposition or in precipitation occurring
either directly onto the water surface or elsewhere within the catchment
area), from the land (either in surface runoff or *via* sub-surface waters)
and directly through effluent discharges (domestic, industrial, or agricultural). The undesirable effects of pollutants in natural water may be due
to:

(a) stimulation of water plant growth—eutrophication—which ulti-
    mately leads to deoxygenation of the water and major ecological
    change;
(b) their direct or indirect toxic effects on aquatic life;
(c) the loss of amenity and practical value of the water body,
    particularly as a source of water for public supply.

Apart from the monitoring of sources of pollutants in liquid effluents (Section 2.1.4) sampling may be carried out:

(a) in rivers, lakes, estuaries, and the sea in order to obtain an overall indication of water quality;
(b) for rainwater, groundwater and runoff water (particularly in the urban environment) to assess the influence of pollutant sources;
(c) at points where water is taken for supply, to check its suitability for a particular use;
(d) using sediments and biological samples in order to assess the accumulation of pollutants and as indicators of pollution.

As well as the measurement of chemical and physical parameters the quantitative or qualitative assessment of aquatic flora and fauna is often used to give an indication of the presence or absence of pollution, and well recognized relationships exist between the abundance and diversity of species and the degree of pollution. This is often used to assess the cleanliness of natural fresh waters and is known as biological monitoring.

Location of sampling sites.   There are two main causes of heterogeneous distribution of quality in a water body. These are

(a) if the system is composed of two or more waters that are not fully mixed (such as in thermally stratified lakes or just below an effluent discharge in a river) and
(b) if the pollutant distributes non-uniformly in a homogeneous water body (for example oil which tends to float, and suspended solids which tend to settle out of the water). Also chemical and/or biological reactions may occur non-uniformly in different parts of the system, so resulting in heterogeneous pollutant concentrations. When the degree of mixing is unknown it is advisable to conduct a preliminary survey before deciding on sampling locations. Rapidly obtainable data of water temperature, pH, dissolved oxygen, or electrical conductance may be used in this respect.

Sampling locations should generally be at points as representative of the bulk of the water body as possible, *e.g.* away from river or lake banks or the walls of channels or pipes, but often it will be desirable (and necessary) to take samples from several locations in order to obtain the required information.

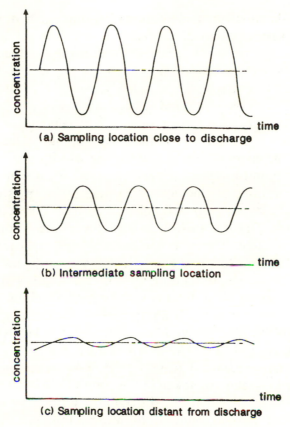

(a) Sampling location close to discharge

(b) Intermediate sampling location

(c) Sampling location distant from discharge

Figure 4 *Schematic diagram of the dependence of pollutant concentration on the distance downstream from a cyclically varying waste discharge*

When sampling from rivers and streams downstream of effluent discharge longitudinal, transverse, and vertical sampling arrays may be necessary to ensure that truly representative data are obtained. Studies of some pollutants require sampling at considerable distances downstream of effluent inputs, *e.g.* in investigating the sag in dissolved oxygen content. When a temporally varying effluent discharge is under study it may be desirable to sample as close to the point of discharge as mixing allows in order to monitor short term variations in concentration. However, if long term average water quality is of interest then sampling should be carried out further downstream where longitudinal dispersion and mixing will have smoothed out the short term variations (see Figure 4).

Sampling in estuaries presents special problems as great spatial and temporal variability may be exhibited. The appropriate locations for sampling will vary from estuary to estuary and will depend on the parameters of interest, but a minimum of 50 samples per survey might be appropriate. If one considers a compound which has an input at only one end of an estuary and which is not removed from or added to solution during its lifetime in the estuary then the concentration of that compound in the estuary will be solely dependent upon the dilution ratio between the river water and the seawater. Thus, for example, the concentration of chloride ion or salinity is dependent only upon the mixing of the fresh and the saline water bodies. This concept of 'conservative' behaviour is an important one which must be taken into account when monitoring estuarine concentrations. If a graph is drawn of the concentrations of the element of interest in an estuary against salinity then the data points will fall on a straight line (the theoretical dilution line) if physical mixing is the only process controlling the concentration of that element in the water. However, if the element of interest is added to or removed from the solution during mixing then the data will not plot on a straight line. In the case of lakes and reservoirs vertical stratification of pollutants may be very pronounced due to a reduction in dissolved oxygen from the surface downwards. A minimum of three samples is then probably necessary, at 1 m below the surface, 1 m above the bottom, and at an intermediate point.

When water is abstracted from a river, lake, reservoir or aquifer, samples should be regularly taken at the point of abstraction and at the point where the water enters the distribution system. Several excellent handbooks with full descriptions of water sampling and analytical methods are available.[9-12]

The determination of concentrations of trace metals in natural waters is a fundamental stage in the calculation of their budgets or cycles, but is subject to the same problems of sample contamination as occur for atmospheric samples from remote areas. All stages of the analysis, from sampling collection, storage, and filtration to actual laboratory manipulation require care to prevent contamination occurring. Indeed for many years measured levels of many trace elements in seawater were purely an artifact of contamination during sampling and analysis.

Case study 3: The behaviour and variations of caesium and plutonium in estuarine waters. Plutonium and caesium isotopes, in addition to other radionuclides, have been discharged into the Irish Sea in effluent from

[9] APHA, 'Standard Methods for the Examination of Water and Wastewater', American Public Health Association, New York, 15th Edn., 1980.
[10] M. J. Suess, 'Examination of Water for Pollution Control—a Reference Handbook', 3 Vols., WHO/Pergamon, Copenhagen, 1982.
[11] J. W. Clark, W. Viessman, and M. Hammer, 'Water Supply and Pollution Control', Harper, New York, 3rd Edn., 1977.
[12] L. G. Hutton, 'Field Testing of Water in Developing Countries', Water Research Centre, 1983.

the nuclear fuel reprocessing plant at Sellafield since 1953. In this study,[13] the plutonium and caesium activity concentrations were monitored in the dissolved phase of estuarine waters, in order to investigate the behaviour of these radionuclides. Water samples were collected by boat at mid-channel, during spring and neap tides, and along the Esk Estuary (10 km from Sellafield discharge point) and its tributaries (Irt and Mite). Samples of 1 or 2 dm^3 were collected using instantaneous isokinetic samplers at near surface (0.5 m depth) on high water surveys and mid-depth on tidal cycle surveys. In addition, a 10 dm^3 sample of river water was collected from the River Esk, well above the tidal limit. Fractionation into particulate and dissolved phases was achieved by filtration with 0.22 mm Millipore filters. Salinities were measured using a Goldberg refractometer and activity concentrations by γ-spectrometry and α-spectrometry following standard preparation and chemical separation procedures.

Figure 5 shows the variation in plutonium and caesium dissolved phase activity (expressed as a percentage of the maximum 'seawater' value for each study) with salinity. Figure 5a clearly indicates that ^{137}Cs activities follow a theoretical dilution line between seawater and river water end members, exhibiting conservative behaviour. Therefore, caesium does not appear to re-equilibrate between the particulate and dissolved phases throughout estuarine waters of different salinities. However, dissolved plutonium activities do not follow a theoretical dilution line (Figure 5b). Higher activities are apparent in the dissolved phase, for salinities below 18‰, than predicted by conservative mixing. In addition, a wide variation of activities were recorded at high salinities. Thus monitoring of $^{239+240}$Pu dissolved phase activity concentrations has shown that plutonium is apparently being released from the particulate phase at low salinities in the Esk Estuary.

Case study 4: Acidification of lakes. The acidification of lakes in North America and Northern Europe (Scotland, Norway, and Sweden) has been directly linked to anthropogenic activities, in particular sulfur dioxide emissions from coal fired power stations. Much of the evidence for increase in acidity has been indirect, such as loss of fish populations, changes in aquatic plant and invertebrate communities, changes in sediment metal concentrations, and results from empirical models of lake acidification. A study of the Adirondack lakes in New York state made direct comparisons of historical and recent lake survey data in order to determine if significant acidification had occurred in that region.[14]

[13] D. J. Assinder, M. Kelly, and S. R. Aston, *Environ. Technol. Lett.*, 1984, **5**, 23–30.
[14] C. E. Asbury, F. A. Vertucci, M. D. Mattson, and G. E. Likers, *Environ. Sci. Technol.*, 1989, **23**, 362–365.

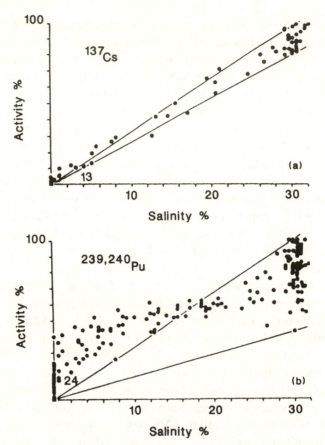

Figure 5 *Variation of (a) ^{137}Cs and (b) $^{239,240}Pu$ dissolved phase activity and salinity in the Esk Estuary, UK (lines refer to level of confidence in theoretical mixing line)* (Reproduced with permission from *Environ. Tech. Lett.*, 1984, **5**, 27, Selper Ltd)

The historical data (1929–1934) included measurements of alkalinity (a measure of the acid neutralizing capacity of the water), pH, and CO_2 acidity. However, the pH data were considered to be unreliable due to the use of colourimetric indicator solutions which can alter the pH of a solution being measured. It was therefore decided to directly compare alkalinity values from this data set with extensive data collected between 1975 and 1985. Alkalinities for each lake were matched on the basis of a unique 'pond number' system adopted by the surveys. Data was excluded where lake identification was ambiguous, the lakes were known to have been treated with lime, or it was suspected that the electrode used for the alkalinity measurements was malfunctioning. The final data set consisted of 274 lakes. Account was taken of the different analytical techniques employed to determine alkalinity. The historic method used titration to a fixed pH end point, determined by the 'faintest pink' colour of the methyl orange indicator. Modern surveys employed the Gran technique to determine the end point. As a result, the historical data set were corrected by an alkalinity subtraction of 54.6 μequiv dm^{-3}.

The historical data had a mean 'corrected' alkalinity of 141 μequiv dm^{-3} compared to a modern alkalinity of 100 μequiv dm^{-3}. This indicates a mean estimated acidification of -41 μequiv dm^{-3}. This loss of alkalinity was found to be significant ($p < 0.01$) by a Wilcoxon signed rank test. It was concluded that significant acidification had occurred in the Adirondack Mountain region since the 1930s. Inter-lake differences in acidification were noted, probably due to factors such as atmospheric deposition, lake hydrology, geology, or soil type. Sensitivity of lakes to acidification has been related to crystalline bedrock and thin acid soils, which are generally associated with high elevation lakes. This study found a direct relationship between the mean lake elevation and mean acidification for lakes in each of the major catchments.

2.2.4 Sediment, Soil, and Biological Monitoring. Soils and sediments may become polluted by a number of routes, including the disposal of industrial and domestic solid wastes, wet and dry deposition from the atmosphere, and infiltration by contaminated waters.

The main pollution hazards on land have been identified as follows:[1]

(a) Harmful substances may get into the soil or plants and so into the food supply.
(b) Substances may wash from the land and so pollute water supplies.
(c) Contaminants may be re-suspended and subsequently inhaled.
(d) Substances polluting the land may make it potentially dangerous or unsuitable for future use (*e.g.* for housing or agriculture).
(e) Ecological systems may be damaged, with consequent loss to conservation and amenity.

Some potentially harmful substances, such as mercury or lead, are naturally present in soils but at concentrations which are not normally deleterious. Some activities, however, can cause elevated levels of these compounds. For example, mining may cause soils to be contaminated by metals, and the dumping of solid wastes on land will invariably introduce a wide variety of pollutants to the soil. On the other hand there are compounds which do not occur naturally, and their presence in soils and sediments is due entirely to human activities. These substances include pesticides, (particularly the organochlorine compounds such as DDT, toxaphene, aldrin, dieldrin) and artificial radionuclides (*e.g.* ^{137}Cs, ^{239}Pu).

As with the other types of media discussed above it is important that the background levels of pollutants be established in soils, sediments, and vegetation. One example of a large-scale investigation of this type is the geochemical survey of stream-bed sediments carried out in England and Wales.[15] Stream sediments are considered to represent a close approximation to a composite sample of the weathering and erosion products of rock and soil upstream of the sampling point and in the absence of pollution provide information on the regional distribution of the elements.

A second type of monitoring programme is required to establish actual levels of contamination in land or sediments known or believed to be affected by pollutants. In this case much more specific and localized monitoring may be required in order to quantify the degree of contamination. The contamination of sites often arises from their previous uses, particularly as coal gas manufacturing plants, sewage works, smelters, waste disposal sites, chemical plants and scrapyards. A typical contaminated site may be found to contain variable concentrations of toxic elements and organic compounds, phenols, coal tars and oils, combustible material from undecomposed refuse, acidic or alkaline waste sludges, and sometimes methane accumulations. In addition, contaminated land is often formed by waste tipping and so may be poorly compacted and very inhomogeneous.

Some sites may contain underground pipework and structures from their previous uses and so present formidable sampling problems. Site investigation of this type is very expensive, involving bore holes or trial pits and, in cases where a detailed site history is unavailable, determination of a large number of pollutants. However, it is most important to be sure that the investigation is sufficiently rigorous, as remedial measures are extremely costly and must be based upon adequate data. Common problems which have been identified are:

— an inadequate number of samples,
— an inadequate range of determinands,
— bulking of samples when individual samples from specific locations are preferable,
— inappropriate analytical methods,
— inadequate referencing of sample locations,
— inadequate descriptions of samples,
— inadequate descriptions of trial pit strata,
— an ignorance of the nature of the required information.

[15] J. S. Webb, I. Thornton, M. Thompson, R. J. Howarth, and P. L. Lowenstein, 'The Wolfson Geochemical Atlas of England and Wales', Oxford University Press, Oxford, 1978.

Monitoring should be carried out following the application of sewage sludge or waste waters to agricultural land. Samples of surface water, groundwater, site soil, vegetation, and sludge applied would normally be tested for faecal coliform, nutrients, heavy metals, and pH. Details of the necessary chemical and physical methods may be found elsewhere.[9] The results from the monitoring exercise may be compared to predicted levels derived from the application rates of sludges to land, soil type, nitrogen, phosphorus, and heavy metal contents of the waste and the nutrient uptake characteristics of the cover crop.

When monitoring background levels and more specific pollution on land or in the sediments of a water body, measurements will often be made of levels in the plants or organisms that the soil or sediments support. In many cases flora or fauna provide excellent indicators of the degree of pollution, as they may act as bioconcentrators (for example of heavy metals from suspended material in shellfish). Furthermore it is obviously important to monitor pollution levels in food and through the food chain. The simultaneous measurements of pollutant levels in soils and plants as well as in water, sediments, and aquatic biota are therefore often carried out. However, the relationship between levels in these various media are often not simple, and sampling and analysis of one of these is no substitute for a comprehensive monitoring programme. For example, laboratory experiments indicate that metals are readily absorbed by some plants but measurements of metals in oil, rainwater, and plants often reveal a lack of correlation between the corresponding sets of concentrations.

An example of the large-scale use of aquatic biota as a medium for the monitoring of pollutants in coastal waters is the US Mussel Watch Program. This programme began in 1976 with the overall aim of providing strategies for pollutant monitoring in coastal waters using mussels and oysters collected on the west, east, and Gulf coasts of the USA. Analyses for trace metals, chlorinated hydrocarbons, petroleum hydrocarbons, and radionuclides (including ^{238}Pu, $^{239+240}$Pu and ^{241}Am) were carried out at three laboratories with extensive intercalibration studies used to ensure compatibility of data. One conclusion of the initial report of this programme was that for metals the frequency of monitoring need not be yearly, but some small multiple of years for a given site. Thus for the same resources, human and financial, greater geographical coverage can be gained by increasing the sampling interval. On the basis of the data obtained in 1977–1978, national baseline concentrations for bivalves in unpolluted waters were suggested. The programme also established the importance of systematic repetitive sampling at the same sites over periods of years, which allowed elevated cadmium and plutonium concentrations in some areas to be attributed to coastal upwelling rather than as a consequence of localized anthropogenic sources.

Monitoring of concentrations of trace metals in crop plants grown on sewage sludge amended soils indicates that the levels found will vary with crop species and properties of the soil substrate on which they have been grown. More recent studies have concentrated on the physico-chemical speciation of the metals in the applied material and the receiving soil and have demonstrated the importance of organic complexes in reducing free metal activity.

Case Study 5: Metals in dusts. A great number of studies have been conducted since the 1970s on dust and soil contamination with heavy metal, in particular lead, due to its predominant source as an anti-knock agent added to petrol. These studies have established the magnitude of different source of lead in the environment. However, it is still unclear whether or not the exposure of young children to relatively small amounts of lead has a detrimental effect on intelligence, behaviour and educational and social attainment.

Although various studies have attempted to assess the significance of different pathways of lead to a child's total lead intake, the study described here[16] was a comprehensive environmental, biological, behavioural, and dietary investigation of two-year-old children in Birmingham. From a set of 183 randomly selected children, 97 completed the study. Various blood samples were taken at the beginning of the study and 56 children also provided a second sample about five months later.

Environmental monitoring was conducted at each of the children's homes. Indoor and outdoor air samples were obtained, over seven consecutive days, using aerosol monitors at a height of 1.5–2 m. House dust samples were collected from the householder's own vacuum cleaner and from the child's playroom and bedroom, using a modified vacuum cleaner. Pavement and road dusts were sampled during dry conditions. A composite soil sample was also collected. On each of the seven days of the study, both hands were wiped thoroughly using a total of three wet wipes. A duplicate of the child's diet over the seven days was provided, along with a dietary record. Finally, mouthing behavioural patterns were analysed using a 30 item questionnaire, or interview of the parents and by recording the child on videotape in set situations. These situations included a period of free play, lunch, being read a 10 minute story and watching a 15 minute video film. Each film was analysed to determine how much time a child touched the floor, objects, and the mouth.

The frequency distributions of lead concentrations in all the samples

[16] D. J. A. Davies, I. Thornton, J. M. Watt, E. B. Culbard, P. G. Harvey, H. T. Delves, J. C. Sherlock, G. A. Smart, J. F. A. Thomas, and M. J. Quinn, *Sci. Tot. Environ.*, 1990, **90**, 13–29.

Table 1 *Lead in blood, environmental samples, handwipes, and water*

Sample	Units	N	Geometric mean	Percentiles 5th	95th
Blood	$\mu g/100$ ml	97	11.7	6	24
Air					
Playroom	$\mu g\ m^{-3}$	607	0.27	0.08	0.88
Bedroom	$\mu g\ m^{-3}$	599	0.26	0.09	0.81
External	$\mu g\ m^{-3}$	605	0.43	0.12	1.53
Dust					
Playroom	$\mu g\ g^{-1}$	97	311	105	1030
Bedroom		96	464	109	2040
Average		94	424	138	2093
Vacuum[a]		92	336	97	1440
Doormat		42	615	120	4300
Pavement		97	360	127	1340
Road		97	527	195	1170
Dust 'loading'	$\mu g\ m^{-2}$	93	60	4	486
Soil	$\mu g\ g^{-1}$	86	313	92	1160
Handwipes	μg	704	5.7	1.9	15.1
Diet (food & beverages)	$\mu g\ week^{-1}$	96	161	82	389
Water	$\mu g\ l^{-1}$	96	19	5	100

After Ref. 16.
[a]Householder's own cleaner.

were found to be approximately log normal, consequently geometrical means were calculated (Table 1). These mean lead concentrations (*e.g.* mean indoor dust lead 424 $\mu g\ g^{-1}$) were similar to those found in previous studies in both Birmingham and elsewhere in the UK, although exposure to dietary lead was believed to be slightly elevated due to higher water lead concentrations. The indoor and outdoor air lead concentrations were strongly correlated, as observed in previous studies, with a mean indoor/outdoor air lead ratio of 0.61. An exception to this occurred in a house in which old paint was being removed by a machine sander. On these occasions the indoor lead concentration exceeded that outdoors. In addition, the dust lead concentration for the householder's own vacuum cleaner was 23 000 $\mu g\ g^{-1}$.

Relationships between blood lead, environmental lead, and behavioural measures were assessed using multiple regression analysis. This indicated the importance of the amount of lead in the house dust combined with a child's rate of hand touching activity to blood lead levels. Water lead and smoking habits of parents are also significant factors where blood lead is concerned. These results were used to predict a mean lead intake of 1 μg day^{-1} via inhalation and 35 μg day^{-1} via ingestion for the Birmingham children.

Case Study 6: Chernobyl-derived radiocaesium in Cumbria, UK. Following the Chernobyl reactor accident on 26 April 1986, a large number of monitoring studies were conducted in the UK and elsewhere in Europe. Radiocaesium contamination of sheep reared on upland fells in Cumbria and North Wales was of particular concern, due to the high levels of fallout received by these areas (20 kBq m^{-2} ^{137}Cs). Restrictions to movement and slaughter of those sheep with a ^{137}Cs muscle specific activity concentrations greater than 1000 Bq kg^{-1} (fresh weight) were imposed by the UK government.

One study[17] conducted at a farm in West Cumbria investigated radiocaesium specific activities in local soils, vegetation, and sheep muscle. Three upland soils and one valley bottom soil were sampled within the same 1 km^2 area, using a 5 cm deep corer. The sampling programme continued from October 1986 to November 1989 with a sampling frequency of every two weeks for the first 5 months. Samples were analysed for total and ammonium exchangeable ^{134}Cs and ^{137}Cs by γ-spectrometry. Vegetation samples were collected from improved pasture at the study farm and also from the nearby open fell, from August 1986 until June 1988 with a frequency of once every two weeks. Live monitoring of the sheep for ^{137}Cs was carried out using a portable γ-spectrometry system which could be taken into the field. A similar monitoring programme to that for the soil and vegetation was adopted.

As a result of this survey it was discovered that a high percentage of the Chernobyl derived ^{137}Cs in upland soils was ammonium exchangeable (86–100%) and thus probably available for uptake by vegetation, compared to 0% for the valley-bottom soil. This was confirmed by the generally higher ^{137}Cs content in vegetation from the open fell (September 1987, 140–2080 Bq kg^{-1} dry weight) compared to the improved pasture (September 1987, < 100–710 Bq kg^{-1} dry weight).

As a consequence of this radiocaesium behaviour, the ^{137}Cs activity concentrations of ewe muscle (Figure 6) declined when the sheep were brought on to enclosed pastures, but rose when they were returned to the open fell. However, despite an overall fall in the ^{137}Cs activity concentration in vegetation with time, the ^{137}Cs levels in sheep were higher in the summer/autumn of 1987 and 1988 than in the autumn of 1986. This appeared to be due, at least in part, to an increase in the availability of radiocaesium originating from Chernobyl fallout, as it had been incorporated into plant material, rather than when it was present as a direct

[17] B. J. Howard, N. A. Beresford, and F. R. Livers, 'Transfer of Radionuclides in Natural and Semi-Natural Environments', ed. G. Desmet, P. Nassimbeni, and M. Belli, Elsevier, New York, 1990.

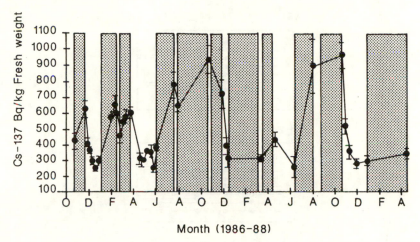

Figure 6 *Changes in the ^{137}Cs activity concentrations of ewe muscle as they are moved between the fell and improved pastures (shading = fell)*
(Reproduced with permission from 'Transfer of Radionuclides in Natural and Semi-Natural Environments', ed. G. Desmet, P. Nassimbeni, and M. Belli, Elsevier Science Publishers Ltd, 1990)

deposit on vegetation surfaces as in 1986. Following the accident the UK Government set up a National Response Plan for dealing with overseas nuclear accidents, which included establishment of a national radiation monitoring network and nuclear emergency response system (RIMNET: Radioactive Incident Monitoring Network).

3 SAMPLING METHODS

3.1 Air Sampling Methods

Sampling systems for airborne pollutants usually consist of four component parts: the intake component, the collection or sensing component, the flow measuring component, and the air moving device. All these (other than the collector or sensor itself) must be constructed of materials which are chemically and physically inert to the sampled air.

3.1.1 Intake Design. The nature of the intake is determined by the type and objective of the sampling technique, and may vary from a vertical opening for the passive collection of dustfall in a deposit gauge to a thin-walled probe used for source sampling of aerosols. Common problems which may require consideration are the non-reproducible collection of the sample portion from the air mass due to poor inlet design, adhesion of aerosols to the tube walls, loss or change of analyte by chemical reaction with inlet materials, adsorption of gaseous compo-

(a)

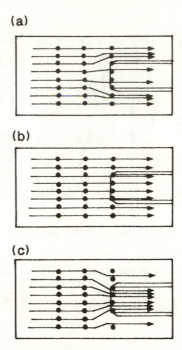

(b)

(c)

Figure 7 *Schematic diagram of (a) under-sampling of suspended particles, (b) isokinetic sampling and (c) over-sampling of suspended particles*

nents on inlet materials, and condensation of volatile components within the transfer lines.

The sampling of aerosols presents particular difficulties in inlet design. A basic requirement is that the velocity of the sample entering the system intake should be the same as the velocity of the gas being sampled. This is necessary because, if the streamlines of the sampled gas are disturbed by the intake probe, particles travelling in the gas flow and possessing inertia directed along the streamlines will continue into the probe while the 'carrier gas' will be diverted away from (if the probe intake velocity is too low) or into (if the intake velocity is too high) the inlet (Figure 7). Thus either a greater number of particles per unit volume of gas than exists in the actual gas flow, or a smaller number, will be collected. Only when the intake velocity at the face of the probe is equal to the approach velocity of the gas stream will the streamline pattern remain unaltered and the correct number of particles per unit volume of gas enter the probe. This is known as isokinetic sampling.

Sampling of ambient air masses is seldom made isokinetically as sophisticated equipment is required to maintain the inlet facing into the wind and to adjust the sampling velocity to match changes in wind speed. This is feasible, but what is difficult is then to interpret the analysis of the collected sample as the flow rate is temporally variable. However, isokinetic sampling is usually practicable, and indeed necessary, when sampling flue gases.

3.1.2 Sample Collection. The methods most commonly used for the collection of atmospheric particulate samples are:

— filtration;
— impingement: wet or dry impingers, cascade impactors;
— sedimentation: by gravity in stagnant air, thermal precipitators;
— centrifugal force, cyclones;

and for gaseous samples:

— adsorption;
— absorption;
— condensation;
— grab sampling.

Filtration. This is by far the most common technique. The type of filter medium chosen will depend upon a number of factors. These include the collection efficiency for a given particle size,[18] pressure drop and flow characteristics of the filter type,[19] background concentrations of trace constituents within the filter medium, and the chemical and physical suitability of the filter with regard to the sampling environment.

Impingement. Impingers consist of a small jet through which the air stream is forced, so increasing the velocity and momentum of suspended particles, followed by an obstructing surface on which the particles will tend to collect. Wet impingers operate with the jet and collection surface under liquid and require high flow rates for optimum collection efficiency.

Cascade impactors use the aerodynamic impaction properties of particles to separate the sample into different size fractions by use of sequential jets and collection surfaces. Increasing jet velocity and/or decreasing gaps between the jet and collection plate fractionates the sample. Figure 8 shows the principle of a commonly used cascade impactor. This consists of up to 15 stages backed by a membrane filter, each stage containing accurately drilled holes which align over a solid portion of the adjacent plates. The holes in each successive stage are

---

[18] M. Katz, 'Measurement of Air Pollutants, Guide to the Selection of Methods', World Health Organization, Geneva, 1969.

[19] B. Y. H. Liu, D. Y. H. Pui, and K. L. Rubow, 'Aerosol in the Mining and Industrial Work Environments', ed. V. A. Marple and B. Y. H. Liu, Ann Arbor Science, Ann Arbor, MI, 1983, Vol. 3, p. 898.

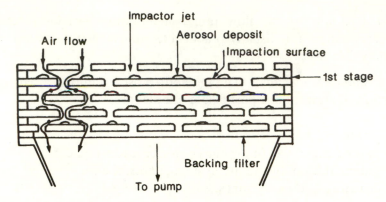

Figure 8 *Schematic representation of a cascade impactor*

smaller than those in the preceding plate, and since air is drawn through the instrument at a constant flow rate the effective velocity at each stage increases. The largest particles are impacted on the first stage and the smallest are collected on the backing filter. The range of particle diameters collected on each stage may be determined by laboratory calibration or by theoretical calculations. However impaction sampling at normal flow rates (about 0.01–0.04 m^3 min^{-1} or 0.6–1.0 m^3 min^{-1} for Hi-Vol cascade impactors) is only efficient for particles with aerodynamic diameters >0.3 mm. Also the collection efficiency of each stage will vary according to particle type, some being very 'sticky', others liable to bounce off. Other problems of cascade impactor sampling include wall losses, the aggregation of particles, and the mechanical breaking of agglomerates which result in inaccurate size distribution measurements.

Sedimentation.   The collection of particulate material by allowing it to deposit into a collection vessel is the simplest of all air pollution measurement techniques. However, the presence of the bowl or cylinder in the path of the falling particles will change their flow pattern and it is not clear whether the collected material is truly representative of actual conditions. The methods are not as widely used now as previously. The British Standard deposit gauge consists of a collection bowl connected to a bottle and supported by a galvanized steel stand. During wet weather dust is washed down from the bowl, but during dry weather high winds may blow dust out of or into the bowl, so producing erroneous dust loadings. At the end of the sampling period (usually one month) a measured volume of water is used to wash any dust in the bowl into the collection bottle, and the pH, total particulate mass, and water volume determined.

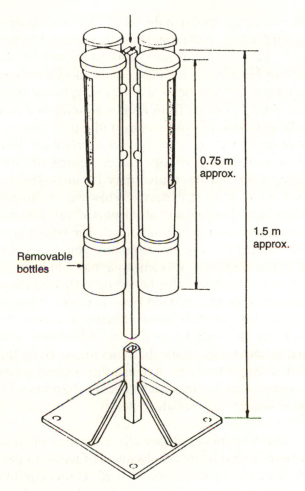

Figure 9 *Directional deposition gauge*

A major disadvantage of the standard deposit gauge is that no directional resolution of the source of particulate material is possible. The directional deposit gauge[20] consists of four cylinders mounted on a common post, with open slots facing the four quadrants of the compass (Figure 9). Each cylinder has a removable collection bottle at its base. After collection a suspension of the dust may be placed in a glass cell and a measure of the dust loading made by the amount of obscuration of a beam of light passing through the cell. Density fractionation of the collected material is also possible by using a mixture of diiodomethane and acetone which allows a density gradient of 0.8–3.3 g cm^{-3} to be achieved. By comparison of the density gradient fractionation of

[20] 'British Standard Method for the Measurement of Air Pollution, Part 5—Directional Deposit Gauges BS 1747 part 5', British Standards Institution, London, 1972.

material collected in a directional deposit gauge with material collected from likely sources (*e.g.* pulverized fly ash storage heaps, slag heaps, cement works, *etc*) simple source apportionment studies may be attempted. Other techniques, including dispersion staining and microscopic examination may also be used in this way, whilst the compilation of a reference library of dusts from known sources in a given area will greatly ease the practical difficulties of dust identification.

The siting of deposit gauges often presents problems. When attempting to monitor emissions from one major source the use of several directional gauges around the source may be successful in confirming that emissions occur from that source. However in areas of multiple sources or where there is significant atmospheric turbulence (*e.g.* in built-up areas) inconclusive data may be obtained. The basic requirement that gauges should be sited at a distance from any object of at least twice the height of the object is often insufficient to avoid significant distortion of the deposition pattern. Siting is further complicated by the need to have tamper-proof sites at or near ground level. It is worth pointing out that the standard deposit gauge is most effective for collection of large particles which readily settle under gravity, whilst the vertical slots of the directional gauge collect smaller particles impacted by the wind more effectively. Source apportionment studies using deposit gauges have now largely been superseded by the specific chemical analysis ('finger printing') of actively sampled particulate material.

Adsorption.   The adsorption of gases is a surface phenomenon. Gas molecules become bound by intermolecular attraction to the surface of a collection phase and so become concentrated. Under equilibrium conditions at constant temperature the volume of gas adsorbed on the collection phase is proportional to a positive power of the partial pressure of the gas, and is also dependent upon the relative surface area of the adsorbent. Materials commonly used as adsorbents include activated carbon, silica gel, alumina, and various porous polymers.

When selecting a suitable adsorbent the relative affinity for polar or non-polar compounds must be considered. For example, activated carbon is non-polar and therefore will adsorb non-polar organic gases, but exclude polar compounds such as water vapour. The wide range of gas chromatographic supports available vary in their degree of polarity and so allow selection of the appropriate type.

The adsorbent used must not react chemically with the collected sample unless chemisorption is used intentionally. Also the analytes must not react with other constituents of the sampled air, either during collection or storage. It has been found for example that some hydrocarbons may decompose on polymers by reaction with atmospheric ozone. This may be prevented by the use of a selective prefilter which removes the oxidant from the air stream but allows the analytes to pass through.

It is important to determine the retention volume of the adsorbent (*i.e.* the volume of air which may be passed without breakthrough of the analyte) with respect to the species being collected. This should be high enough to allow sufficient of the analytes to be collected for analysis. The desorption properties of the material are also important to ensure quantitative recovery of the sample, preferably with regeneration of the adsorbent for subsequent use. Activated carbon is a very efficient adsorber, so making quantitative desorption difficult. Steam stripping may result in hydrolysis reactions with the analytes and vacuum distillation and solvent extraction are not without their problems. Compounds on support bonded porous polymers may conveniently be thermally desorbed by flushing with an inert carrier gas. In the case of reactive hydrocarbons on polymer a two-stage thermal desorption system utilizing an intermediate cryogenic trap cooled with liquid N_2 followed by flash heating allows quantitative recovery as well as direct injection of the sample in a very small volume of gas onto the gas chromatograph column.

Since adsorption is temperature dependent, collection efficiency and an increase in retention volume may be achieved by cooling the adsorbent. However, problems with blockages by ice may then occur. With the increase in sophistication of detection systems in recent years, particularly by the interfacing of chromatographic separation techniques with mass selective detectors, the use of adsorbents as preconcentrators is also increasing. However, care must always be exercised to avoid non-quantitative collection, breakthrough effects due to exceeding the retention volume of the system, decomposition, and non-quantitative recovery of the sample.

Absorption. Gases may be collected by being dissolved in a liquid collection phase or by chemical reaction with the absorbent. The simple Dreschel bottle may be used or may be modified by the inclusion of a fitted diffuser to create small bubbles and so enhance the collection efficiency.

An example of a simple absorption technique which allows an estimate of NO_2 concentration to be made with relatively little capital outlay is the use of triethanolamine diffusion tubes.[21] A small acrylic tube is fitted with a fixed cap at one end. A fine wire mesh coated in triethanolamine is placed in the closed end of the tube and absorbs NO_2 as it diffuses from the open end. The NO_2 is determined spectrophotometrically at the end

[21] M. R. Heal and J. N. Cape, *Atmos. Environ.*, 1997, **31**, 1911–1923.

of the sampling period. These passive samplers are very cheap to construct and analyse and have been used as an effective primary survey technique before embarking upon a more expensive monitoring exercise based upon the standard chemiluminescent technique.

Condensation. By cooling an air stream to temperatures below the boiling point of the substance of interest it is possible to condense gases from the air and so concentrate them. However, a limitation of the method is that water vapour present in the air will also freeze and so progressively block the trap. This may be overcome by using a first trap of large volume designed to collect water and a second trap at a sufficiently low temperature to collect the analytes. Coolants of temperatures $-183\,°C$ or lower (*e.g.* liquid N_2) should not be used for this purpose as they will condense atmospheric oxygen and result in a serious combustion hazard.

Grab sampling. Rather than utilizing a concentration technique in the field, samples may be collected in an impermeable container and returned to the laboratory for analysis. Grab samples of this type have been collected in FEP-Teflon bags or specially treated stainless steel cans for hydrocarbon determination by GC, for example. In the bag technique the deflated container is housed within a rigid box which is slowly evacuated. Air is thus drawn into the flexible bag which may be sealed when inflated. Samples can then be drawn at a later stage from the bag by hypodermic gastight syringe. Pumps constructed of inert material may be used to fill rigid cans to a high pressure, allowing a large volume of air to be sampled.

Specific techniques. Two air sampling techniques are in such widespread use that they require separate consideration. These are:

(a) the UK National Survey of Air Pollution smoke and SO_2 sampling apparatus,
(b) the US National Air Sampling Network Hi-Vol method for total suspended particles (TSP) measurement.

(a) NSAP smoke and SO_2 method. The equipment used in the National Survey of Air Pollution in the UK consists of a pump which draws about $2\,m^3$ of air per day through a filter paper held in a brass clamp. This removes particulate material. The air then passes through dilute H_2O_2 which removes SO_2, converting it to H_2SO_4, which is determined by titration. An eight-port sample changer allows eight 24 h samples to be collected sequentially in an array of eight filter papers and bottles, and so requires operator attendance only once a week. Although

the H_2O_2 method will actually measure the net gaseous acidity of the air it is considered that for ordinary urban situations it will give a good estimate of the SO_2 concentration and hence results are expressed in terms of $\mu g\ SO_2\ m^{-3}$. The particulate loading of the exposed filter paper is estimated by measuring the reflectance of the paper, so obtaining a measure of the staining property of the air. A calibration curve is then used to convert the darkness of the stain to concentrations of equivalent standard smoke.

Inaccuracies in these methods may become pronounced when there are acidic or basic gaseous components other than SO_2 present in the air (HCl giving a positive interference and ammonia a negative interference) or when particulates of 'non-standard' staining properties (*e.g.* light coloured ammonium compounds) are present. When the filter paper is extremely heavily loaded the reflectometer method may severely underestimate the actual smoke level. Particulate losses may also occur at the inlet to the apparatus and in the inlet tube. The smoke stain method gives concentrations of Standard Smoke which are not directly comparable with TSP levels determined by the US Hi-Vol method.

Although still widely used, the titration method has been superseded by the preferred technique of gas phase pulsed fluorescence, which allows the determination of SO_2 concentrations with one or two minute averaging times with very high precision and accuracy at low concentrations. Similarly, the smoke stain method has now been superseded by the state-of-the-art oscillating microbalance method for determination of particulate loadings in ambient air. Both these sophisticated methods require substantial capital investment and demand skilled operator attendance for calibration and maintenance purposes.

(b) Hi-Vol method for suspended particulates. This method is the current US EPA reference method for total suspended particulates (TSP) and is used in the US National Air Sampling Network,[22] although it is being superseded by the use of size selective inlets which excludes particles of aerodynamic diameter $> 10\ \mu m$ (the PM_{10} instrument) or $> 2.5\ \mu m$ ($PM_{2.5}$). A high flow rate blower draws the air sample into a covered housing and through a 20×25 cm rectangular glass fibre filter at 1.1–$1.7\ m^3\ min^{-1}$. The mass of particles collected on the filter is determined gravimetrically and extraction techniques may then be used to remove the material for chemical analysis. However, when glass fibre filters are used reaction with acidic components may result in artifact formation, for example sulfate from gaseous SO_2. Although 24 h

[22] Environmental Protection Agency, 'Reference Method for the Determination of Suspended Particulates in the Atmosphere (High Volume Method)', US Federal Register 36, No. 84, 1971.

sampling periods are commonly used, timer devices are available to switch on and off the blower at predetermined intervals. However, passive sampling by the settlement of particulates onto the filter during periods when the pump is not operating may cause a positive error in the determination. Recent modifications include size selective inlets which will exclude particles of greater than a given size. However, the cutoff efficiency is usually dependent upon wind speed and may not be sufficiently selective.

3.1.3 Flow Measurement and Air Moving Devices. In order to measure the concentration of an airborne constituent it is necessary to know the volume of air sampled. This may be achieved by measuring the rate of flow with a rate meter or by directly measuring the volume of air passed with a dry or wet test gas meter, a cycloid gas meter, or by the use of a mass flow controller. All these devices require calibration and regular checking for leaks. As air volume is dependent upon both temperature and pressure the measurement of these two parameters at the meter inlet is essential so that volumes may be expressed at standard temperature and pressure.

Many different types of pump are available for air sampling, both mains and battery operated, but the precise type chosen will depend upon the required flow rate, the availability of power, and whether continuous or intermittent flow is required.

3.2 Water Sampling Methods

For many applications no special water sampling system is required as an appropriate sample container immersed in the water may be adequate. The main requirement is that a portion of the material under investigation small enough in volume to be transported and handled conveniently but still accurately representing the bulk material should be collected. Typically a $0.5–2$ dm^3 volume is sufficient. When samples are required from depth, two types of collection vessel may be used.[9] The first consists of a cylinder with hinged lids at both ends. The container is lowered into the water with both lids opened and at the desired depth a messenger weight sent down the wire which closes them. This type is not suitable for trace metal work as contaminated surface waters may result in contamination of the vessel and the messenger may scour metallic particles from the wire. The second type consists of a sealed container filled with air which is lowered to the required depth. A messenger is again sent down to open, and another to close, the lid. Alternatively, pressure sensors may activate the lid.

Automatic sequential samplers are available which will collect a given volume of water into an array of bottles. They have been used, for example, in collecting stormwater runoff from roads and the sampling sequence may be triggered when the flow in a flume reaches a certain height.

Adsorption or filtration media may also be used to concentrate the species of interest *in situ*. Using a completely self-contained sealed unit of inert material housing a peristaltic pump and power supply with only the adsorption tube inlet and outlet open to the water, contamination of the sample can be completely avoided. Very large volumes of water may be processed. Another ongoing development is the use of passive permeation devices which may be immersed in water for long periods of time, so giving time-averaged concentrations.

When considering sampling methods for use on inland waters or in coastal waters and estuaries the sophisticated techniques developed for use at sea may be found to be impracticable due to their need for heavy lifting gear on the sampling vessel. Monitoring work must often be undertaken on such waters using small boats without such equipment. One ingenious method of collecting water samples at different depths using very limited resources on a small boat is to lower a weighted plastic tube to the desired depth and to use a small peristaltic pump to draw water up and into a collection bottle. In this way completely uncontaminated samples may easily be obtained from depths of 30 m or more. Whether the particulate fraction of material present in the water can be quantitatively collected in this manner is not clear.

Whichever method of collecting samples is used care must be taken to ensure that neither the sample storage containers nor any collecting vessels used contaminate or alter the sample. Contamination may occur by:

(a) Leaching of contaminants from the surface of imperfectly cleaned containers.

(b) Leaching of organic substances from plastics, or silica and sodium or other metals from glass.

(c) Adsorption of trace metals onto glass surfaces or organics onto plastic surfaces. In the case of metals this may be avoided by prior acidification of the container, but this may in turn exacerbate problem (a).

(d) Reaction of the sample with the container material, *e.g.* fluoride may react with glass.

(e) Change in equilibrium between pollutants in particulate and solution phases.

If a solvent extraction technique is used to concentrate the analytes prior to analysis, care must be taken to ensure that the reagents and containers used are themselves sufficiently clean. Some commonly used materials and techniques have been shown to cause severely elevated metal levels in water.

Different determinands require different methods of preservation in order to prevent significant changes between the time of sampling and of analysis. Generally acidification to pH 2 and refrigeration to 4 °C will be adequate although complete stability of every constituent can never be guaranteed and, at best, chemical, physical, and biological processes affecting the sample can only be slowed down. Samples may be filtered directly after collection in the field to separate the particulate and solution phases. The solution phase may then be acidified to prevent adsorption of the pollutant to the container wall.

3.3   Soil and Sediment Sampling Methods

Soils and sediments are typically very inhomogeneous media and large lateral and vertical variations in texture, bulk composition, water content, and pollutant content may be expected. For this reason large numbers of samples may be required to characterize a relatively small area. Although surface scrapings may be taken it is often necessary to obtain cores so that vertical profiles of the determinands may be obtained or cumulative deposition estimated. Plastic or chromium plated steel tubing of ~ 2.5 cm internal diameter is often suitable, and if the samples are sealed into the tubes and air excluded they may be satisfactorily stored at low temperatures until required. Otherwise they may be extruded in the field and stored in plastic bags. Various core sampling devices are available for obtaining cores of bottom sediments from lakes *etc* (*e.g.* the Jenkin corer).

Grab samples of soils are easily obtained manually and stored in pre-cleaned plastic bags. Sometimes composite samples formed by the bulking together of a number of individual samples may be sufficient, but generally analyses of individual samples is to be preferred. In the case of sediments, grab samplers are available for operation at considerable depths, examples being the Ponar, Orange-peel and Peterson grabs. Alternatively a dredge may be used to obtain a composite sample along a strip of the sediment surface.

Wet soils and sediments that are to be analysed while still wet should not be collected or stored in bags, but in rigid containers. The vessel should be filled as completely as possible leaving no airspace at the top and a bung inserted so as to displace excess water without admitting air.

Some determinands in soils and sediments are liable to change during storage and require the use of preservation techniques. For example

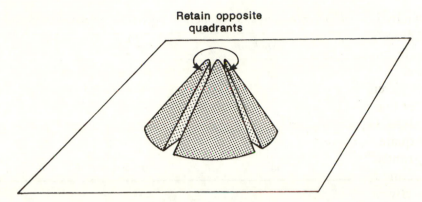

Figure 10 *Schematic diagram of the sub-sampling of dried soil or sediment using the technique of coning and quartering*

nitrate in soil can be extracted into potassium chloride solution and preserved with toluene. Usually, however, air-dried soils and sediments may be disaggregated, sub-sampled by coning and quartering (see Figure 10) and stored in suitable containers, but as always sample contamination must be avoided at each stage.

There are important effects associated with grain size which should be considered in the analysis of soils or sediments. First, many pollutants are associated with particle surfaces and therefore occur in highest concentrations in the smaller grain sized material. Secondly sub-sampling from a bulk sample may be very difficult due to size segregation effects and it may be necessary to grind the sample to a very fine powder to ensure homogeneity prior to division of the sample.

4   MODELLING OF ENVIRONMENTAL DISPERSION

A characteristic feature of environmental monitoring studies is that substances may be found over very large ranges of concentrations, and therefore the analytical techniques employed must be extremely flexible. Some typical concentrations of substances in polluted environmental media are given in Table 2. Not only will large differences of concentrations be found from area to area but even small temporal and lateral changes can result in large changes in pollutant concentrations. The temporal and spatial variability of lead in air and dust samples is illustrative of the way in which pollutant concentrations vary and have been discussed in relation to the design of monitoring programmes.[23] In this study it was shown by collating data collected at several sites in London, UK, that short term concentrations of atmospheric lead (of

[23] M. J. Duggan, *Sci. Total Environ.*, 1984, **33**, 37–48.

Table 2   *Typical concentrations of substances in polluted environmental media*

Pollutant	Medium	Typical ranges
Cadmium	air	$0.1–10 \text{ ng m}^{-3}$
Lead	seawater	$2–200 \text{ pmol dm}^{-3}$
Lead	soil	$5–5000 \text{ mg kg}^{-1}$
Lead	air	$1–100 \text{ ng m}^{-3}$
Sulfur dioxide	air	$0.5–200 \text{ ppb } (10^{-9} \text{ v/v})$
Sulfate	air	$0.1–20 \text{ } \mu\text{g m}^{-3}$
Benzo(a)pyrene	freshwater	$0.1–10 \text{ ng dm}^{-3}$
Carbon monoxide	air	$0.01–50 \text{ ppm } (10^{-6} \text{ v/v})$

sampling period less than one week) can vary considerably at the same site, with a factor of up to 4 between the highest and lowest weekly averages at any one site. Traffic flow differences cannot entirely account for these variations, and local weather conditions are probably the dominant factor. However, when longer averaging periods are used the temporal variability at any one site is much reduced, but a second effect becomes apparent. Thus monitoring periods of at least a month or two are required in order to obtain representative results with possibly some adjustment made for the season. The spatial variability of urban lead-in-air concentrations was found in this study to be rather slight and although lead concentrations decreased away from the carriageway small changes in sampling position (*e.g.* at the ground floor as opposed to the third floor windows of a building facing the street) made up to 30% difference to lead concentration. Thus the precise choice of sampling position at a site may not be critical unless the concentrations are very close to some pre-determined guideline or limit value, in which case compliance may be found in one position but not at another close by.

Dust samples collected weekly at the same pavement sites were found to have lead concentrations which varied by about the same relative amount as air-lead concentrations (*i.e.* weekly samples had a coefficient of variation of *ca.* 30%). However, spatial variability of lead-in-dust concentrations was found to be high over short distances (*i.e.* a few metres). It was suggested therefore that this can be overcome by taking dust samples over a large area (*e.g.* 5 m^2) and that in this way representative dust-lead values may be obtained.

In order to appreciate the variability of pollutant levels, and hence appreciate the complexity of designing an adequate monitoring programme, it is necessary to have some understanding of environmental dispersal, mixing, and sink processes and of the time-scales on which these processes act. Rather than discussing in detail the physical, chemical, and biological processes responsible for changes in pollutant concentrations after discharge to the environment, which have been extensively presented elsewhere, the salient features of some of the techniques by which these changes can be anticipated will be shown. A summary of some of the computer codes which are currently available for modelling environmental dispersion is provided in Table 3.

4.1 Atmospheric Dispersal

Material discharged into the atmosphere is carried along by the wind and mixed into the surrounding air by turbulent diffusion. In the vertical plane the dispersion continues until the turbulent boundary layer is uniformly filled whilst in the horizontal plane dispersion is theoretically unlimited and usually proceeds more rapidly than in the vertical plane. In the simplest models of plume dispersion the degree of turbulence, and hence of mixing, is described by an atmospheric stability classification that is dependent upon the amount of incoming solar radiation, wind speed, and cloud cover, but surface roughness is also important in producing turbulence, especially in the case of large buildings or topographic features.

The most commonly used model of plume dispersion[24] is that described by a Gaussian distribution characterized by standard deviations σ_y and σ_z in the vertical and horizontal directions respectively.

The basic equation for Gaussian plume dispersion is:

$$\chi_{(x,y,z)} =$$

$$\frac{Q}{2\pi\sigma_y\sigma_z u_{(z)}}\exp\left[-\frac{1}{2}\left(\frac{y}{\sigma_y}\right)^2\right]\left\{\exp\left[-\frac{1}{2}\left(\frac{z-H_e}{\sigma_z}\right)^2\right]+\exp\left[-\frac{1}{2}\left(\frac{z+H_e}{\sigma_z}\right)^2\right]\right\}$$

where χ is the pollutant concentration at point (x,y,z) (μg m^{-3}); Q is the pollutant emission rate (μg s^{-1}); $u_{(z)}$ is the wind speed (m s^{-1}); at the effective emission height, H_e; σ_z is the standard deviation of the plume concentration in the vertical at distance x; σ_y is the standard deviation of the plume concentration in the horizontal at distance x; x,y,z are the lateral, transverse, and vertical directions (m), downwind with the base of the stack as the co-ordinate origin; H_e is the effective height of the plume (m).

Assumptions made are that:

— no deposition occurs from the plume at the ground surface;
— pollutant levels are not altered by chemical processes in the plume;

[24] F. Pasquill, 'Atmospheric Diffusion', Ellis Horwood, Chichester, 1974.

Table 3   *Examples of environmental dispersion modelling tools*

Environmental media	Code	Overview	Availability
Air	ADMS	ADMS is a PC-based model for atmospheric dispersion of passive, buoyant or slightly dense releases from single or multiple sources	Cambridge Environmental Research Consultants, 3, Kings's Parade, Cambridge, CB2 1SJ, UK
Freshwater	HSPF	HSPF is a comprehensive modelling package for the simulation of the quantity and quality of runoff from multiple-use catchments, and of processes occurring in streams of fully-mixed lakes receiving catchment runoff.	US Environmental Protection Agency, Centre for Exposure Assessment Modeling, Athens, Georgia, United States
	OTTER	OTTER is a model for the transport of heavy metals and radionuclides through the freshwater environment, typically for simulating the after effects of deposition from the atmosphere.	AEA Technology, Thomson House, Risley, Warrington WA3 6AT, United Kingdom
	PRAIRIE	PRAIRIE is a risk assessment software tool for predicting the risks associated with accidental releases of hazardous materials into rivers or estuaries.	AEA Technology, Thomson House, Risley, Warrington WA3 6AT, United Kingdom
	QUAL2E	QUAL2E is a stream water quality model designed primarily to simulate conventional constituents (*e.g.* nutrients, algae, dissolved oxygen) under steady-state conditions, both with respect to flow and input waste loads.	US Environmental Protection Agency, Centre for Exposure Assessment Modeling, Athens, Georgia, United States
Multi	MMSOILS	The MMSOILS model is a methodology for estimating the human exposure and health risk associated with releases of contamination from hazardous waste sites.	US Environmental Protection Agency, Centre for Exposure Assessment Modeling, Athens, Georgia, United States
Multi	MULTIMED	The Multimedia Exposure Assessment Model (MULTIMED) simulates the movement of contaminants leaching from a waste disposal facility or contaminated soils	US Environmental Protection Agency, Centre for Exposure Assessment Modeling, Athens, Georgia, United States

— there is no effect from surface obstructions (*e.g.* buildings);
— dispersion by diffusion in the downwind direction is negligible compared with bulk transport by the wind;
— the constituents are normally distributed vertically and horizontally across the plume.

The effective height of emission H_e (which is greater than the stack height due to the momentum and buoyancy of the plume) may be calculated from Holland's equation:[25]

$$H_e = H + \left\{ \frac{V_s d}{u} \left(1.5 + 2.68 \times 10^{-3} pd \frac{T_s - T_a}{T_s} \right) \right\}$$

where V_s = stack gas exit velocity (m s^{-1})
d = internal stack diameter (m)
u = wind speed (m s^{-1})
p = atmospheric pressure (mb)
T_s = stack gas temperature (K)
T_a = air temperature (K)
H = stack height (m)

and 2.68×10^{-3} is a constant with units of mb^{-1} m^{-1}.

The wind speed at height z (u_z) may be obtained from the power law

$$u_z = u_{10} (z/10)^n$$

where u_{10} is the speed at the reference height of 10 m, and n varies according to surface roughness, but may be taken to equal 0.2.

The next step is to determine the appropriate Pasquill stability class from Table 4 and then evaluate the estimate of σ_y and σ_z as a function of downwind distance.[24] In order to obtain the ground level concentration below the plume centre line (*i.e.* $y = z = 0$) the general equation reduces to:

$$\chi_{(x)} = \frac{Q}{\pi \sigma_y \sigma_z u} \exp \left[-\frac{1}{2} \left(\frac{H_e}{\sigma_z} \right)^2 \right]$$

or for a ground level source with no effective plume rise ($H_e = 0$)

$$\chi_{(x)} = \frac{Q}{\pi \sigma_y \sigma_z u}$$

[25] J. Z. Holland, 'A Meteorological Survey of the Oak Ridge Area, Report ORO-99', Atomic Energy Commission, Washington, DC, 1953, p. 554.

Table 4 *Pasquill's Stability Categories*

Surface wind speed (at 10 m) (m s^{-1})	Day			Night	
	Incoming solar radiation			Thinly overcast or ⩾4/8 low cloud	⩽3/8 cloud
	strong	moderate	slight		
<2	A	A–B	B		
2–3	A–B	B	C	E	F
3–5	B	B–C	C	D	E
5–6	C	C–D	D	D	D
>6	C	D	D	D	D

After Ref. 24.
The neutral class D should be assumed for overcast conditions during day or night.

Full treatments of these equations and their applications are available elsewhere, including modifications to include line and area sources.[26,27]

A frequently required type of calculation is to determine the minimum height of a chimney which will give adequate dispersion to ensure that critical ground level concentrations are not exceeded. A simple algorithm has been developed which allows this estimate to be easily made.[28]

All attempts to model atmospheric diffusion and mixing processes are liable to be, at best, only good estimates of the real situation and care must be taken to:

(a) understand the physical, chemical, and mathematical limitations of the model, and
(b) avoid treating the output from models as providing definite answers.

Although accurate measurements may be made of wind speed it is in the determination of the turbulence characteristics of the atmosphere that uncertainties arise, which in turn lead to uncertainties in the model. At short distances downwind, with steady winds, uniform terrain, and no local obstructions the Gaussian plume model may be expected to give good estimates, but in any but ideal conditions they soon become order-of-magnitude estimates only.

[26] D. B. Turner, 'Workbook of Atmospheric Dispersion Estimates', US Environmental Protection Agency, Research Triangle Park, NC, 1970.
[27] R. H. Clarke, 'A Model for Short and Medium Range Dispersion of Radionuclides Related to the Atmosphere', Report NRPB-R91, National Radiological Protection Board, Harwell, 1979.
[28] HM Inspectorate of Pollution, 'Technical Guidance Note (Dispersion) D1. Guidelines on Discharge Stack Heights for Polluting Emissions', HMSO, London, 1993.

4.2 Aquatic Mixing

The physical transfer and transport of pollutants in the aquatic environment is determined by the same two processes that determine the mixing of pollutants in the atmosphere. These are:

(a) advection, caused by the large-scale movement of water, and
(b) mixing or diffusion, due to small-scale random movements which give rise to a local exchange of the pollutant without causing any net transport of water.

The combined effect of advection and diffusion is known as dispersion. These two processes occur over a very wide range of scales, both of spatial extent and frequency, which necessitates the use of averaging procedures when defining their role in pollutant transfer. As in the lower atmosphere, there is usually a constraint on the vertical dispersion component, induced either by water depth, or in deeper waters by thermal or density stratification. On a large scale in the oceans there will be a vertical circulation driven by density differences but, on a small scale, vertical motion due to turbulence or eddy diffusion will tend to be suppressed by stratification.[29] Similar restrictions on vertical mixing occur in rivers due to limited depth, but here horizontal mixing in the cross-channel direction is also constrained.

A large number of models have been developed to describe the movement of pollutants in the aquatic environment, many being analogous to those used in air pollution studies. The fact that such models are necessary is due to limitations in our detailed knowledge of the velocity field, and this in turn may lead to uncertainties in the prediction of dispersion patterns.

The dispersion of pollutants in rivers has attracted a great deal of study, particularly in the context of effluent discharges and the ability of rivers to dilute them. Unlike the oceans, which have traditionally, but unreasonably, been considered to have an infinite capacity for dilution, the deterioration in water quality due to pollutant discharges is often manifestly apparent in rivers.

Various models have been applied to the problem of modelling river dispersion with varying levels of complexity. An extremely simple model of change in water quality downstream from a discharge into a river would be to assume that an exponential decay in concentration occurs, *i.e.*

$$C_x = C_0 \, e^{-kt}$$

[29] G. Kullenberg, 'Physical Processes in Pollutant Transfer and Transport in the Sea', ed. G. Kullenberg, CRC Press, Boca Raton, 1982, Vol. 1.

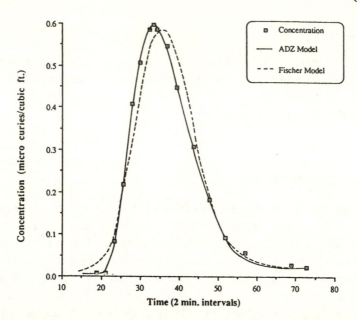

Figure 11 *Modelling of Copper Creek dispersion data: comparison of ADZ model fit with the results obtained by Fischer using the conventional Fickian diffusion model* (Reproduced with permission from 'Pollution Causes, Effects and Control', ed. R. M. Harrison, Royal Society of Chemistry, Cambridge, 2nd Edn., 1990)

where C_x is the concentration at point x, C_0 is the concentration at the point of discharge, k is the decay rate and t is the time taken for flow from the point of discharge to point x.

An alternative approach to river modelling is to consider the river to be divided into a number of reaches, with each reach being considered as a continuously stirred tank reactor, and to place an appropriate time delay between each reach. It is assumed that the major dispersive mechanism can be explained by the aggregated effect of all 'dead zone' phenomena in the river between the two sampling points. This 'aggregated dead zone' model is thus a combination of the continuously stirred tank reactor and a factor to account for the advection component of dispersion. The two approaches to modelling river dispersion represented by the Fickian Diffusion analysis and the Aggregated Dead Zone (ADZ) model have been compared with data obtained from a tracer experiment.[30] Figure 11 compares the monitored concentrations with those predicted by the two models, and in this example the ADZ model is able to explain the data better than the conventional diffusion model.

---

[30] P. C. Young, 'Pollution: Causes, Effects and Control', ed. R. M. Harrison, Royal Society of Chemistry, Cambridge, 2nd Edn., 1990.

More recent models[31] take account of the various physico-chemical processes a pollutant may be subjected to following a spill into a river (*e.g.* sorption, volatilization, hydrolysis, photolysis, oxidation, and biological degradation). The effects of weirs on pollutant removal may also be modelled.

4.3 Variability in Soil and Sediment Pollutant Levels

Obviously, the same mechanisms of dispersion as operate in fluid or gaseous media do no occur in soils and sediments. Physical mixing may occur, such as during agricultural practices, dredging of estuaries, or bioturbation by burrowing organisms, but usually only on a fairly limited scale. The level of contamination in a soil or sediment will depend upon the deposition rate of the pollutant and its subsequent rate of movement through the soil or sediment column. The rate of movement of a contaminant through these solid media is dictated by the degree of adsorption to or leaching from the particles and the flux rate of pore water, transferring the pollutant to deeper horizons. The physico-chemical properties of the water, soil or sediment particles and the pollutant all influence the rate of adsorption or leaching. For example, conditions that favour the adsorption of radiocaesium and lead in soils include low rainfall and high clay (particularly illite) content, whereas mercury and copper tend to accumulate in soils with a high organic content. Contaminant concentrations are generally always higher in soils or sediments with finer grain sizes, due to the increased total surface area available for adsorption. These effects are illustrated in Figure 12 for trace metal concentrations in sewage sludge in treated and control soil plots. Copper concentrations in sludge-treated plots show a strong correlation with organic carbon content, while zinc concentrations in both treated and control plots increase with decreasing particle size, and increase with increasing organic carbon content. All these factors lead to great variability in pollutant concentrations, which simply emphasises the need for carefully designed monitoring programmes.

5 DURATION AND EXTENT OF SURVEY

5.1 Duration of Survey and Frequency of Sampling

The duration of a pollution monitoring programme is entirely dependent upon the purpose of the study, and can vary from the time taken to collect a limited number of, for example, street dust samples in an urban

---

[31] A. Watson, L. S. Fryer, and J. W. Clark, 'Aqueous Pollution Modelling—Approaches used by PRAIRIE™, AEAT-0843', AEA Technology, Harwell, 1990.

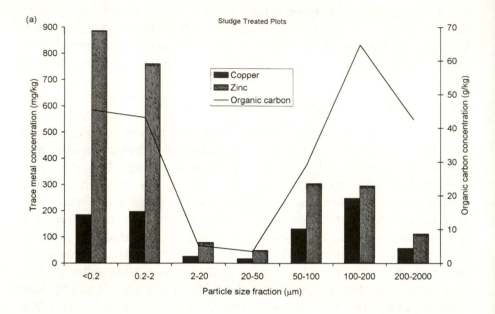

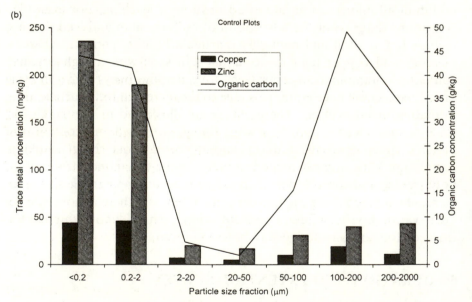

Figure 12   *The distribution of copper and zinc concentrations in (a) sludge-treated soil*
plots and (b) control soil plots
(After J. Ducaroir and I. Lamy, *Analyst*, 1995, **120**, 741–745)

area to tens of years for long term surveillance projects such as the US EPA Ambient Air Monitoring Program. The choice of the frequency of sampling, *i.e.* the duration of each sample period and the interval between successive measurements, is also dependent upon the objectives of the study. Pollutant concentrations in air and water fluctuate with varying degrees of rapidity and in order to characterize their behaviour it is necessary to measure these changing levels; long term mean data may be sufficient for some purposes but will not be adequate where information of short term high level episodes is required. Generally, if random sampling techniques are used, the number of samples required will increase as the standard geometric deviation of the pollutant concentrations increases, *i.e.* the greater the fluctuation of the pollutant level, the more numerous the samples that must be taken to assess the variations accurately. If the variations in levels during the period of interest are essentially random, independent, and normally distributed then the number of samples which must be taken in order to estimate the period mean within certain limits and prescribed confidence limits may be calculated. However, in order to do this a reasonable estimate of the standard deviation of the data is required: it is assumed that the random variations follow a normal distribution and that the results of successive samples are not serially correlated but are independent. These criteria are rarely met.

As a more general guideline it may be assumed to be necessary to have a sampling interval at least ten times shorter than the fluctuation cycle time. For example, the entire variance structure of a diurnally fluctuating pollutant concentration profile may be obtained from roughly 12 samples, each with a two-hour averaging period. If the annual trend of levels is required then probably 12 monthly samples each year would be adequate. Thus, although a continuous and instantaneous record of pollutant level may be required for some purposes (*e.g.* to monitor very short-term changes in air quality due to the oscillating passage of a plume over the sampling station), it is not always necessary. Further, if a short sampling period is chosen it rapidly becomes necessary to record and store efficiently the large volumes of data generated, which will themselves often be averaged over a longer period for analysis. A case in point is the UK automatic air pollution monitoring network where one minute average readings are themselves averaged to give one-hourly values.

Fast-response continuous monitors are now available for the more common gaseous pollutants and for many determinations of water quality. The term 'fast-response' implies a response time, measured as a 90% rise, of less than about 2 min, but in most cases it is of the order of seconds. Thus very rapid temporal variations are measurable and with

Table 5 *Summary of commonly employed methods for measurement of gaseous air pollutants*

Pollutant	Measurement technique	Sample collection period	Response time[a] (continuous techniques)	Minimum concentrations
Total hydrocarbons	NDIR		5 s	1 ppm (as hexane)
	Flame ionization analyser		0.5 s	10 ppb (as methane)
Specific hydrocarbons	GC-FID	[b]		1 ppb
Carbon monoxide	NDIR		5 s	0.5 ppm
	Catalytic methanation/FID	[c]		10 ppb
Sulfur dioxide	Absorption in hydrogen peroxide/titration	24 h		2 ppb
	Fluorescence analyser		2 min	0.5 ppb
Oxides of nitrogen	Conversion to nitrite/azo dye formation	days		5 ppb
	Chemiluminescent reaction with ozone		1 s	0.5 ppb
Ozone	UV absorption		10 sec	1 ppb
Peroxyacetyl nitrate	GC/electron capture detection	[c]		1 ppb

[a] Time taken for a 90 per cent response to an instantaneous concentration change.
[b] Grab samples of air collected in an inert container and concentrated prior to analysis.
[c] Instantaneous concentrations measured on a cyclic basis by flushing the contents of a sample loop into the instrument.

the advent of on-line microprocessor data handling and reduction systems the large amounts of data produced are more easily handled. Fast-response continuous monitors do not generally have a predictable response so they cannot be calibrated solely in terms of chemical stoichiometry and hence need calibration with a standard atmosphere or solution. Some of the commonly employed fast-response methods of gaseous air pollutant analysis are shown in Table 5, and some of the methods used for water analysis are shown in Table 6.

One further consideration when deciding on sampling frequency is that if the measurements are being made in order to assess whether a given environmental quality standard is being satisfied, then the data resolution must be sufficient for that purpose. As an example the US Ambient Air Quality Standard for sulfur dioxide specifies that 3 hourly and 24 hourly values are required. It would be pointless to collect SO_2 data with a continuous fluorescent analyser with a response time of two

Table 6 *Summary of some methods of analysis of water*

Pollutant or determinand	Measurement technique	Response time
pH	Electrometric	10 s
Biochemical oxygen demand	Dilution/incubation	5 days
Chemical oxygen demand	Dichromate oxidation	2 h
Metals	AAS	–
Organometallics	GC–AAS	–
Nitrate	Colourimetric	5 min
Nitrate	UV spectrophotometric	1 min
Formaldehyde	Photometric	6 min
Phenols	GC	30 min

minutes, if the only purpose of the exercise is to see whether this standard is being met.

5.2 Methods of Reducing Sampling Frequency

Once the desired sampling frequency has been selected it may be found to be impracticable with the resources available and some means of reducing the number of samples will then be required. This may be done by:

(a) reducing the number of sampling locations,
(b) reducing the sampling frequency, or
(c) reducing the number of determinands.

It has already been shown above that there are several methods available for the rationalization of an existing monitoring network where the quality at one location is correlated sufficiently well with that at another location, and these methods of analysis may be applied to all types of monitoring networks. Similarly, statistical analysis of past data may show that one determinand is sufficiently well correlated with one or more others, such that it may be used as an indicator of quality.

When sampling water, soils, sediment, and flora and fauna the use of composite samples may be of value in reducing sample numbers. Composite samples are formed by mixing together individual samples to give an indication of the average quality over an area or during a sampling period. They may also be formed by continuous or intermittent collection of samples into one container over a given period, as, for example, the collection of atmospheric particulate material on a filter. Individual samples collected at different locations may be mixed together in proportion to the volumes of the sampled bodies, as in the case of a non-homogeneous water body, and so give a better indication of average quality.

In the case of atmospheric pollutants it is often desirable to estimate the likely daily maximum as well as the daily mean concentration, and several methods have been proposed that allow this to be done on the basis of a few discrete samples of short duration. For example, the same statistical concentration frequency distribution of SO_2 levels as is provided by continuously recorded data can be obtained from a limited number of randomly collected short term samples.

5.3   Number of Sampling Sites

The choice of the number of sampling sites to be used in a particular survey is very dependent, as are so many other design parameters, on the objectives of the study. In the simplest case of source-orientated monitoring of atmospheric emissions from a single stack or site then 4–6 sampling locations might be considered sufficient. This may be thought to be too few but operational constraints may prevent this number being increased. In the case of the UK National Survey of Air Pollution it was previously thought necessary to obtain daily data from about 1200 sites, although this number has now been greatly reduced.

When too few sampling sites are used in a source-orientated monitoring programme it is possible that atmospheric (or aqueous) emissions may pass between them without being detected at all. The probability of a fixed number of sample stations detecting a release is a function of the quantity released, the number of samplers, the distance of the samplers from the source, the plume dimensions, and the height and duration of the release. This type of analysis has been applied to the environmental monitoring of atmospheric releases.

A recent innovation is the use of hand-held Global Positioning Systems for accurately determining the longitude and latitude of a sampling point on the Earth's surface. This is particularly useful when it is necessary to return to the same point for further sampling.

6   PREREQUISITES FOR MONITORING

Before monitoring begins, certain information, techniques, and methodologies must be available in order for the survey to be successfully carried out. As has already been stressed, a prime requirement is that the objectives of the study should be defined, but the following also need consideration:

— definition of a monitoring protocol;
— availability of meteorological or hydrological data;
— availability of emission data;

— likely pollutant concentrations;
— availability of suitable monitoring equipment;
— availability of sensitive and specific analytical techniques;
— definition of suitable environmental quality standards.

6.1 Monitoring Protocol

Monitoring is a complex task and carefully planned and documented procedures are necessary to ensure that reliable and comparable results are obtained.[32] The documented procedure by which this is achieved is known as a Monitoring Protocol. The main components of a Monitoring Protocol are as follows:

(a) A reference measurement method—standard methods may not always be available so some reference methods may actually be non-standard.
(b) A standard methodology which sets out for each process:
   (i)   frequency of sampling;
   (ii)   duration of sampling and the number of samples to be taken;
   (iii)   pollutants to be measured;
   (iv)   accuracy required (to some extent this is dictated by the pollutant and the sampling method used);
   (v)   variability of the pollutants under investigation;
   (vi)   analytical methods;
   (vii)   analytical resolution;
   (viii) general principles of sampling (*e.g.* position, isokinetic);
   (ix)   plant operating conditions (for source monitoring);
   (x)   health and safety considerations of sampling staff (and plant operators for source monitoring);
   (xi)   procedures to be adopted when meaningful measurements by standard procedures are not possible.
(c) Quality control procedures which define requirements for:
   (i)   calibration of measurement devices;
   (ii)   maintenance of instruments;
   (iii)   sample storage and transport to ensure that the sample is identifiable throughout the sampling, sample preparation and analysis, and to ensure the sample integrity is maintained;
   (iv)   data handling and reporting.

---

[32] HM Inspectorate of Pollution, 'Technical Guidance Note (Monitoring) M2: Monitoring Emissions at Source', HMSO, London, 1993.

(d) A quality assurance programme to ensure that:
  (i)   measurements are made in accordance with a standard methodology;
  (ii) the correct quality control procedures are in place;
  (iii) the quality control procedures are being adhered to;
  (iv) sample identification and routing procedure is well documented;
  (v)   the correct reporting procedures are used.

It is most important to have a documented Monitoring Protocol when source monitoring is being carried out for comparison against emission limits or Environmental Quality Standards. Regulatory Authorities may wish to agree such a protocol prior to monitoring being carried out. Since each monitoring programme is different both in the way in which it is operated and in the pollutants which are monitored, separate Protocols will be needed for each programme. Measurement teams, operators, and contractors should be instructed in the use of the Protocols.

6.2   Meteorological Data

When carrying out air pollution measurements it is desirable, and often essential, to have access to meteorological data. Care must be taken to ensure that the data used are meaningful and representative of the area of study, wind data in particular being very susceptible to local interference. Light, robust anemometers (*e.g.* hog wire) and wind vanes are now generally available and when mounted on a dismountable 10 m mast may be used in the field, thus obviating the need to rely on data obtained from another, possibly less representative, source. Less easily obtained parameters that may be required are the lapse rate and the height of any atmospheric temperature inversions. These are rather difficult to measure, requiring accurate measurements of temperature at increasing heights or acoustic radar observations, but such data are usually obtainable from large meteorological stations.

Measurements of low-level atmospheric turbulence are made using a bivane and anemometer on a 10 m mast. This instrument measures the horizontal and vertical fluctuations of the wind; siting of the mast is obviously critical and a data logging system is required to cope with the large amounts of data generated.

Meteorological data may be required for several reasons. For instance, when fast response measurements are unavailable it may be desirable to construct time-weighted pollution roses, which show how pollutant concentrations vary with surface wind directions. For this the hourly wind direction at 10 m height is determined, the record is divided into sixteen 22.5° sectors, and the duration in each sector is tabulated. The rose may then be calculated using:

$$(\text{TWMC})_n = \frac{\sum_{i=1}^{m}(t_{i,n}c_i)}{\sum_{i=1}^{m}t_{i,n}}$$

where $(\text{TWMC})_n$ is the time-weighted mean concentration of the pollutant for the nth sector, $t_{i,n}$ is the number of hours for which the wind was in sector n during the ith sampling period, c_i is the concentration of the pollutant for the ith period, m is the number of sampling periods and n takes values from 1 to 16 (sector $n = 1$ being 0–22.5°, sector $n = 2$ being 22.5–45° *etc.*).

Alternatively the trajectory of a parcel of air over synoptic-scale distances may be required for source-apportionment or dispersion studies. For this the surface pressure field over a very large area is needed and the geostrophic wind vector estimated from the isobar spacing and alignment.[33] The analogous hydrographic data may be required for dispersion studies in the sea, and flow data for rivers and lakes may be needed although these may not be so readily available as meteorological data.

6.3 Source Inventory

One, often very cost-effective, method of identifying the likely occurrence of pollutants prior to actual monitoring is by collecting and collating detailed information on the pollution emissions in a given area or to a particular river. Such an emission or source inventory should contain as much information as possible on the types of source as well as the composition of emissions and the rates of discharge of individual pollutants. Supplementary information describing the raw materials, processes, and control techniques used should also be collected. Detailed discussion on the test procedures to adopt in compiling an emission inventory is available,[34] and the application of the completed inventory were identified by these authors to be:

(a) guiding emission–reduction efforts;
(b) helping to locate monitoring stations and alerting networks;

[33] R. I. Sykes and X. X. Hatton, *Atmos. Environ.*, 1976, **10**, 925–934.
[34] A. T. Rossano and T. A. Rolander, 'Manual on Urban Air Quality Management', World Health Organization, Copenhagen, 1976.

(c) indicating the seasonal and geographic distribution of the pollu-
    tion burden;
(d) assisting in the development of implementation strategies;
(e) pointing out the priority of air or water quality problems;
(f) aiding regional planning and zoning;
(g) air and water quality diffusion modelling;
(h) predicting future air and water quality trends;
(i) determining cost-benefit ratios for air and water pollution control;
(j) community education and information programmes.

Although a very useful tool, the emission inventory is no substitute for
actual measurements of pollutant levels and should be considered as a
complementary technique to monitoring, not as an alternative.

6.4   Suitability of Analytical Techniques

As was shown in Table 2, pollutants may be found in environmental
media over very wide concentration ranges. Often there are several
procedures available by which a pollutant may be analysed, but they
may have widely differing sensitivity and specificity. As an example,
sulfur dioxide in air may be determined by absorption in hydrogen
peroxide and determination of the resultant acid by acid–base titration or by conductivity measurement. Although acidic or basic
compounds interfere, these techniques can yield useful results in
urban areas where sulfur dioxide concentrations are high and the
levels of interfering compounds are low. In rural areas this is not the
case due to natural production of ammonia, and misleading results
are obtained.

  Very fast-response analysis of sulfur dioxide is possible using the flame
photometric sulfur analyser in which gaseous sulfur compounds are
burned in a reducing hydrogen–air flame and the emission of the S_2
species at 394 nm is measured. The method is very sensitive but unless
used with gas chromatographic separation is not specific to sulfur
dioxide, but rather gives a measure of the total volatile sulfur content.
Alternatively SO_2 may be analysed with a fast response time by excita-
tion at 214 nm and the resultant fluorescence measured. This is now the
standard and recommended method for measuring SO_2 concentrations
in air.

  Some of the considerations that will affect the choice of an analytical
method may thus be summarized:

— sensitivity (depends on detection limit and pollutant levels);
— specificity (to allow unequivocal determinations);

Table 7 *Instrumental analytical methods*

Method	Sample[a]	Specificity	Detection limit[b]
Gravimetric	SLG	good	$>1\ \mu g$
Titrimetric	SLG	good	$>10^{-7}$ M in solution
Visible spectroscopy	SL	fair	>0.005 ppm in solution
Ultraviolet spectroscopy	SLG	fair	>0.005 ppm in solution
Flame emission spectroscopy	SL	good	>0.001 ppm in solution
Atomic absorption spectroscopy	SL	excellent	>0.001 ppm in solution
Gas chromatography	GL	excellent	>10 ppm
Liquid chromatography	SL	good	>0.001 ppm
Polarography	L	good	>0.1 ppm
Anodic stripping voltammetry	L	good	>0.001 ppm
Spectrofluorimetry	SL	good	>0.001 ppm
Emission spectroscopy	SL	excellent	>0.1 ppm
X-ray fluorescence	SL	good	>10 ppm
Neutron activation	SL	excellent	>0.001 ppm
Mass spectrometry	SLG	good	>0.003 ppm

[a] S = solid, L = liquid, G = gas.
[b] Approximate only, depending upon the particular element being analysed.

— response time;
— response range (particularly linearity of continuous monitors);
— ease of operation;
— ease of calibration;
— cost and reliability;
— precision and accuracy.

Some of the more commonly used instrumental methods are shown in Table 7 and described comprehensively elsewhere.[35]

Figure 13 highlights the difference between precision and accuracy for an analytical technique. Precision may be defined as the reproducibility of analyses, whereas accuracy is a true measure of the determinant present. In order to assess the precision of an analytical method it is necessary to analyse separate representative sub-samples of the sample under investigation. These sub-samples should be included at intervals during the analysis. In this way the drift in the precision of a technique may also be detected. It is not possible to know how accurate a series of determinations have been, even if they are precise. However, a best estimate of the accuracy of an analytical technique may be found from the regular analysis of national and international standard reference materials. These have undergone inter-laboratory analyses, using different techniques and calibration standards, for a variety of species.

[35] 'Instrumental Analysis of Pollutants', ed. C. N. Hewitt, Elsevier, London, 1991.

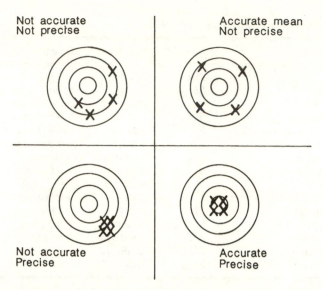

Not accurate
Not precise

Accurate mean
Not precise

Not accurate
Precise

Accurate
Precise

Figure 13   *Schematic illustration of precision and accuracy*

6.5   Environmental Quality Standards

Of the six possible reasons for carrying out a monitoring programme outlined in Section 1, four rely upon the prior formulation of a standard of environmental quality. Only in the case of source apportionment and pollutant interaction and pathway studies or when monitoring is carried out with the intention of obtaining a historical record of environmental quality is this not a prior requirement. There is little point in monitoring in order to pinpoint pollutant health effects or to assess the need for legislative controls on emissions, for example, unless a certain pollutant level has been defined as being undesirable or likely to cause damage. Environmental quality standards have been devised and adopted for many atmospheric and water pollutants, some of which are shown in Tables 8 and 9.

Having adopted a standard for environmental quality there may be great difficulty in ensuring compliance. In the case of, for example, lead in drinking water, various reduction strategies are possible, culminating in the wholesale removal of lead pipes (although this may not entirely solve the problem when lead-based solder is used with copper pipe) and in the case of primary air pollutants similar 'simple' remedies are possible. The difficulties arise in the case of secondary pollutants (*i.e.* those formed within the atmosphere itself) or for pollutants with both primary and secondary origins. In the case of atmospheric suspended particles (smoke) both primary and secondary sources are important. Primary emissions have in recent years been greatly reduced by the use of

Table 8 *US ambient air quality standards*

Pollutant	Measurement period	Standard
PM_{10}	24 h average	$150~\mu g~m^{-3}$
	Annual arithmetic mean	$50~\mu g~m^{-3}$
$PM_{2.5}$	24 h average	$65~\mu g~m^{-3}$
	Annual arithmetic mean	$15~\mu g~m^{-3}$
Sulfur dioxide	Annual arithmetic mean	$80~\mu g~m^{-3}$
	24 h average	$365~\mu g~m^{-3}$
	3 h average	$1300~\mu g~m^{-3}$
Carbon monoxide	8 h average	$10~mg~m^{-3}$
	1 h average	$40~mg~m^{-3}$
Ozone	8 h average	$157~\mu g~m^{-3}$
Nitrogen dioxide	Annual arithmetic mean	$100~\mu g~m^{-3}$
Lead	Quarterly average	$1.5~\mu g~m^{-3}$

Table 9 *EC water quality standards for inland waters (e.g. rivers, lakes)*

Pollutant	Country	Measurement period	Concentration
Aldrin, dieldrin, endrin, and isodrin	EC	Annual mean	0.03 µg/litre (total) 0.005 µg/litre endrin
Cadmium and its compounds	EC	Annual mean	5 µg/litre (total)
Carbon tetrachloride	EC	Annual mean	12 µg/litre
Chloroform	EC	Annual mean	12 µg/litre
DDT (all isomers)	EC	Annual mean	0.025 µg/litre
p,p'-DDT	EC	Annual mean	0.01 µg/litre
Hexachlorobenzene	EC	Annual mean	0.03 µg/litre
Hexachlorobutadiene	EC	Annual mean	0.1 µg/litre
Hexachlorocyclohexane (all isomers)	EC	Annual mean	0.1 µg/litre
Mercury and its compounds	EC	Annual mean	1 µg/litre (total)
Pentachlorophenol and its compounds	EC	Annual mean	2 µg/litre

efficient control techniques on industrial sources and substantial change in domestic fuel usage from coal towards cleaner fuels. However secondary particles, produced in the atmosphere by formation of the ammonium salts of strong acids from industrial emissions of SO_2, NO_x and HCl, together comprise a substantial proportion of the atmospheric aerosol. Thus reduction of primary emissions of particles does not necessarily ensure a reduction in atmospheric concentrations or compliance with a standard. One secondary air pollutant that is likely to prove difficult to control adequately in the next decade or so is ozone. This is formed in the lower atmosphere by reactions involving several primary pollutants, and our present understanding of the chemistry of the atmosphere is probably insufficient to accurately predict the effect of control strategies.

7 REMOTE SENSING OF POLLUTANT

Many highly sophisticated techniques are now available for the remote sensing of atmospheric and water pollutants. However, their use is almost exclusively restricted to specialized monitoring exercises due to the very considerable capital cost of the instrumentation. Probably the cheapest and most widely used methods are those of aerial photography, including infrared sensing and optical correlation spectrometry. Uses of aerial photography include the monitoring of liquid effluent dispersion using dye tracers and conventional colour film. Infrared photography has been used for monitoring the condition of crops and forests. Airborne heat-sensing infrared linescanning equipment has been routinely used to monitor thermal plumes in waters receiving industrial effluents and also for the detection and mapping of oil spills at sea using thermal infrared data from satellites.

The most common use of the correlation spectrometer in air pollution analysis is for the determination of sulfur dioxide and nitrogen dioxide concentrations in plumes from tall stacks, and this provides a good technique for studying the transport and dispersion of a plume. The instrumentation may be ground based in a mobile laboratory or airborne, in which case the plume is viewed from above.

The use of tunable lasers allows long-path absorption measurements of a range of gaseous pollutants such as SO, NO_2, SO_2, CO and O_3 and minor reactive species such as $OH^{\bullet}$. Reliable measurements of this latter species are of great importance because of its dominant role in the chemistry of the troposphere. In long-path laser adsorption methods a detector is used to monitor adsorption of specific wavelengths in the light path. In lidar techniques, however, the back-scattered radiation from a laser is monitored. By using a pulsed system the time taken for receipt of back scatter can be related to the distance of travel, allowing spatial resolution of pollutant concentration data within the light path. By monitoring back-scatter intensity at two close wavelengths, one strongly absorbed by the species of interest and one unabsorbed, the species' concentration may be inferred as well as its spatial distribution. Care is required to avoid spectral interferences but this method has been successfully used for measurement of sulfur dioxide up to a range of *ca.*

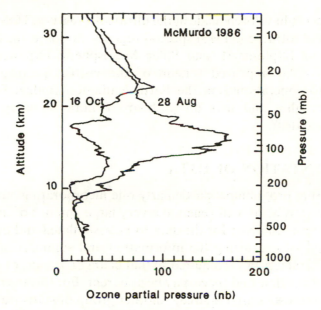

Figure 14   *Ozone profiles from balloon soundings over McMurdo (South Pole) during 1986*
(After Hoffman *et al.*, *Nature*, 1987, **326**, 59–62) (reproduced with permission from 'Stratospheric Ozone', Department of the Environment, HMSO, London, 1987)

2 km. Further significant developments of laser methods using the Raman back scatter, which is highly characteristic of the scattering molecule, are likely.

Case study 7: Measurement of stratospheric ozone.[36]   There are several techniques by which stratospheric ozone may be measured. From the ground, the total ozone in a column extending vertically into the atmosphere can be measured passively with a Dobson spectrophotometer. This spectrophotometer is also used to estimate the vertical distribution of ozone, by observing the zenith clear blue sky, while the sun traverses a range of solar zenith angles. Scattering from different altitudes in the atmosphere allows the height profile of ozone to be deduced. Greatly improved vertical resolution of ozone concentrations has been achieved using the differential absorption laser (DIAL) technique. Ozone profiles have been measured from spectrophotometers carried aloft by balloon or rocket. The balloon-sondes provided evidence of a reduction in stratospheric ozone over Antarctica during October 1986 (Figure 14).

A major achievement in the remote sensing of ozone has been the verified measurements of ozone profiles obtained from satellites.

[36] UK Stratospheric Ozone Review Group, 'Stratospheric Ozone', HMSO, London, 1987.

Measurements have been made in the ultraviolet (SBUV/TOMS), visible (SAGE) and infrared (LIMS) spectral regions on board the Nimbus 7, Applications Explorer II, and Solar Mesospheric Explorer satellites respectively. An improved version of the visible spectrophotometer (SAGE-2) is operational on the Earth Radiation Budget Experiment (ERBE) satellite and uses an on-board calibration lamp to check instrumental drift.

8 PRESENTATION OF DATA

A monitoring programme, particularly one incorporating automatic or fast-response systems, can generate a very large amount of information in a short time. In order for the data to be assimilated and understood, some means of organizing the information and summarizing its most essential characteristics is required so that changes, trends, or patterns in behaviour over time and space may be apparent. For this the methods of descriptive statistics are required. Another group of statistical methods, those of inferential statistics, are used when information of the relationships and processes operating between measurements is required. Details of these methods are available from standard texts and from books devoted to the environmental sciences (*e.g.* References 37–38).

Many statistical tests depend upon having data that are normally distributed, but often environmental analytical data do not satisfy this criterion. In a normal distribution the arithmetic mean and median are the same, but in log normally distributed data the *geometric* mean and median are the same. This is the situation that applies to many environmental data sets and comes about from a few results having high values whilst the majority of results are closely grouped together. If such data are treated as belonging to a normal distribution too much weight will be applied to the outlying values and the wrong deductions may be made. In this case some method of transforming the data is required before statistical analysis is carried out. For example, it might be appropriate to use the logarithms of the data, or the square or cube roots. Similarly it is often better to quote the 95 percentile range of values, excluding the extreme 5%, again in order to avoid giving prominence to a few outliers.

Monitoring data may be incorporated in Geographical Information Systems (GIS) which are specialized database management systems for handling geographical data—*i.e.* data whose key characteristic is 'place'. These manage data on land use, industry, roads, habitations, and so on.

[37] C. Chatfield, 'Statistics for Technology', Chapman & Hall, London, 3rd Edn., 1983.
[38] R. M. Haynes, 'Environmental Science Methods', ed. R. Haynes, Chapman & Hall, London, 1982.

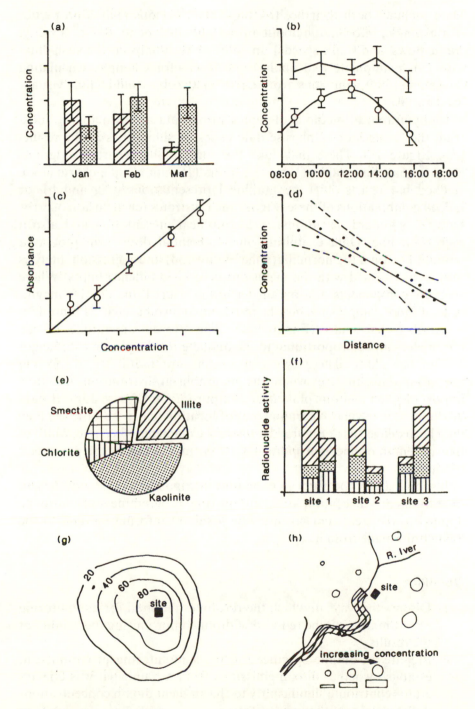

Figure 15 *Examples of different approaches to presenting data*

Data includes both their positions and their various related properties (for example, information about a road may include its traffic capacity, traffic flows, and traffic speeds). In addition to simply storing such data, GIS can manipulate and analyse the data—for example, calculating contours of NO_x emissions around road networks—and include visualization tools.

Many graphical methods of representing data are available to illustrate the changes or emphasise differences or similarities in the results (*e.g.* Figure 15). These may take the form of bar charts/histograms (Figure 15a), line graphs (15b), X Y plots (15c and d), pie charts (15e), stacked bar charts (15f), geographical presentations (15g and h), or indeed combinations of these. Figure 15c illustrates the inclusion of error bars (or standard deviations) in a graphical presentation and also a regression line. Data exhibiting an exponential decay will plot as a straight line on a log normal graph (Figure 15d) and regression analysis may be performed with the inclusion of 95% confidence limits. Where separate components of a parameter are measured, the relative magnitude of those components may be represented on a pie chart (Figure 15c) or stacked bar chart (Figure 15f). The overall magnitude of the parameter can be proportional to the diameter of the pie chart or height of the bar chart. These types of graphs are useful for displaying speciation data. In cases where the geographical distribution of data is important then contour plots may be appropriate (Figure 15g). If data are limited or specific to certain types of location (*e.g.* road or river) they may be displayed as in Figure 15h where the diameter of a circle, width of bar, or length of line is proportional to the magnitude of the parameter of interest.

It should always be borne in mind that the rigour applied to the design, operation, and execution of a monitoring programme must also be applied to the treatment given to the resultant information and to the deductions made from it.

Questions

1. Discuss the ways in which the resources required for a monitoring programme could be reduced without compromising the validity of the results.
2. Prepare an outline specification for a monitoring programme to establish whether dioxin emissions from a municipal waste incinerator contribute significantly to the ambient dioxin concentrations in the vicinity of the incinerator.
3. Compare the advantages and disadvantages of remote sensing with conventional *in-situ* environmental monitoring.

4. Describe how the results of regulatory compliance monitoring for emissions from a sulfuric acid plant should be presented to the Environment Agency and the general public. Reasons for the differences in the approach should be highlighted and discussed.
5. Describe, with examples, the differences between planned, fugitive, and accidental emissions of pollutants. Using a hypothetical industrial process as an example, outline how a monitoring network could be used to detect all three types of emissions.
6. Describe how mathematical modelling techniques can be used to determine the optimum siting of a limited number of monitoring devices in the vicinity of an industrial plant emitting pollutants to the atmosphere.
7. A cement kiln company has been asked by the US EPA to provide reassurance that the short term emissions of sulfur dioxide during kiln start up do not breach the 3 hour air quality standard for sulfur dioxide ($1300 \ \mu g \ m^{-3}$). As the company Environmental Manager you decide to estimate the likely concentration of sulfur dioxide at the point where the plume grounds.

Given that the height of the stack is 90 m, the stack gas velocity is $17 \ m \ s^{-1}$, the internal stack diameter is 2.5 m, the mean windspeed at a height of 10 m above ground-level (u_{10}) is $1 \ m \ s^{-1}$ and the stack gas temperature is 90 °C calculate the effective stack height.

The concentration of sulfur dioxide in the stack gas is 260 mg m^{-3} during kiln start-up. The plume grounds most quickly in Pasquill stability category A at a distance of 650 m from the base of the stack and the concentration is greatest at this point on the plume centre line. Given that the standard deviation of the plume concentration in the vertical direction (σ_z) is 97 m and the standard deviation of the plume concentration in the vertical direction (σ_y) is 240 m (both at a distance of 650 m from the release point, for a 3 h release and in Pasquill stability category A), assess whether it is likely for the 3 hour air quality standard to be breached.

Discuss what action you should take to address the concerns of the US EPA.

CHAPTER 8

Ecological and Health Effects of Chemical Pollution

S. SMITH

1 INTRODUCTION

This chapter is concerned with the hazards of chemical pollution to human health and ecological systems, and an evaluation of these requires knowledge of the effects of pollutants, the levels which cause such effects, and the likely occurrence of hazardous levels in the environment. Clearly in tackling or preventing pollution hazards the aim is to ensure that environmental levels are below those at which pollutants have known effects.

There are, on the one hand, substances which we recognize as pollutants and, on the other, organisms on which pollutants exert their effect. A basic maxim of toxicology that can be extended to pollution is that all substances administered at sufficiently high doses are harmful to biota. Conceivably, biota may be able to cope with a small amount of a very toxic substance, whereas an otherwise essential one may reach overwhelming proportions and adversely affect many organisms.

The impact of pollution on biota depends on the amount entering the environment and its fate in the environment. In the process of dispersion in air, water, or soil, pollutants may undergo transformations to more innocuous forms or to ones that are more hazardous. In the latter case the impact is due to secondary forms of pollution. Dispersion in the environment leads to dilution of pollution; however, certain environmental processes can concentrate pollutants and where this occurs the impact may be greater. On the other hand they may become bound up or immobilized in some way and in this form they are less available to biota and may be considered less hazardous. The fate of pollutants in the environment is covered in some detail in the preceding chapters.

Pollutants encompass a broad range of chemical and physical properties which strongly influence both their fate in the environment and their effects on biological systems. Particularly during this century vast quantities of many different chemical substances have been released into the environment; the majority of these substances are waste products generated by industry and society consuming manufactured goods. Synthetic fabrics and fibres, pharmaceuticals, fertilizers, pesticides, paints, and building materials, as well as chemicals for industrial processes, are just some of the products of the chemical industry that are integral to almost every aspect of modern living. Many such substances are natural constituents of the environment, others are synthetic chemicals. Inevitably wastes generated during the manufacturing process, the substances themselves, and perhaps their degradation products, are released into the environment. Gases (*e.g.* sulfur dioxide, carbon dioxide, and oxides of nitrogen) and particulates from combustion processes are discharged into the atmosphere. Liquids and solids containing inorganic and organic substances are discharged into water and hazardous chemical wastes are buried on land, incinerated, or dumped at sea. Fires, explosions, and tanker accidents can result in sudden unintentional but devastating pulses of pollutants into the environment, and pesticides are intentionally released at certain times of the growing season. It has to be said that there are now a number of controls regulating such discharges and that these have arisen from actual and perceived chemical pollution problems in the past.

Pollutants exert their effect on individuals, but recognition of an effect is only seen when large numbers and even whole communities are affected. The rate of supply or the amount of pollutant reaching a receptor (organism, population, or community) is referred to as the *exposure* and an *effect* is a biological change caused by an exposure. Directly toxic substances must gain access to the exterior membranes or interior components of tissues and cells in order to exert their effect. The amount that is taken into an organism is referred to as the *dose*; essentially it is a function of concentration and the period of exposure, although dose is perhaps more correctly defined as the amount of substance received at the site of effect. Dose can be a difficult parameter to measure for pollutants, and exposure has become the standard means of assessing how much is received by a receptor. Therefore, a fundamental goal of pollution studies is to establish the relationship between exposure and effect, so that a level at which no effect occurs can be identified and this can be set as an objective to work towards in controlling pollution.

A pollutant may be supplied in one large pulse (acute exposure) or an equivalent amount may be supplied at a lower concentration over an extended period (chronic exposure). Similarly, the effects caused by a pollutant are categorized as being acute or chronic. Acute effects are observed immediately following an exposure; they are rarely reversible and are very often fatal. Chronic effects or damage follow a period of prolonged exposure which results in biochemical or physiological disturbances. Outwardly they may be manifested as visible or clinical symptoms of damage, *e.g.* chlorotic regions on plant leaves, and inability to maintain homeostatic balance, co-ordinate activities or breed. These symptoms are often quite diffuse and not specific to a particular pollutant. Following cessation of exposure, chronic damage is frequently reversible, although continued exposure may prove fatal. However, it may be argued that any severe disturbance will impair the efficiency of an individual and so shorten its life.

As well as the direct impact of pollutants on living species, other secondary effects require careful consideration. These relate to community and habitat changes that may follow initial pollution damage. Obvious examples include disturbance of predator–prey relationships following a dramatic decline of one or more species in a food web and reduced turnover in biogeochemical cycling of nutrient elements if, say, decomposer organisms are affected. These secondary impacts are a feature of the ecosystem level of organization.

Pollution studies are very wide ranging and so extend over a multitude of situations involving interactions of pollutants with living organisms. Damage to certain groups of organisms arouses more concern than others. Clearly humans and their resource species—livestock, crops, and fisheries—for obvious social and economic reasons are the two groups which engender most concern. Surveying the range of effects of pollutants, it is evident that these also can be arranged on a scale of increasing concern, with cellular disruption arousing less concern than an acutely toxic effect.[1]

2 TOXICITY: EXPOSURE–RESPONSE RELATIONSHIPS

The toxicity of a substance or industrial discharge is estimated by determining the concentration or dose which is lethal to a particular organism, and the most commonly used yardstick is the LC_{50} or LD_{50}, the concentration or dose which results in 50% mortality of test organisms.

Because chemicals discharged into the environment largely end up in aquatic systems, toxicity tests are commonly based on aquatic organisms. There is an extensive database on the acute toxicities of a variety of

[1] M. W. Holdgate, 'A Perspective of Environmental Pollution', Cambridge University Press, Cambridge, 1979.

chemicals, although by no means all potential contaminants. The organisms selected as test organisms are those that are easy to obtain and maintain in the laboratory. In the UK toxicity tests are commonly undertaken with the water flea (*Daphnia* sp.) and trout; in the US the fathead minnow is the fish most frequently used.

Mortality is only one of many end-points of toxicity; others relate to effects on the general wellbeing and survival of an organism, *e.g.* growth, reproduction, and feeding rate. The parameter used in these cases is the effective concentration or EC_{50}, the concentration that results in 50% reduction in growth or some physiological function. At the biochemical level the emphasis is on determining the mode of toxic action. Chronic toxicity tests are based on the same principle. Test organisms are subjected to a range of concentrations of pollutants over extended periods and their effects on growth and reproduction are determined. Toxicity tests are examples of dose or exposure–response relationships, in that the proportion of test organisms showing the damaging effect is measured. These relationships are fundamental to all forms of pollution; they provide a framework for estimating exposures associated with the onset of effects and in turn those that have no effect. The no-effect exposure is the one which is striven for in the environment. To get to this stage a large number of tests on a variety of organisms are required to ascertain the most sensitive species and the most sensitive end-point.

Many thousands of chemicals have been and currently are discharged into the environment, and it is difficult to provide full toxicological data on all of them. Many belong to chemical groups or homologous series which have related physico-chemical properties and it is possible to use these to predict their toxicological characteristics and fate in the environment. Examples of homologous series are the chlorophenols and chlorobenzenes. These are based on phenol and benzene respectively and in each case chlorine can be added to form mono- or polychlorinated compounds. The end member of the chlorophenols is pentachlorophenol and that of the chlorobenzenes is hexachlorobenzene. Pentachlorophenol has been used as a wood preservative and hexachlorobenzene has pesticidal properties. Toxicity increases progressively through each of these series such that phenol and benzene are the least toxic, while the most toxic ones are the end members of each series, pentachlorophenol and hexachlorobenzene.

This progression can be related to the accumulative capacity of the compounds and to one particular parameter, the octanol–water partition coefficient (K_{OW}). This parameter is a measurement of the tendency of organic compounds to partition between water and an organic solvent. Taking equal volumes of water and octanol, a water soluble compound will reside mainly in the water, whilst a non-polar organic compound

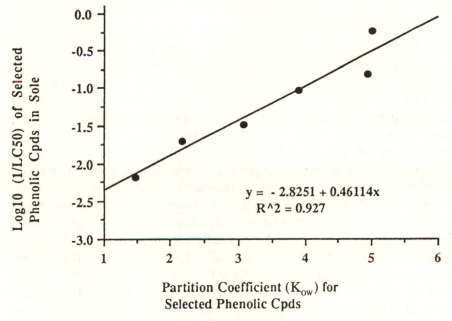

Figure 1   *Relationships between octanol–water partition coefficient K_{OW} and toxicity of phenol and five chlorophenols to sole,* Solea solea
(Reproduced with permission from Furay, PhD, University of London, 1994)

with low water solubility will be mainly in the octanol. The ratio of the amount of a compound in octanol to that in water is the octanol–water partition coefficient (K_{OW}) and clearly hydrophobic substances will have large values for these coefficients. This physico-chemical system is analogous to the situation where a pollutant in water is exposed to an organism, the lipid in the cell membranes being equivalent to octanol. Therefore substances with a high K_{OW} (hydrophobic and low water solubility) will partition from the water to the lipid component and enter the cells of the organism and such substances will bioconcentrate to a significant extent.

Figure 1 shows the relationship between the toxicity and the K_{OW} of chlorophenols, essentially the most toxic members are the ones with the highest K_{OW}. These relationships are known as 'Quantitative Structure Activity Relationships' or QSAR. They are useful for predicting the toxicity of chemicals with related properties. They can also give some insight into different modes of toxic action. The gradient of a similar relationship with chlorobenzenes tends towards unity and it is typical of a large number of organic hydrophobic chemicals that have a general narcotic effect. Chlorophenols on the other hand form a different gradient and these chemicals are known to be potent electron uncouplers in aerobic respiration.

3 EXPOSURE

Exposure is the amount of pollutant reaching a receptor, and is commonly expressed as the concentration of pollutant in the media (*e.g.* air, water, soil) and in food that is available to the receptor (individual, population, community). It is important to take into account the period of exposure to estimate the total or rate of supply to a receptor and in turn the intake (or dose). In the case of the more persistent pollutants, the amount or concentration in an organism, or some part of an organism, is a function of exposure. Consequently measurements of such concentrations can be used to indicate exposure.

Experimental studies aim to establish exposure–response relationships and measurements of exposure in the environment indicate whether levels are hazardous. Whilst this is a simple concept, in practice it is not quite so easy to translate experimentally derived data into the ambient environment. Exposure can arise from a single source of supply, as in the transfer of gaseous air pollutants such as sulfur dioxide (SO_2), nitrogen dioxide (NO_2), or ozone (O_3) from the atmosphere to the leaf surfaces of plants. Alternatively more than one pathway may contribute to the total exposure, for example human exposure to toxic metals such as lead and mercury and many pesticides is from water, air, and food.

General measurements of substances in air, water or soil may not be representative of the actual amount that reaches a receptor. This may be a consequence of how, when, and where measurements are taken, or because a pollutant exists in more than one form, all of which may not be equally available to biota. Exposure can vary in space and time; in rivers the concentrations of pollutants vary with season, time of day, magnitude of freshwater runoff, depth of sampling, intermittent flow of industrial effluent, and hydrological factors such as tides and currents. Human exposure to atmospheric pollutants such as lead (Pb), SO_2, and airborne particulate matter can be equally if not more variable and therefore difficult to estimate accurately. Many samples are required to iron out variation due to methods of collection and analysis as well as actual spatial and time differences. In practice, however, exposure estimates are based on data from a small number of monitoring stations (perhaps only one) representing a relatively large area.

There are further difficulties in taking account of the movement and different activities of people. Many adults spend a part of their day at work and children at school, and consequently such groups may experience different exposure regimes during these times. We all spend extended periods indoors removed from the outside air, and certain susceptible groups such as the aged spend most of their time indoors. Furthermore, such habits as tobacco smoking, which is a very direct form of air pollution, quite obviously increases a person's exposure to an array of substances and it is well known that cigarette smoking provides the major supply of cadmium (Cd) to habitual smokers; smoking 20 cigarettes can result in a daily intake of 2–4 μg Cd.[2,3]

Much of the research on the effects of gaseous air pollutants on plants has been done in the controlled environment of growth chambers. The concentration of the gas in question in the inlet or outlet of the chambers has been used as an estimate of plant exposure. These values, however, can be quite different from those at the leaf surface. Design and size of growth chamber, density of plants, and air velocity are just three of many plant and environmental factors that can influence the concentration gradient between ambient air around the plants and that in contact with the leaf surfaces. These and other factors have a marked effect on the thickness of the boundary layer of air that surrounds the leaves. Gases must pass through this almost laminar flow of air by the relatively slow process of molecular diffusion to gain access to the leaf surfaces. In situations where varied responses have been reported for seemingly equivalent exposures, it is apparent that at least a part of the discrepancy can be explained by the experimental conditions and the way in which exposure was quantified. Open top chambers permit plants to be exposed to natural precipitation regimes and their design is such that the microclimate inside is similar to that in the surrounding environment. Nevertheless, the forced ventilation within these chambers can have an effect on pollutant deposition and over a growing season fluxes of pollutants may be enhanced over that in ambient air, which means that these chambers may overestimate the sensitivity of vegetation to air pollutants.[4]

The nature of vegetation can affect the flux of air pollutants. The canopy of a forest creates a larger resistance than say cereals or grassland with the result that there is more turbulence and larger inputs or fluxes of pollutants to forests. The deposition of O_3 to forests is influenced by the opening and closure of stomata; O_3 uptake is known to be less in tree species that close their stomata during the night-time.[4]

The physico-chemical properties of pollutants and the nature of the environment into which they are discharged dictate their fate and therefore the exposure levels encountered by organisms. This can be illustrated by a simple example in which equilibrium partitioning of organic chemicals between air, water, and biota and organic matter, can

[2] 'Biological Monitoring of Toxic Metals', ed. T. W. Clarkson, L. Friberg, G. F. Nordberg, and P. R. Sager, Plenum, New York and London, 1988.
[3] 'Handbook on the Toxicology of Metals', ed. L. Friberg, G. F. Nordberg, and V. B. Vouk, Elsevier, 2nd Edn., 1986, Vol. 2.
[4] United Kingdom Critical Loads Advisory Group, 'Critical Levels of Air Pollutants for the United Kingdom', Department of the Environment, London, 1996.

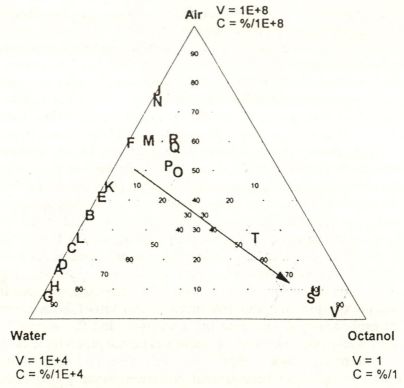

Air V = 1E+8
C = %/1E+8

Water
V = 1E+4
C = %/1E+4

Octanol
V = 1
C = %/1

Figure 2 *Locations of phenol, alkylphenols, and chlorophenols on a triangular diagram of volume ratios V_A:V_W:V_O of 10^8:10^4:1. The arrow depicts the effect of increasing substitution of chlorine. A, phenol; B, o-cresol; C, m-cresol; D, p-cresol; E, 2,4-dimethylphenol; F, 2,6-dimethylphenol; G, 3,4-dimethylphenol; H, 3,5-dimethylphenol; I, 2,4,6-trimethylphenol; J, 2-chlorophenol; K, 3-chlorophenol; L, 4-chlorophenol; M, 2,4-dichlorophenol; N, 2,6-dichlorophenol; O, 2,3,4-trichlorophenol; P, 2,3,5-trichlorophenol; Q, 2,4,5-trichlorophenol; R, 2,4,6-trichlorophenol; S, 2,3,4,5-tetrachlorophenol; T, 2,3,4,6-tetrachlorophenol; U, 2,3,5,6-tetrachlorophenol; V, pentachlorophenol*
(Reproduced with permission from *Environ. Toxicol. Chem.*, 1995, **14**, 1839)

be represented by three properties; water solubility, vapour pressure, and the octanol–water partition coefficient. Water soluble substances will tend to remain in water, substances with a high vapour pressure will evaporate, and hydrophobic or lipophilic ones will partition to lipids in living and dead organic matter. K_{OW} values can be used to assess this last tendency. Figure 2 shows the relative mass distribution of phenol, alkylphenols, and chlorophenols between water, air, and 'octanol' compartments. Phenol and some of the alkylphenols tend to remain in the water compartment, mono-, di-, and tri-chlorophenols are relatively volatile, and the more highly chlorinated members, with higher K_{OW} values, accumulate in organic matter. Benzene and the lower chlorobenzenes have high vapour pressures and therefore escape to the air; the higher members of the series are lipophilic and bioconcentrate in biota.

Clearly the actual environment is more complex. Organic matter in soil and sediment is an important substrate for the adsorption and accumulation of lipophilic substances. Once bound to organic matter in particles such lipophilic substances are less available to organisms, including bacteria. There is evidence that the bioavailability of these bound forms can decline with time as the compounds become more tightly sequestered or less accessible within the organic matter.[5] A further level of complexity is that organic chemicals are susceptible to chemical or biological degradation. Some are broken down quite quickly whilst others, such as the larger polycyclic aromatic hydrocarbons (PAHs) and PCBs, are resistant to breakdown and are persistent in soil and sediment for many years.

Metals exist in different forms. In soils and aquatic systems, metals are partitioned between solid and liquid phases and within each, further partitioning or speciation occurs among specific ligands, determined by ligand concentration and the strength of each metal–ligand association. Consequently, at any one time the amount of free ion available for uptake by organisms is less, and in many instances much less, than the total concentration. As a rule the free ion (*e.g.* Cd^{2+}, Cu^{2+}, Pb^{2+}) is the most bioavailable and therefore the most toxic, although organometallic compounds such as methylmercury and tributyltin are more available and therefore more completely absorbed than their free ion counterparts.

Organisms with different feeding habits can have very different exposures in the same ecosystem. Those foraging in sediments may be exposed to elevated levels of metals or lipophilic organic compounds whereas those swimming in the water may have a much lower exposure. The amount accumulated depends on the bioavailability of the substances in different parts of the system and the integrated level in biota is one way of assessing this. Increasingly, exposure assessment is carried out using computer models which simulate environmental processes and pollutant inputs to predict levels of pollutants reaching different types of receptors.

4 ABSORPTION

Absorption is the process whereby a substance traverses the body membranes. By and large, lipid soluble substances are absorbed more efficiently than polar water soluble substances. Lipid soluble substances diffuse across the lipid bilayer of membranes and those that are more

---

[5] M. Alexander, *Environ. Sci. Technol.*, 1995, **29**, 2713–2717.

lipid soluble (*i.e.* those with a larger octanol–water partition coefficient) tend to be absorbed more efficiently. Complex carrier systems involving the membrane proteins serve to transport polar substances such as metals, although certain forms which are more lipid soluble or of reduced polarity traverse membranes by passive diffusion, and this is the most likely explanation for the greater accumulation of such compounds as methylmercury and tributyltin.

The characteristics of the interface between the environment and an organism strongly influences the absorption of a substance. In mammals the main routes of entry for environmental chemicals are through the lungs, the skin, and the gastrointestinal tract. The physico-chemical properties of a pollutant and the nature of exposure strongly influence the amount presented at each portal of entry. In terrestrial mammals, including humans, penetration through the skin is an unimportant route of entry for environmental pollutants. The skin is relatively impermeable to aqueous solutions and most ions, although many synthetic organic chemicals, because of their lipid solubility, can penetrate the dermal barrier. Such situations tend to be confined to occupational exposures.

Atmospheric pollutants occur as gases or as particulate matter. The site and extent of absorption of inhaled gases are, for the most part, determined by their water solubility. For instance, sulfur dioxide (SO_2) is a soluble gas, which is mainly absorbed in the upper respiratory tract whereas a less soluble gas such as nitrogen dioxide (NO_2) reaches the lower airways. Inhalation is the most important route of uptake for elemental mercury vapour, and the major site of absorption is alveolar tissue where about 80% of inhaled mercury vapour is absorbed.[2,3,6]

The fraction of inhaled particulate material deposited in the various parts of the respiratory tract is a function of particle size, and the fraction absorbed from the tract is dependent on the chemical nature of the aerosol. The upper limit for respirable particles is of the order of 10 μm diameter and modern air monitoring equipment has specially engineered intakes to sample particles of less than this diameter as a means of estimating the respirable fraction. This is now generally referred to as PM10.[7] Only particles less than 2 μm diameter penetrate as far as the alveolar region. Larger particles are trapped in the upper tracheobronchial and nasopharyngeal regions where they may be either absorbed or transported up the pharynx entrained in mucus propelled by ciliary action. Subsequently, these particles are swallowed and become available for absorption in the gastrointestinal tract. Particles taken up from the alveoli (the most important site of absorption) may pass directly into the

[6] 'Environmental Health Criteria, 1, Mercury', World Health Organization, Geneva, 1976.
[7] Quality of the Urban Air Review Group, '3rd Report, Airborne Particulate Matter in the United Kingdom', Department of the Environment, London, 1996.

bloodstream or be retained in lung tissue. For respirable particles of lead, at a particle size of 0.05 μm about 40% may be retained but for larger sizes, *e.g.* 0.5 μm, only about 20% is deposited and probably only about 50% of the deposited lead is absorbed into the blood, although this will depend on the solubility of lead compounds in the particles. Deposition in the alveolar region of inhaled cadmium (Cd) aerosols up to 2 μm diameter is 20–35% of which less than 50% is absorbed; the rest is exhaled or swallowed.[2,3]

In humans, absorption through the gastrointestinal tract is an important route of entry for many environmental chemicals. The circulatory system is closely associated with the intestinal tract, and once toxicants have crossed the epithelium, entry into capillaries is rapidly effected. Venous blood flowing from the stomach and intestine introduces absorbed materials to the hepatic portal vein, resulting in transport to the liver, the main site of metabolism of foreign compounds. Exposure to metals in the general environment is usually greater *via* food and drink than *via* air. However, the absorption of ingested lead and cadmium in adults is normally relatively low, being about 10% for lead and 5% for cadmium. Absorption is influenced by various dietary factors, and increased absorption of lead has been found in cases of low dietary calcium. Absorption of dietary lead is much higher in young children. In the case of inorganic mercury compounds absorption from foods is about 7% of the ingested dose; in contrast, gastrointestinal absorption of methylmercury is practically complete.[2,3,6]

The gills of aquatic animals present another type of interface between the external medium and an organism and are an important route of entry for water-dispersed pollutants. They present a large surface area of diffusion, and at the same time continual circulation of water across the gill filaments ensures maximum exposure.

One other type of external surface membrane worthy of note is that of the leaves of plants. Leaf surfaces are covered by a waxy cuticular layer, interspersed with stomatal pores which may occur on both leaf surfaces or on a single surface (usually the lower). Stomata, as well as regulating uptake of carbon dioxide and water loss, are the major route of entry for gaseous pollutants. Rate of uptake of gaseous pollutants into a leaf is a function of several physical factors; one such factor is the resistance to diffusion caused by the boundary layer, which in turn depends on the velocity and turbulence of airflow over the surface, and another factor is the stomatal resistance. Before a pollutant can gain access and cause injury within a plant cell it must first enter into solution in the extracellular water enveloping the cell wall. In solution, SO_2 is active in the form of either HSO_3^- or SO_3^{2-}. Ozone (O_3) is less soluble, but since it is a highly reactive molecule, it is thought that decomposition products such as hydroxyl radicals and other free radicals are important reactive species produced from reactions involving organic compounds.

Changes in stomatal aperture induced by air pollutants has attracted considerable attention. This is not surprising since any change in gaseous flux to and from metabolic sites in mesophyll tissues may eventually affect the overall growth and yield of plants. The effect of SO_2 on stomatal aperture is complex; initial studies found enhanced stomatal opening in *Vicia faba* plants exposed to SO_2. It was also shown that relative humidity strongly influenced the direction of the response; humidities above 40% enhanced stomatal apertures, whereas at humidities below 40% apertures decreased. A host of different species have now been investigated and it would appear that there is no uniformity of response between species; both concentration of pollutant gas and duration of exposure influence the outcome.[8]

5   INTERNAL PATHWAYS

Once inside an organism, a pollutant may follow a number of different pathways. In general terms four main routes can be identified (see Figure 3): some molecules are metabolized (converted) into other compounds which are frequently less toxic to the organism than the parent compound, although some are more toxic. A second pathway involves storage in certain tissues, *e.g.* lead in bone, cadmium in kidney, and DDT in fatty tissues. Thirdly, a pollutant and its metabolites may be excreted from an organism. Because metabolites are, as a rule, more water soluble they are more easily excreted than the original pollutant, and metabolism can be seen as an essential preliminary to excretion. On this rather simplistic view, the remaining fraction can be regarded as being available to exert an effect at a site of action.

The fundamental effect occurs at the molecular level; for example, lead inhibits the action of ALA-D, an enzyme in the haem synthesis pathway, although the significance of this to the overall toxicity of lead is not clear. Organophosphorus insecticides specifically inhibit the action of acetylcholinesterase. Perhaps surprisingly, the primary toxic lesion of many pollutants is unknown. A large group of non-polar organic substances are thought to have a general narcotic action, whereby metabolism is seen to slow down in proportion to exposure and eventually it ceases. In recent years there has been considerable attention directed at finding fundamental modes of toxic action. The realization that several pollutants can mimic natural hormones, and as a result disrupt the endocrine

---

[8] J. Wolfenden and T. A. Mansfield, 'Acid Deposition—Its Nature and Impacts', ed. F. T. Last and B. Watling, *Proc. Roy. Soc. Edinburgh*, 1996, Section B, **97**, 117.

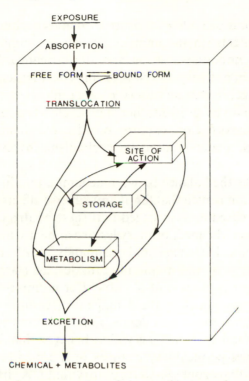

EXPOSURE

ABSORPTION

FREE FORM ⇌ BOUND FORM

TRANSLOCATION

SITE OF ACTION

STORAGE

METABOLISM

EXCRETION

CHEMICAL + METABOLITES

Figure 3 *Diagrammatic representation of the possible internal pathways followed by pollutants*
(Adapted with permission from T. A. Loomis, 'Essentials of Toxicology', Lea and Febiger, Philadelphia, 2nd Edn., 1974)

system of animals, has raised the possibility that this is a basic event that could explain widespread reproductive problems caused by many pollutants in wildlife and possibly humans (this is discussed more fully below). Another area of growing importance is the possible role of highly reactive free radicles in causing molecular disruption which may be linked to short term damaging effects such as inflammation of lung surfaces caused by air pollutants or to longer term damage to DNA and the induction of cancers. Free radicals are atoms or molecules which are capable of independent existence and which contain one or more unpaired electrons, examples include $H^{\bullet}$, $O_2^{\bullet-}$ and $OH^{\bullet}$. They are produced in the body all the time and there are several scavenging mechanisms for removing them and so preventing damaging effects. Several pollutants have free radical properties or may be converted into species with such properties, for example primary metabolites of PAHs. Put simply, if the production of free radicals exceeds the body's capacity to scavenge them, damage to molecular systems will occur and, in particular, this may include damage to DNA. There is also a growing number of reports providing evidence of metals such as Cd, PAHs and PCBs having effects on the immune system of animals. The nervous, immune, and endocrine systems are all intricately linked with one another and it is not easy to establish in which system a fundamental lesion takes place, in that an observed effect in one system may actually be a response to an effect in another. This is clearly a rapidly developing area and in the next five or six years considerable advances will be made in identifying and understanding the mechanisms involved in the toxicity of pollutants.

At low intakes the balance between accumulation in tissues, metabolism, and excretion is such that few if any adverse effects are observed. As exposure is increased it is envisaged that these controlling processes are progressively overwhelmed, primary lesions appear which in turn lead to major physiological damage and ultimately death of an organism. Internal pathways of the major phytotoxic gases are short, *i.e.* they exert their effect near the point of entry. For mesophyll tissues of plant leaves, for example, much of the damage due to sulfur dioxide involves disruption of chloroplasts and depression of photosynthesis along with changes in stomatal aperture. It is widely believed that much of the effect of ozone involves permeability changes in the membranes of palisade cells. Pollutants like ozone are so transitory that it is virtually impossible to detect them within tissues; their presence is detected from the effects induced. Exposure to sulfur dioxide can increase plant sulfur content three- to four-fold; it is rapidly incorporated into organic molecules, notably the amino acids, glutathione and cysteine. In fact the balance between incoming sulfur dioxide and its destruction is probably a critical factor in resistance to acute injury in different varieties and species of plants. Nitrogen dioxide is similarly incorporated into amino acids such as glutamine and asparagine. The incorporation of SO_2 and NO_2 into metabolic pathways in this way explains why these pollutants can stimulate growth of plants where S and N are limiting nutrients.

For many metals, storage or accumulation within a tissue or an organ is essentially a detoxication step. Inorganic lead is transported in the bloodstream attached to red blood cells (erythrocytes) but a major proportion, about 90% in adults and 70% in children, accumulates in bone. The biological half-life in this tissue is about 10 years but turnover of lead in the bloodstream and soft tissues is rapid and responds quite quickly to changes in lead intake and exposure.

In normal healthy humans about 50% of the cadmium in the body is in the kidneys and liver with about one-third in the kidneys alone. The biological half-life in these tissues is probably more than 10 years and as a consequence cadmium accumulates with age. A significant proportion occurs bound to metallothioneins, which are low molecular weight proteins rich in sulfhydryl groups. In this bound form it is believed that cadmium is less available to exert its toxic action. A number of other trace metals, including zinc, copper, mercury, and silver also induce and bind to metallothioneins. The precise role of metallothioneins is unclear and the regulation of intracellular concentrations of metals also involves other mechanisms; for example, metals may be incorporated into extracellular structures such as carbonate granules, or bound to intracellular components such as nuclei, mitochondria, lysosomes, and phosphate granules.

DDT, PCBs, and other chlorinated organic compounds because of their lipophilic properties, are concentrated in the fatty tissues of animals and those with the largest amount of fat accumulate the highest concentrations. In normal circumstances there is a slow turnover of these compounds in the body's fat reserves. However, during periods of stress and starvation mobilization of the fat deposits can release the stored organochlorines into the bloodstream which in turn may induce toxic effects and ultimately death. Laboratory studies have shown that starved animals exposed to organochlorines such as DDT are critically affected at lower exposures than well fed ones. A large number of guillemot deaths were recorded in September 1969 from the coasts around the Irish Sea and it is believed that they were due to a combination of stress and mobilization of organochlorine residues, perhaps PCBs, into the blood stream.[1]

A feature of most, if not all, vertebrate and invertebrate animals is a built-in capacity to metabolize a wide range of foreign compounds to less toxic entities that are more easily eliminated from the body. In higher organisms metabolism mainly takes place in hepatic tissues and it essentially consists of two phases; phase 1 consists of the monooxygenase system which is responsible for metabolizing foreign compounds by such mechanisms as hydroxylation, dealkylation and epoxidation; phase two involves conjugation reactions in which glucuronide and sulfate derivatives are formed from the oxidized products of phase one. The net effect is to convert lipophilic foreign compounds to water soluble metabolites which facilitates their elimination from the animal. Numerous hydrocarbon compounds, including various phenolic and benzene compounds, are known to undergo biotransformation to water soluble conjugates. The influence of metabolism in regulating tissue concentrations and hence on the toxicity of a substance can be deduced from studies that have used inhibitors of the biotransformation pathways. Salicylamide is a potent inhibitor of the phase II glucuronide conjugation reactions and it has been shown that pretreatment of fish with salicylamide significantly increases the toxicity of phenols as a consequence of preventing the production of the conjugated metabolite, phenylglucuronide.

Organophosphorus insecticides on the other hand need to be metabolized to become active, so inhibiting this step results in the pesticide being less toxic. In animal tissues DDT is rapidly converted to the more persistent DDE and this explains why DDE, rather than DDT, is the most widespread and abundant form in wildlife.[9]

A fundamental relationship in pollution and ecotoxicology studies is that which relates exposure, body burden or accumulation, and toxicity of a substance. Residues in an organism (or some part of it) form an important link between exposure and biological response or damage. For the more persistent pollutants, such as metals and organochlorine compounds, enhanced exposure from food or the surrounding media generally results in a greater concentration in an organism. When uptake (absorption) of a substance exceeds elimination (including metabolism) accumulation occurs in the whole organism or some part of it. Eventually a steady state may be attained when uptake and elimination are equal.

For many organic chemicals, the bioconcentration factor (*i.e.* the ratio of the concentration of the chemical in the organism to the exposure) is proportional to the lipid solubility and inversely related to the water solubility of the compounds; it can therefore be estimated from the *n*-octanol–water partition coefficient.

Many studies have examined the relationship between blood lead concentration and concentration of lead in air, water and diet. A curvilinear plot is obtained for the total intake *versus* blood lead concentration, indicating that successive increments in intake or exposure result in progressively smaller contributions to blood lead concentrations. This relationship can form the basis of predicting one parameter from the other. The accumulation and elimination rates of methylmercury in humans conform to a single exponential first order function (Figure 4). The interesting feature of this model is that, assuming a constant hair-to-blood ratio, mercury analysis of segments of hair can be used to predict both the body burden and blood concentration of mercury at the time the hair segment was laid down.

Organisms in aquatic systems can acquire residues from both food and the surrounding medium. From laboratory based studies comparing equivalent exposures from both routes, uptake directly from the water medium would seem to be more important. However, in reality lipophilic substances partition to living and dead organic matter in the system and the concentration in water would be very small, therefore exposure to such substances would be mainly from ingesting contaminated food. In

[9] F. Moriarty, 'Ecotoxicology: The Study of Pollutants in Ecosystems', Academic Press, London, 1998.

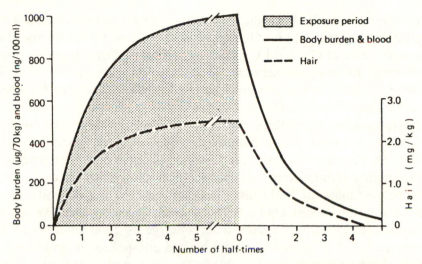

Figure 4 *The changes in the body burden and hair and blood concentration of mercury during constant daily exposure (shaded area) and after exposure. This calculation was based on a daily intake of 10 µg of methylmercury during the exposure period, an elimination half-time of 96 days, and a hair-to-blood concentration of 250*
(Reproduced with permission from WHO, Environmental Health Criteria, 'Mercury', 1976, **1**)

terrestrial ecosystems, most if not all lipophilic substances, including pesticides, are transferred via the food chain. Seed corn dressed with dieldrin undoubtedly killed many pigeons in England in the 1950s and 1960s; the poisoned pigeons were responsible for killing predators such as foxes and perhaps peregrine falcons. It is common to find that residues of organochlorine compounds are highest in carnivorous birds, followed by insectivorous and then herbivorous birds. Pollution of the St. Lawrence–Great Lakes basin with persistent organochlorine substances first became apparent in the 1960s and it has been estimated that there is a 100 million-fold bioaccumulation of PCBs between water, through the food chain and into bald eagle's eggs.[10]

6 ECOLOGICAL RISK ASSESSMENT

Ecological risk assessment is concerned with estimating whether current or predicted environmental levels of contaminants have the potential to cause harmful effects on populations and communities. Such assessments combine estimates of hazard and exposure. Hazard is defined in terms of toxicity and is based mainly on single species tests that quantify

[10] M. Gilbertson, *Environ. Toxicol. Chem.*, 1997, **16**, 1771–1778.

responses to acute and chronic exposures. The general effects measured are related to survival, growth, and reproductive performance and they are expressed as LC_{50} and EC_{50} values (as discussed in Section 2). From such tests the lowest observable effect concentration (LOEC) can be estimated and at some concentration below this is the No Observed Effect Concentration (NOEC).

Extrapolation of NOECs based on laboratory test organisms to wildlife populations is a contentious issue. Tests are based on a few species that are used as surrogates for wildlife, and they are carried out under highly standardized conditions. Uncertainty factors are usually built into estimates of NOEC and the magnitude of these depend on how well effects have been characterized on species representative of the ecosystem in question, ranging from as large as 1000 where tests are restricted to a few species and with limited end-points, to as low as 10 where chronic toxicity is well established for test species representative of the different levels of organization in the system and especially where data are available for the most sensitive species. Hence NOEC values are a synthesis of many toxicity tests with different organisms and a range of end-points.

Exposure is measured or predicted from models as discussed in Section 3. The ratio of NOEC to the measured or predicted environmental concentration (PEC) is taken to indicate the degree of potential hazard. If the ratio is less than one, then environmental concentrations are expected to equal or exceed the NOEC value and the chances of harmful effects occurring in wildlife are high, but if it is less than one, or less than the built-in uncertainty factor, then the likelihood of adverse effects is low. Figure 5 highlights the concept, and also shows that whilst from initial screening tests the ratio may be less than unity there is uncertainty because the estimates are based on very few measurements and so more tests with different species and additional end-points are required, and perhaps more sophisticated experiments that closely simulate the ambient environment. At the same time more concrete information is required on the behaviour or fate of the toxicant in the actual environment. A stage is reached where a no-effect level can be defined with some degree of confidence and it can be used as an objective for environmental levels, either as an upper level for new chemical inputs or as a target to aim for in reducing the levels of existing chemicals.

A related concept of critical concentrations and critical loads has been developed for controlling the impacts of air pollutants on the environment. Critical loads refer to the total deposition of sulfur, nitrogen, or hydrogen ions, and they are primarily aimed at preventing acidification or eutrophication of sensitive aquatic or terrestrial ecosystems. Critical levels refer to the direct effects of atmospheric pollutants on vegeta-

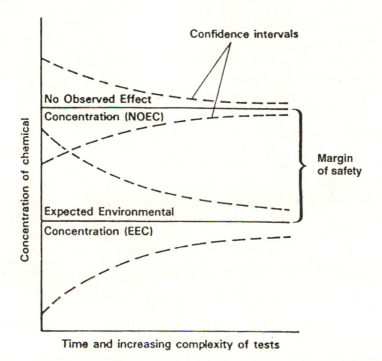

Figure 5 *Qualitative representation of the relationship between No Observed Effect Concentration and Exposure with progressively increasing complexity of tests*

tion.[4,11] As shown in Figure 6 critical levels are the concentrations in the atmosphere of pollutants above which direct adverse effects on receptors, such as plants, ecosystems, or materials, may occur according to present knowledge. Across a region such as Europe the number of receptors is very large and therefore the emphasis is on determining exposure–response relationships for the most sensitive receptor at a particular location.

7 INDIVIDUALS, POPULATIONS, AND COMMUNITIES AND THE ROLE OF BIOMARKERS

As a general rule, increasing pollution exposure results in progressively more severe and damaging effects. Figure 7 provides a useful model summarizing the interaction of pollution with receptors.[1] The so-called 'cascading effect' suggests that some degree of homeostatic control or carrying capacity operates at each level of organization. As exposure increases in magnitude, the capacity of cells, tissues, whole organisms, populations, and communities are successively overwhelmed and

[11] K. R. Bull, *Environ. Pollut.*, 1991, **69**, 105–123.

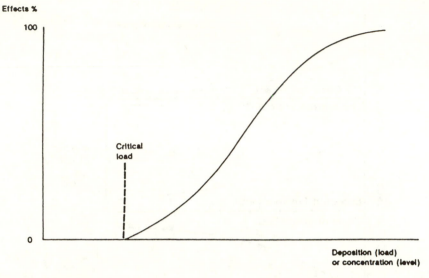

Figure 6 *Theoretical dose–response curve showing a threshold, or 'critical load' at which effects are observed*
(Reproduced with permission from *Environ. Pollut.*, 1991, **69**, 105–123)

damaged. Disruption at the biochemical level may be quite specific but at higher exposures, as physiological systems are affected, more general symptoms of injury become apparent; reproductive capacity and recruitment may be affected and behavioural responses may be evident. At still higher exposures fatalities start to occur and as these become as significant as losses due to natural causes, noticeable reductions in population size result. The cascade model is generally accepted; however, in practice the relationships between the different levels of organization have yet to be firmly established. Moreover, the model is useful for explaining the role of biomarkers and bioindicators.

A biomarker can be defined as 'a biological response that can be related to an exposure to, or toxic effect of, an environmental chemical or chemicals'.[12] The biological response can range from one at the biomolecular level to one that indicates changes in ecosystems. Most examples come from the lower levels of organization. Their potential lies in their ability to signal the onset of adverse exposures or effects at a higher level of organization. Biomarkers can be categorized according to their specificity. Only one or two biomarkers are sufficiently well developed or understood to be used for quantifying exposure and or toxicity. The inhibition of acetylcholinesterase by organophosphate and carbamate insecticides is specific and reproducible. The inhibition leads to disrup-

---

[12] D. B. Peakall, *Toxicol. Ecotoxicol. News*, 1994, **1**, 55–60.

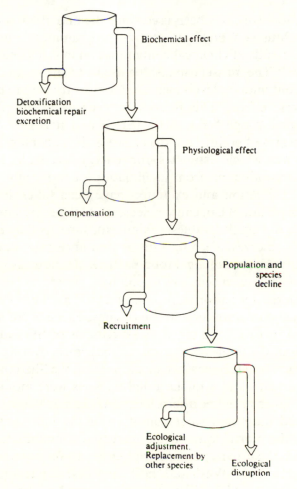

Figure 7   *Crude model of 'cascading' effect of pollution, as a series of reservoirs. Effects do not spill over from one level to another unless inputs exceed capacity of relieving systems depicted on left of each reservoir. Variation between individuals can be thought of in terms of variation in capacity of these systems, which may also be increased by evolution, producing more resistant organisms*
(Reproduced with permission from M. W. Holdgate, 'A Perspective of Environmental Pollution', Cambridge University Press, Cambridge, 1979)

tion of the nervous system; 20% inhibition has been used as a criterion of exposure and 60% inhibition as a criterion of death caused by these compounds. The monooxygenase system represents a broader category and the induction of these oxidases has been related to exposures to dioxins, PCBs, and PAHs in both field and laboratory studies (an example is given below).

Case Study 1. Scope for growth in mussels, an example of the practical use of a biomarker in the assessment of biological effects. The common mussel, *Mytilus edulis*, has been used since the mid-1970s as a 'sentinel' organism in 'Mussel Watch' monitoring programmes to assess spatial and temporal trends of chemical contamination in estuarine and coastal environments.[13] The mussel can also be used as a means of demonstrating whether contaminant levels measured in their tissues are high enough to have adverse effects. This is done by measuring the physiological response termed 'scope for growth' (SFG) and it is one of the most sensitive measures of pollution-induced stress. It is an integrated physiological response which estimates the energy available for activity, growth, and reproduction from assimilation of food, after the energy expended in respiration and excretion have been taken into account. Exposure to a pollutant can cause a decrease in the energy available and it may decline to negative values at high exposures as the animal draws on the body's reserves. Feeding rate or clearance rate, defined as the volume of water cleared of algal food particles per hour, is generally the component of SFG most sensitive to pollutant stress.

One of the largest monitoring programmes to have been undertaken in which chemical contamination and biological effects were measured at the same time involved combined measurements of SFG and chemical contaminants in the tissues of mussels at locations covering 1000 km in the North Sea.[14] In all, 26 coastal locations from the Shetland Islands to the Thames estuary and 8 offshore light vessels were monitored. SFG values were higher at the northern locations than at the southern ones, which reflected the clean water condition of the inflow of the North Atlantic into the North Sea and the increased contamination of water associated with the urban and industrial locations further south. Coastal regions of the Humber–Wash and the Thames estuary recorded some of the lowest values for SFG (Figure 8). For several of the contaminants detected in the mussel, experimentally derived tissue concentration–response relationships were available and these were used as a basis for estimating the additive contribution of the different contaminants to the overall decline in SFG. It was found that toxic hydrocarbons (mainly PAHs) accounted for most of the reduced SFG at the majority of sites, whilst tributyltin made a significant contribution to the toxic effect at a number of locations. Metal concentrations were generally below those associated with a reduction in SFG. This example demonstrates the practical use of biomarkers in the assessment of the biological effects of pollution and in the detection of the toxic agents concerned with these effects. Other examples of biomarkers such as eggshell thinning in birds caused by exposure to organochlorine pesticides and vitellogenin induc-

[13] J. Widdows and P. Donkin, 'The Mussel *Mytilus*', Elsevier, Amsterdam, 1992, Ch. 8.
[14] J. Widdows *et al.*, *Mar. Ecol. Prog. Ser.*, 1995, **127**, 131–148.

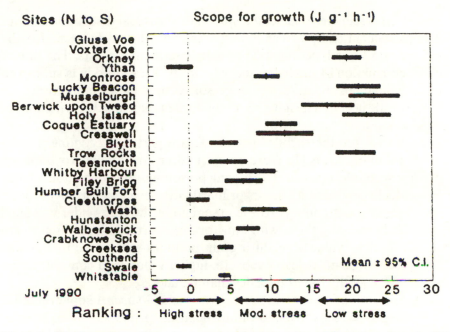

Figure 8 Mytilus edulis. *Scope for growth of mussels collected from sites along the UK North Sea coastline (mean ±95% CI, n = 16)*
(Reproduced with permission from *Mar. Ecol. Prog. Ser.*, 1995, **127**, 131–148)

tion in relation to effects on the endocrine system in fish are discussed below.

An important aspect of pollution studies is the diverse response of organisms to equivalent exposures. Variation in sensitivity is apparent between:

(1) Individuals;
(2) Groups or populations differentiated with regard to sex, stage of development, nutritional well-being, *etc*;
(3) Varieties of strains;
(4) Species.

In general, differences between individuals are much less evident than those between different species. Differences in sensitivity to pollution may be viewed broadly in terms of differences in homeostatic capacities between various groups of organisms *i.e.* differences in rate of absorption, metabolism, storage and excretion.

(1) In a set of individuals a measured response will reveal a group which is sensitive, another which is moderately sensitive and one

that is relatively resilient; a spread or statistical distribution about a median value is usually apparent and this is the basis for the response of experimental populations in toxicity tests. The rate of elimination of methylmercury from the human body is subject to individual variation and a person with a slow elimination rate would accumulate a higher body burden than another with a more rapid rate.

(2) Very often the early stages of life are particularly vulnerable to pollution stress. For example, children are known to be more susceptible than adults to lead poisoning, and depletion of fish stocks in acidified lakes is primarily due to high mortality among eggs, larvae, and fry. At the other end of the scale, elderly people, and especially those with a history of heart and lung disease, are the most vulnerable during air pollution episodes. In humans receiving similar lead exposure, higher blood lead concentrations are generally found in males than females.

(3) Certain varieties of tobacco (*Nicotiana tabacum*) show different degrees of sensitivity to photochemical oxidants. Three in particular, which are described as supersensitive, resistant, and intermediate, have proved useful biological indicators of elevated ozone levels.

Prolonged exposure to certain pollutants can generate resistant or tolerant varieties. This is especially true of fast breeding forms of life. Examples include:

(a) Genotypes tolerant of one or more trace metals (*e.g.* Cu, Zn, Pb, Ni) have been identified in many plant species growing on metal-rich substrates. Such substrates include soils of mineralized areas, metal mine wastes, and soils contaminated by metal-rich fumes originating from smelters, refineries, and automobiles.[15]

(b) A genotype of the grass *Lolium perenne*, known as Helmshore (named after the district 15 miles north of Manchester where it was first discovered), has a higher resistance to sulfur dioxide than the cultivated variety, S23.[16]

(c) An enormous number of insect pests have developed resistance to various insecticides and in the late 1970s resistant

[15] M. H. Martin and P. J. Coughtrey, 'Biological Monitoring of Heavy Metal Pollution – Land and Air', Applied Science, London and New York, 1982.

[16] J. N. B. Bell, M. R. Ashmore, and G. B. Wilson, 'Ecological Genetics and Air Pollution', ed. G. E. Taylor *et al.*, Springer-Verlag, New York, 1991.

varieties had been recorded in over 400 species of insects and mites.[9]

(d) In the lead-, copper- and zinc-rich sediments of old metal mine workings in estuaries of south-west England, tolerant varieties of the polychaete *Nereis diversicolor* and asellus among others, can be found.[17]

(e) Darker, melanic forms of many species of the larger moths, *e.g. Biston betularia*, occurred with increasing frequency in areas within and adjacent to the industrial regions of Britain. The distinctive feature of these darker varieties is the greater preponderance of melanic granules in the cuticle and wing scales. The significance of the adaptation would seem to lie in concealment from bird predation on darker surfaces which are typical of areas affected by air pollution.[9]

(4) Species, even closely related ones, can differ quite markedly in their response to one form of pollution or another. This is an important consideration in the assessment of the ecological effects of pollution since it is common to find that communities that are subject to pollution stress undergo a reduction of species diversity as the most susceptible ones disappear first, to leave behind fewer and more resilient types. Some notable examples include forest ecosystems affected by air pollution and freshwater systems subject to biodegradable wastes, artificial eutrophication, or acidification. In certain situations changes in community structure can be used as a sentinel of the degree of pollution stress, *e.g.* biotic indices that are used to indicate water quality.

Bryophytes (mosses and liverworts) and lichens (fungal–algal symbioses) are among the most sensitive receptor organisms to several atmospheric pollutants. Corticolous lichens (*i.e.* those growing on tree trunks) along a gradient of sulfur dioxide as say from a rural area to an industrial centre, reveals that the lichens become progressively impoverished as higher sulfur dioxide concentrations are encountered and commonly within urban–industrial centres there is a complete absence of lichens. Hawksworth and Rose[18] devised a scale composed of 11 zones based on presence and abundance of corticolous lichens, and a map of England and Wales depicting these lichen zones shows them to be closely related to the winter mean sulfur dioxide concentration for the particular year of the survey (1972–73).

[17] G. W. Bryan, 'Pollution and Physiology of Marine Organisms', ed. W. B. Vernberg and F. J. Vernberg, Academic Press, XXX, 1974, p. 123.
[18] D. L. Hawksworth and F. Rose, *Nature*, 1970, **227**, 145.

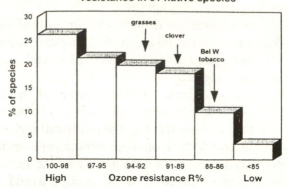

Figure 9   *Frequency distribution of the ozone resistance of 61 native species, compared with that of the sensitive bioindicator,* Nicotiana tabacum *cv. Bel-W3. Ozone resistance is expressed as growth in ozone as a percentage of that in filtered air* (Data provided by Prof. A. W. Davison)
(Reproduced with permission from United Kingdom Critical Loads Advisory Group, 'Critical Levels of Air Pollutants for the United Kingdom', Department of the Environment, London, 1996)

It is increasingly recognized that the most important impact of O_3 on semi-natural plant communities is through shifts in species composition, loss of biodiversity, and changes in genetic composition. Figure 9 shows the frequency distribution of resistance to O_3 of 61 plant species native to Derbyshire, together with the resistance of the tobacco variety, Bel W3. All the species were subjected to the same O_3 exposure over a period of two weeks, and whilst between 25 and 45% of them showed no detectable effect, over 25% of species showed a marked reduction in growth and they were found to have a similar sensitivity to the tobacco cultivar.[19] A similar experiment in which populations of the broad leaved plantain (*Plantago major*) collected from different geographical areas of the UK were exposed to O_3 demonstrated that there was as much variation in resistance within this one species as there was between the 61 species. The O_3 resistance of the populations was significantly correlated with the O_3 concentrations prevailing in the areas from which they were collected. The most resistant populations were from southern England, indicating that these populations had evolved resistance to O_3.[20]

In some cases, differences in species sensitivity may be more apparent than real because certain organisms may experience greater exposure as a consequence of pollutants becoming more concentrated in specific parts

[19] K. Reiling and A. W. Davison, *New Phytol.*, 1992, **109**, 1–20.
[20] K. Reiling and A. W. Davison, *New Phytol.*, 1992, **122**, 699–708.

of an ecosystem such as the upper soil layers or the sediments at the bottom of lakes and estuaries. Organisms which inhabit these areas and in particular those that do so by virtue of their feeding habits, *e.g.* earthworms or filter feeds such as mussels, will experience greater exposures from ingesting contaminated particles.

Fatalities due to pollution do not necessarily affect the overall size of a population. Organisms die from a variety of natural causes, and unless pollution is on a par with these its influence on population size is unlikely to be significant. In broad terms a population is controlled by birth rate, death rate, and the balance between immigration and emigration. For many wildlife populations, recruitment (*i.e.* the number of 'recruits' produced annually) has a strong influence on population size. This is certainly the case for most commercially important fish species in which fluctuations of the early life stages account for much of the fluctuation in overall population size, and environmental factors, rather than spawning stock, control the number of early larval and juvenile stages.

Natural populations of animals and plants are subject not only to short term changes but also long term fluctuations of population size which are often large-scale and irregular events that may take place rapidly. These changes are often due to changes in weather patterns such as periods of lower or higher temperatures. Clearly for disturbances caused by pollution to be apparent in a population, they must be on a scale over and above fluctuations due to natural causes. In addition, other human activities, *e.g.* overexploitation of a natural resource, can cause significant perturbations.

Whether or not pollution exerts a permanent effect on the population size will depends on the persistence of the polluting agent. Once the stress has been withdrawn, populations have a remarkable ability to recover from large-scale fluctuations, due to whatever cause. The rapid recolonization of the tidal river Thames is a striking example of recovery of an ecosystem from pollution. However, to conclude that recovery is complete assumes a precise knowledge of the situation prior to pollution onslaught and this is rarely known.

In general terms, the ecological effects of pollution are reductions in species diversity, productivity, and biomass. These trends have been described as a retrogression to an earlier successional stage of ecological development. However, perception and evaluation of such changes at a community level is difficult, since even the simplest of ecosystems contain a considerable number of species, population interactions are complicated, and measurement of pollutant exposure in the various niches of the system can be very problematic.

8 HEALTH EFFECTS OF THE MAJOR AIR POLLUTANTS

Health effects of air pollution in the general population are associated with an increase in mortality and worsening of lung and heart conditions during pollution episodes. The potential of smoke laden air from coal fires to increase incidences of respiratory disease in London was recognized as long ago as 1661 by Evelyn.[21]

Case Study 2. The London Smog 1952. The most notorious incident occurred in London in early December 1952, when 3500–4000 deaths above the norm were recorded during an exceptional smog episode that lasted unremittingly for five days.[22,23] At the time, much of the air pollution was due to coal combustion, with numerous near-ground sources generating considerable quantities of smoke and sulfur dioxide. The fog developed across the capital as a consequence of a very stable high pressure zone and an inversion layer. This prevented the dispersal of pollutants, and concentrations built up to very high levels, the particles of smoke acting as condensation nuclei which added further to the density of the fog. The epidemic went largely unnoticed by the population of London and it was only when death certificates for the whole of the London area were later examined that the sudden upsurge in the number of deaths became apparent. Deaths were mainly confined to the elderly and those people with a history of heart and lung diseases. The central areas of London, where the fog was at its densest and most persistent, had the greatest increases in mortality, some 200% more than the average for that time of year. Estimates indicate that sulfur dioxide and smoke concentrations (48 h mean) during the London episode attained very high values and were in the region of 3.7 mg m^{-3} (1.3 ppm) for SO_2 and above 4.5 mg m^{-3} for smoke. The episode was also at the time of the Smithfield Agricultural Show at Earls Court and it is interesting to note that a number of cows also died during the course of the show. The London smog episode was very important in establishing the strong link between air pollution and respiratory health.

It is generally accepted that the combined effects of sulfur dioxide and smoke particles on the respiratory tract were responsible for the excess deaths and for exacerbating heart and lung disease in susceptible people. To the healthy majority of Londoners the pollutant laden fog was merely an inconvenience. Epidemiological studies established that the spatial

[21] J. Evelyn, 1661, reprinted in 'The Smoke of London: Two Prophecies', ed. J. P. Lodge, Maxwell Reprint, New York, 1969.
[22] Ministry of Health, 'Report on Public Health and Medical Subjects No 95', HMSO, London, 1954.
[23] 'Environmental Health Criteria, 8, Sulphur Oxides and Suspended Particulate Matter', World Health Organization, Geneva, 1979.

and temporal trends of sulfur dioxide and smoke correlated closely with the distribution of the enhanced mortalities, but other pollutants would also have been elevated, *e.g.* carbon monoxide, sulfates, and sulfites. However there are insufficient data on such substances to allow exposure–response relationships to be established.

Follow-up studies in other large cities concentrated on more moderate day-to-day variations of mortality and morbidity in relation to pollution levels. A WHO Task Group provided a summary[23] of the salient features of many of these studies and the collated data have formed the basis for developing short and long term exposure–response relationships. It is important to emphasise that concentrations of sulfur dioxide and suspended particulate matter vary considerably from place to place and from one time period to the next. Also, most monitoring data relate to levels prevailing in the outdoor environment and take little account of indoor exposure, which is usually lower but is where most people spend much of their time. The elderly and the chronically sick spend most of their time indoors. Also there is no consideration of occupational exposure, which can be significant, and smoking which is the most direct form of air pollution. Notwithstanding, short term exposures of 500 μg m^{-3} (24 h mean) for both sulfur dioxide and smoke can be expected to result in excess mortality among the elderly and the chronically sick. Exposures of 250 μg m^{-3} (24 h mean) to both pollutants are likely to lead to worsening of the condition of patients with existing respiratory disease. With regard to long term exposure, increased prevalence of respiratory symptoms among both adults and children and increased frequency of acute respiratory illnesses in children are more likely to occur when annual mean concentrations of sulfur dioxide and smoke exceed 100 μg m^{-3}. Based on these relationships, and incorporating a margin of safety, WHO advocate, as a guideline, that 24 h mean values of sulfur dioxide and smoke should remain below 100–150 μg m^{-3}, with an annual mean below 40–60 μg m^{-3}.

The relationship between photochemical smog episodes and human health is less clear cut. Early epidemiological studies, based mainly on Los Angeles, failed to establish direct relationships between increased mortality rates and the frequency of smog episodes, although eye irritations (perhaps due to peroxyacyl nitrate compounds) and upper respiratory tract discomfort are common complaints during smog episodes. Increased breathing difficulties in heavy smokers and asthmatics and reduced performance in people indulging in physical activity are associated with periods of high oxidant concentrations.[24]

Chamber studies with short term but realistic exposures to O$_3$ and

[24] 'Environmental Health Criteria, 9, Photochemical Oxidants', World Health Organization, Geneva, 1979.

healthy adult subjects have consistently shown effects such as reductions in lung function (lung volume and expiratory flow-rate), increases in lung reactivity to other irritants, and pulmonary inflammation.

Field studies of adults who exercise heavily for short periods have shown short term reversible decreases in pulmonary function in association with ozone concentrations at or near the previous US National Ambient Air Quality Standard of 120 ppb. Much of the field evidence comes from children attending summer camps in North America where ambient O_3 concentrations, particularly above 120 ppb have been associated with short term declines in the average lung function.[25,26] However, other studies have failed to reproduce this relationship. A recent study re-analysed the data from six summer camp studies on lung function and O_3 exposures. They were all based in North America with two in north-west New Jersey, two in southern California and two in Ontario. Data were available for forced expiratory volume in 1 second (FEV_1) and peak expiratory flow rate (PEFR) as well as O_3 exposure. Statistically significant positive relationships were found between FEV_1 and O_3 exposure but those with PEFR were inconsistent.[27] The 1997 revised US ambient air quality standard for ozone is 80 ppb daily maximum eight hour average over three years.

In London, daily hospital admissions for respiratory disease over a five year period (1987–92) were examined in relation to air pollution (NO_2, SO_2, O_3, and smoke) and it was found that O_3 above all the others had a small yet significant effect on admissions.[28]

One of the key issues at the present time is whether the lower concentrations of respirable particles encountered in towns and cities today are having an effect on respiratory health, and much has been made of the possibility that they are responsible for the recent increase in asthma among children.

The primary source of particulate matter in towns and cities is traffic and so with the ever increasing volume of traffic respiratory health has to be assessed in relation to this source. Traffic-related air pollution, however, is not restricted to particulate matter, the other main pollutants being nitrogen dioxide, ozone, carbon monoxide, lead and hydrocarbons.

Over the last seven years or so, a number of studies have demonstrated an association between elevated PM_{10} concentrations and hospital

[25] M. Berry, P. J. Lioy, K. Gelperin, G. Buckler, and J. Klotz, *Environ. Res.*, 1991, **54**, 135.
[26] D. M. Spektor, G. D. Thurston, J. Mao, D. He, C. Hayes, and M. Lippmann, *Environ. Res.*, 1991, **55**, 107.
[27] P. L. Kinney, G. D. Thurston, and M. Raizenne, *Environ. Health Perspect.*, 1996, **104**, 170–174.
[28] H. R. A. Ponce de Leon, R. Anderson, J. M. Bland, D. P. Strachan, and J. Bower, *J. Epidemiol. Comm. Health*, 1996, **50**, S63–S70.

admissions for respiratory disease.[29] In particular, admissions for chronic obstructive pulmonary disease (COPD) shows the largest increase during periods of high PM_{10} concentrations (>50 μg m^{-3}).

In one localized study, centred on a steel mill in Utah, it was shown that hospital admissions of children with respiratory disease in winter in the Utah Valley were associated with emissions of particulates from the nearby steel mill. In one particular winter a strike by the workforce resulted in much reduced PM_{10} concentrations and fewer hospital admissions.

Other time-series and cross-sectional studies have shown similar associations between airborne particulate matter concentrations and respiratory disease and mortality. The association is strongly linked to respirable particles regardless of weather conditions (winter or summer pollution episodes), confounding influences of other pollutants such as SO_2 and O_3, different geographical areas, and different types of sources of particulate matter (*e.g.* wood smoke *versus* traffic emissions). Estimates indicate that for a 100 μg m^{-3} rise in PM_{10} the relative risk factor of increased admission to hospital with respiratory disease is of the order of 1.1–1.3; although this is a relative small degree of risk, in large cities it does mean that tens of thousands are likely to be affected.[29]

There is therefore strong evidence that current levels of air pollution in the towns and cities of Europe and North America are having an effect on the respiratory and perhaps the cardiovascular health of the vulnerable and those with a history of these diseases. The balance of evidence suggests that respirable particulate matter is the major cause, although some recent studies have suggested that the effects can also be related to O_3.

The possible role of air pollution in the increased incidence and prevalence of asthma in countries such as Europe, North America, Australia, and New Zealand is controversial and far from clear. It has to be stressed that while asthma prevalence and hospital admissions for this disease have shown a dramatic increase in the last two decades in the USA and elsewhere, the trends in the major air pollutants have been distinctly downwards. There is no obvious correlation between air pollution and asthma hospital admissions in the UK where the prevalence of childhood asthma has increased by about 50% over the past 20–30 years. During this period SO_2 and particulate matter concentrations have declined markedly. In Britain and elsewhere there is little correspondence between the geographical distributions of childhood asthma and air pollution, and in particular there appears to be no difference in the prevalence of the disease between children living in rural and urban areas.

---

[29] J. Schwartz, 'Health at the Crossroads: Transport Policy and Urban Health', ed. T. Fletcher and A. J. McMichael, John Wiley & Sons, Chichester, 1997, Ch. 2.

9 EFFECT OF AIR POLLUTION ON PLANTS

The main pollutants which affect vegetation are SO_2, NO_x, O_3, fluorides, and ethylene. The effects on vegetation include reduction in yields, visible leaf damage, loss of sensitive species and therefore reduction in diversity of plant communities, and changes in sensitivity to other stresses. In the UK, concentrations of SO_2 have decreased in recent decades, NO_x concentrations are elevated but steady, and there would appear to be increased incidences of photochemical pollution producing elevated O_3 concentrations. Fluorides and ethylene are more localized pollutants. Oxides of nitrogen and sulfur are ultimately metabolized by the endogenous nitrogen and sulfur biochemical pathways and thus can contribute to the normal N and S metabolism in plants. At relatively low exposures ecosystems absorb or act as a sink to the influx of pollutants and little or no harm can be detected. Indeed, there are reports that growth can be stimulated in N and S deficient soils.

In the past the main concern was focused on visible damage and impaired growth due to SO_2 (and acid aerosols) and smoke around point sources and in cities. One of the earliest works in urban areas was in the early twentieth century in Leeds and a nearby industrial area where air pollution was shown to cause reduction in yields of lettuces by up to 70% and extensive damage to evergreen privet during autumn and winter months. These changes were correlated with soot and sulfite deposits. Certain conifer species are now known to be especially sensitive to sulfur dioxide and other pollutant gases. For a number of years, attempts to establish conifer plantations in southern parts of the Pennine Hills in the North of England generally proved unsuccessful. At one stage Scots Pine (*Pinus sylvestris*) was either absent or very sparse over a 50 km wide corridor, downwind of the major conurbations of Greater Manchester and Merseyside. The frequency of the species inversely corresponded to mean winter sulfur dioxide concentrations in the area, and no trees occurred wherever SO_2 concentrations exceeded 0.076 ppm. Indigenous grassland has also been shown to be tolerant of SO_2 pollution.[16] With the decline of sulfur dioxide concentrations there has been marked improvements in conifer plantation in the area.[30] General improvement in urban air quality can be seen in a number of areas of the UK from the gradual recolonization by lichens on the bark of trees, including in parts of London.[31]

Growth chamber experiments which subject plants to unfiltered (ambient) air and filtered (control) air provide strong supportive evi-

[30] J. N. B. Bell, 'Gaseous Air Pollutants and Plant Metabolism', ed. M. J. Koziol and F. R. Whatley, Butterworth, London, 1984, Ch. 1.
[31] C. I. Rose and D. L. Hawksworth, *Nature*, 1981, **289**, 5795, 289–292.

dence of urban air pollution damage to vegetation. One of the earliest studies (1950–51), examined the effect of pollutants in the air of a Manchester suburb on the growth of perennial ryegrass, *Lolium perenne*. Significant reductions in growth were found in the grass exposed to the ambient air which had a mean SO_2 concentration of 0.07 ppm. Later investigations in the Sheffield area recorded reduced yields in plants exposed to much lower mean SO_2 concentrations. In other studies using more realistic open-top chamber designs, growth reductions only occurred at higher SO_2 exposures. The differences may be due to the different design of the chambers, although other factors may be involved. For instance, chronic SO_2 injury can be enhanced if plants are exposed to SO_2 during winter, when they are growing more slowly. Furthermore in these studies, only SO_2 was measured and it is now evident that other phytotoxic pollutants would have been also present, *e.g.* NO_x and O_3. The Sheffield studies coincided with some of the highest O_3 concentrations yet recorded in Britain.[30]

Vegetation injury caused by photochemical smog was first reported in the Los Angeles basin in 1944 and it has continued to be a chronic problem in southern California. It was established that O_3 was the main phytotoxic agent in the smog complex. Ozone has caused widespread injury to agronomic and horticultural crops and natural and managed forest ecosystems, not only in California but also in many other states where meteorological conditions and primary pollution concentrations were favourable. Ozone has been the most economically damaging air pollutant to vegetation in the USA. Large scale injury to tobacco crops in the eastern USA has been reported over a number of years. Extensive pine needle damage to Ponderosa and Jeffrey pines in western locations and white pine in the east were due to atmospheric oxidants. In the mixed conifer forest ecosystems in the mountains of Southern California, the dominant tree species, Ponderosa and Jeffrey pines, for many years, suffered annual mortalities of about 3%, with the result that hundreds of thousands of trees have died. In the 1970s it was estimated that in the San Bernardino forest 46 230 acres had suffered severe ozone-type injury, 53 920 acres moderate injury, and 60 800 acres light or no injury.

Evidence that ambient ozone concentrations in Britain during the summer can inflict deleterious effects on vegetation comes from a series of experiments, using open-top chambers, at a rural site near Ascot in Berkshire, some 32 km west of London. At this location SO_2 and NO_x concentrations are generally low, but episodes of high ozone concentrations have been regularly recorded. During three such incidents between 1978 and 1983, *Pisum sativum*, *Trifolium repens*, and *T. pratense*, grown in open-top chambers and receiving unfiltered air, developed visible leaf necrosis typical of O_3 damage. Concentrations of O_3 during the course of

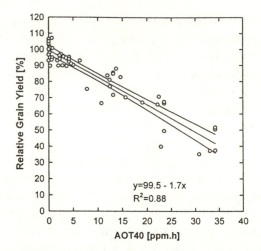

Figure 10 *The relationship between relative grain yield of spring wheat and ozone exposure expressed as AOT40 for 3 months, based on data from eight European open-top chamber experiments*
(From Fuhrer, J. (1994) in: Fuhrer, J. and Ackermann, B., Critical levels for ozone; a UN-ECE workshop report. FAC Report No. 16, Swiss Federal Research Institute for Agricultural Chemistry and Environmental Hygiene, Liebefeld-Bern)

these experiments exceeded 0.1 ppm, which is considered to be the threshold above which visible leaf necrosis appears in these sensitive plants. Field observations in the surrounding areas revealed symptoms typical of O_3 injury in a variety of crop plants. It has since been realized that *Pisum sativum* crops grown in the UK for a number of years, frequently developed necrotic lesions typical of ozone injury without the cause being known.[32] Field observations have established that many commercially grown crops in Europe develop visible injury, such as chlorosis or bronzing, due to ambient levels of O_3 and that these symptoms commonly develop in response to short term episodes rather than longer term average concentrations.[4]

Studies have shown that increasing the concentration of O_3 has a greater impact than increasing the period of exposure and that levels over and above a threshold concentration give a better fit to reductions in growth. The European open-top chamber experiment (EOTC), conducted in 9 countries over a five year period, was a major international research study on the effects of O_3 over the complete live-cycle of several crop species. Figure 10 shows the linear relationship between O_3 exposure, expressed as the accumulated exposure over a threshold of 40 ppb (AOT40), and the relative yield of spring wheat. Similar relation-

[32] M. R. Ashmore, Proceedings of an International Workshop on 'The Evaluation and Assessments of the Effects of Photochemical Oxidants on Human Health, Agricultural Crops, Forestry, Materials and Visibility', Swedish Environmental Research Institute, (IVL) Goteborg, 1984, p. 92.

ships have recently been established for tree species such as beech and Norway spruce. Beech is more sensitive to O_3 with an AOT40 value of 10 000 ppb h associated with a 10% reduction in growth. Recent work with semi-natural vegetation suggests that changes in species composition are more important than either growth reduction or visible injury. In grassland communities a marked shift from forb to grass species has been observed at comparable O_3 exposures to those which produce yield reduction in spring wheat.[4]

Other effects of air pollutants include damage to the epicuticular wax layer of leaves and needles which can lead to increased water loss and there are reports that air pollutants can delay the onset of winter hardening. There is also evidence that air pollutants acting singly and in combination (SO_2, SO_2/NO_2 and O_3) have an effect on the relative distribution of growth above and below ground, with root growth being more severely affected than shoot growth. This may be a consequence of interference with phloem translocation and assimilation in the plant. There are also reports that plants in polluted areas are more susceptible to attack by insect pests and exposures of plants to small or medium doses of SO_2 and/or NO_2 result in increases of the population growth of aphids feeding on the plants. This in turn can result in significant increases in pest damage to plants.[8,33]

The past two decades has seen a change in emphasis from the impact of local sources of air pollution to effects across national boundaries. Thus extensive areas of Europe and North America are subject to elevated levels of gaseous air pollution and in such instances a polluted air mass rarely contains a single phytotoxic agent but a complex gas–aerosol mixture that includes varying proportions of SO_2, nitrogen oxides (NO_x), ozone (O_3), and acid aerosols. This has led to the application of the critical level concept whereby each air pollutant for a defined set of environmental conditions is given a limiting value, above which available evidence predicts that adverse effects are likely to become apparent. The receptors are classified according to the type of vegetation—crops, forests and woodlands, and semi-natural vegetation. Modifying factors such as winter stress, or nutrient deficient soils can be taken into account. Taking semi-natural vegetation as one category, the critical levels for SO_2 and NO_x are 20 and 30 μg m^{-3} respectively and that for O_3 is expressed as 3000 ppb h (ppb O_3 above a threshold of 40 ppb accumulated during daylight hours). In the UK it has been estimated that whilst the land area of semi-natural vegetation where the critical levels of SO_2 and NO_x are exceeded is in the region of 1%, that of

[33] 'Air Pollution and Forest Ecosystems in the European Community', ed. M. R. Ashmore, J. N. B. Bell, and I. J. Brown, Air Pollution Research Report 29, Commission of the European Communities, 1990.

O_3 is of the order of 70%, highlighting the current importance of this pollutant.

These critical levels are based on individual pollutants and do not take account of pollutant combinations. This is a very complicated subject, as research has shown that responses of vegetation to pollutant mixtures is not simply the sum of the responses to individual pollutants. For example SO_2 and NO_2 can act synergistically and there is evidence that NO_2 and O_3 may behave antagonistically. Thus the exposure–response relationship for one pollutant may be modified by the presence of another and the exposure at any one location and time is a complex function of source strength, meteorology, and atmospheric chemistry.

10   ECOLOGICAL EFFECTS OF ACID DEPOSITION

Over large parts of western Europe and north-eastern North America streams, rivers, and lakes have become progressively more acidic. It is now generally accepted that acid deposition with major acidic, or acidifying ions is the cause of freshwater acidification in geologically sensitive areas. Such areas are those with slowly weathering granites and gneiss rocks that support acidic and weakly buffered soils. Freshwater acidification is typified by a loss of acid neutralizing capacity (essentially the carbonate buffering system), decrease in pH by as much as 1–2 pH units, increases in sulfate, nitrate, ammonium ions, aqueous aluminium, and metals such as manganese and zinc. Acidification is responsible for the loss and depletion of fish populations from numerous freshwater ecosystems in parts of Norway, Sweden, UK, Canada, and USA.

Freshwater organisms generally maintain their internal salt concentration by active uptake of ions (in particular Na^+ and Cl^-) from water against a concentration gradient. Substantial experimental evidence and field data indicate that fish mortality at low environmental pHs is primarily caused by a failure to regulate internal salt concentrations at gill surfaces. A characteristic symptom of acid-stressed fish is therefore a reduction in the salt or electrolyte ion concentration of blood plasma.[34] At an early stage it was realized that mortalities were not simply a function of acidity. Field data indicated that fish mortality occurred in waters of pH 5, whereas laboratory experiments with purely acid solutions failed to reproduce these results and significant mortalities only occurred at pH 4. It was later shown that the dissolved concentration and chemical species of aluminium was the important factor influencing toxicity in acidified rivers and lakes. This common element

[34] H. Leivestad and I. P. Muniz, *Nature*, 1976, **259**, 391.

in rocks is more soluble in acidic environments, and various surveys of lakes and rivers have revealed that concentrations of aqueous aluminium increase in acidified waters. Because of reactions with hydroxide ions, aluminium toxicity to fish is pH dependent and the most bioavailable forms of aluminium occur at pH 5. Several reports have shown the effects of the interaction of acidity and soluble aluminium on fish survival. The survival of brown trout fry is reduced in concentrations of aluminium of $250 \ \mu g \ l^{-1}$ and at lower concentrations if calcium is low. Effects on growth have been found at much lower concentrations, for example the growth rate of brown trout is reduced at a concentration greater than $20 \ \mu g \ l^{-1}$, a concentration much lower than that found in acid waters. The most toxic combinations were found with elevated aluminium concentrations at pH 5. Increasing calcium concentrations from 0.5 to $1.0 \ mg \ l^{-1}$ offset the effect of increasing aluminium and acidity on the survival of brown trout. Sulfates, fluorides, and soluble organic matter in acidified waters which form complexes with aluminium can also have a significant ameliorating influence on aluminium toxicity.[35]

In summary, loss of fish populations can be expected in acidified clear water lake ecosystems, low in organic matter and low in calcium. Such environments tend to be characteristic of mountain areas and in southern Norway in particular these are the regions where the loss of fish populations was first recorded.

The effects of acidification are not restricted to fish; all facets of freshwater ecosystems are affected. The assemblages of all the major groups, micro-organisms, plants and animals, alter considerably and there is a general tendency for a reduction in species diversity at all trophic levels. For many groups this is accompanied by a reduction in productivity and a loss or decline of populations, *e.g.* several species of crustacea (snails and crayfish), amphibians and fish.

The following summary of the effects of acidification on the major trophic levels in freshwater systems is based on a review of the subject by Muniz.[36]

Decomposers. Accumulation of undecomposed and partly decomposed organic matter is taken to indicate reduced activity of bacteria, fungi, and protozoans in acidified waters. Some field and experimental studies have shown decreased rates of decomposition, but others have found no evidence of any changes.

Primary Producers. Reductions in the numbers of species of several

[35] D. J. A. Brown, 'Acid Deposition, Sources, Effects, and Controls', ed. J. W. S. Longhurst, British Library Technical Communication, 1989, p. 107.
[36] I. P. Muniz, 'Acid Deposition—Its Nature and Impacts', ed. F. T. Last and B. Watling, *Proc. Roy. Soc. Edinburgh*, 1991, Section B, **97**, 228.

phytoplanktonic algae, especially the green algae (Chlorophyceae) have been reported. Dinoflagellates dominate the plankton of many acid lakes. Fewer species of periphyton occur on submerged surfaces and these habitats are commonly superseded by mass encroachments of filamentous algae (*e.g. Mougeotia, Zygogonium,* and *Spirogyra*). In many lakes in Sweden macrophyte populations of *Lobelia* have declined and have been invaded by *Sphagnum* species. Synoptic surveys generally indicate that biomass and primary productivity are less in acidified lakes, although experimentally acidified lakes have shown no change and even increases in biomass.

Zooplankton. Species diversity decreases with increasing acidity, particularly below pH 5.5–5.0 and this is usually followed by a reduction in biomass. Acid sensitive species of *Daphnia* and *Cyclops* frequently disappear below pH 5.

Benthic macroinvertebrates. Many species of mayflies, amphipods (freshwater shrimps), crayfish, snails, and clams are very sensitive to acid conditions. In Norway, snails were found to be absent in waters below pH 5.2 while mussels disappeared at pH 4.7. Recruitment failure and reduced growth are important causes of the elimination of species. Crayfish decline has been due to the effects of acidity on the eggs and juvenile stages, and the moulting stages are particularly sensitive. A similar situation has been found for freshwater shrimps, mayflies, and snails.

Fish. There are numerous examples for the loss of individual species and changes in the composition of fish communities in acidified waters. Minnows seem to be some of most sensitive fish and population declines often start below pH 6. An approximate order of sensitivity to acid stress is as follows (beginning with most sensitive): roach, minnow, Arctic char, trout, European cisco, perch, white pike and eels are consistently the most tolerant species. The younger life stages, particularly newly hatched or 'swim-up' fry just starting to feed, are the most sensitive stages to acid stress. As a consequence many acid-stressed populations are dominated by older fish due to recruitment failure.

Amphibians. Reproductive failure is the main effect of acid stress on amphibians. In Sweden, populations of the frog *Rana temporaria* have been lost and reproductive failure of the toad *Bufo bufo* has occurred.

Birds. Studies have shown that the breeding density of dippers in Wales (songbirds that feed almost exclusively on aquatic invertebrates) has decreased as acidity has increased. This has been related to the decrease in abundance of mayflies and caddis fly larvae in acidified streams. In general, some species of songbirds that breed alongside acidified lakes and streams produce smaller clutches with thinner eggshells, and breed later with fewer second clutches. Evidence also indicates that their young grow less rapidly and suffer more mortality.

Although the loss of decline of populations of animals and plants is largely attributable to changes in water chemistry, indirect effects may also cause additional stress. These essentially arise out of changes in predator–prey relationships caused by a decline of food resources. For example, instances of increases in phytoplankton biomass may be due to a decrease in the performance of herbivores. The decline of the dipper is strongly associated with the disappearance of its prey from acidified streams.

Case Study 3. Lake Acidification; an example of a large scale field experiment. Experimental acidification of entire lakes has yielded very valuable information and increased understanding of the effects of acidification on communities and ecosystem function. Lake 223, in the Experimental Lakes Area of Canada, was acidified over an 8 year period during which time the pH was gradually reduced from 6.8 to 5.0. One of the surprising features of the study was that primary production and decomposition showed no overall reduction. However, the composition of the phytoplankton and zooplankton communities showed distinct changes. At pH 5.9 several key organisms in the lake's food web were severely affected; the opossum shrimps (*Mysis relicta*) declined from almost 7 billion to only a few animals and fathead minnows (*Pimephales promelas*) failed to reproduce. At pH 5.6 the exoskeleton of crayfish (*Oronectes virilis*) hardened more slowly after moulting and remained softer. The animals later became infected with a microsporozoan parasite.

When the acidification of Lake 223 was reversed, several biotic components recovered quickly. Fish resumed reproduction at pHs similar to those at which it failed on acidification. The condition of lake trout improved as the small fish on which they depended for food returned. Many species of insects and crustaceans returned as the pH was raised.[37,38]

Diatom communities in lakes have provided valuable information on the history of aquatic systems and in particular on their acidity. These organisms have characteristic siliceous scales which are preserved in lake sediments. Species of diatom can be classified according to their

[37] D. W. Schindler, K. H. Mills, D. F. Malley, D. L. Findley, J. A. Shearer, I. J. Davis, M. A. Turner, G. A. Linsey, and D. R. Cruikshank, *Science*, 1985, **228**, 1395.
[38] D. W. Schindler, T. M. Frost, K. H. Millerin, P. S. S. Chang, I. J. Davis, D. L. Findley, D. F. Malley, J. A. Shearer, M. A. Turner, P. J. Garrison, C. J. Watras, K. Webster, J. M. Gunn, P. L. Brezonik, and W. A. Swenson, 'Acid Deposition—Its Nature and Impacts', ed. F. T. Last and B. Watling, *Proc. Roy. Soc. Edinburgh*, 1991, Section B, **97**, 193.

relative tolerance of acid or alkaline conditions and assemblages of different species can be used to indicate the pH of lake water. Palaeo-limnological data from sediment cores have been used to construct the pH histories of lakes and in relation to recent changes they have provided evidence of the major causes. It is clear that sensitive lakes in the UK with moderate to high sulfur deposition have been acidified since the middle of the nineteenth century. Figure 11 shows the diatom assemblages at different depths from Lochnagar in the north-east of Scotland, the onset of acidification is evident from about 1890 when the acid neutralizing capacity had been exceeded and the pH of the lake decreased by 0.5 pH units.[39]

Evidence for widespread acidification of soils in sensitive areas of catchments showing surface water acidification was slow to change. Many such soils are naturally quite acidic and it was difficult to envisage what further changes would occur. However, the acidifying ions of sulfate and nitrate are important in bringing about soil chemical changes. In particular they promote the leaching of base cations, impoverishing the nutrient status of such soils further, creating stronger acidic conditions and promoting mobilization of aluminium. In less acidic soils with a higher base cation status and ones in which cation exchange processes act as a buffer, the first signs of acidification are increased leaching of base cations as hydrogen ions replace them on exchange sites. Ultimately the soil becomes more acidic and deficient in base cations, aluminium is mobilized and in turn also displaces further cations from the exchange sites.

Long term changes in the pH of soils of south-west Sweden have been shown to have occurred by repeating measurements that were done over 50 years previously on exactly the same soil plots. Figure 12 shows that in the surface litter layer of forests of equivalent ages, pH levels in 1984 were lower than in 1927; in lower horizon, pH of soil was independent of stand age and recent soil samples had a lower pH. Although slightly smaller in scale, similar results have been reported for soils in northern Sweden. The falls in pH are thought to be caused mainly by acid deposition although biological processes also could have been involved.

The critical loads approach is being applied over large areas of Europe and North America in an attempt to reverse surface water acidification. The objective is to reduce sulfur dioxide emissions such that the resulting acidic input to aquatic systems is equal to or less than the production of base cations from catchment weathering processes. A critical alkalinity

[39] V. J. Jones, R. J. Flower, P. J. Appleby, J. Natkanski, N. Richardson, B. Ripley, A. C. Stevenson, and R. W. Batterbee, *J. Ecol.*, 1993, **81**, 3–24.

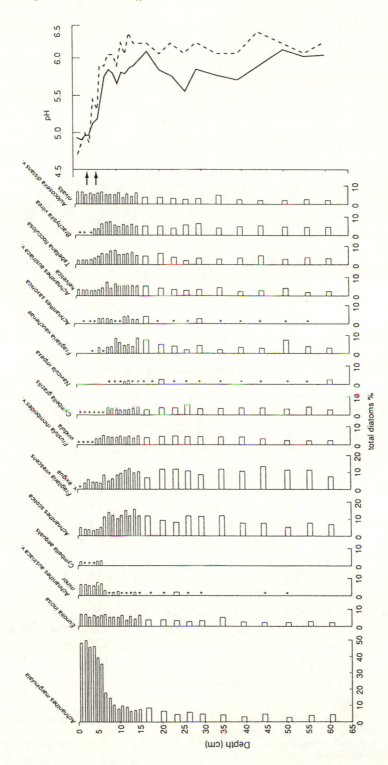

Figure 11 *Summary diatom total percentage frequency diagram (left) and pH reconstructions (right), based on weighted averaging (——) and multiple regression (– – –) methods from Lochnagar. Arrows show 1900 (lower) and 1950 (upper) levels at each site based on* ^{210}*Pb dating (Reproduced with permission from J. Ecol., 1993,* **81***, 3–24)*

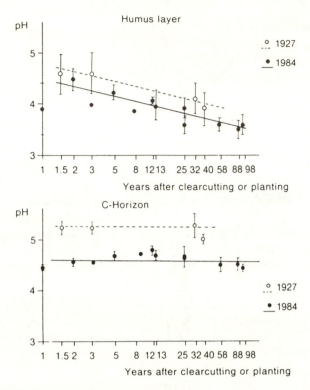

Figure 12   *pH in water suspensions of fresh soil as a function of the logarithmic stand age of Norway spruce stands at the Tönnersjöheden experimental area in southwest Sweden. Data are shown for the humus layer (O horizon) and the C horizon. Vertical bars show the standard deviation of the mean for* n > 3
(Hallbäcken, L. & Tamm, C. O., *Scand. J. Forest Res.*, 1986, **1**, 139–143)

value is chosen to protect vulnerable aquatic species or as a critical point to aim towards in reducing the acid inputs. These reductions in acid load are translated into reduction of SO_2 emissions which are needed to meet the critical load. Recently it has been argued that whilst a fixed alkalinity value may define the critical load for a particular species, it does not represent the overall alkalinity of a lake. Palaeolimnological methods using diatoms to reconstruct historical pH values of lakes show that the pH has varied by at least one pH unit throughout its history. It has been suggested that palaeolimnological information obtained from diatom studies provide a useful means of setting critical loads for a particular system.[40] Critical loads with respect to soil acidification are based on ratios of base cations (*e.g.* Ca^{2+}) relative to Al concentrations in the soil solution.[41]

[40] R. W. Batterbee, T. E. H. Allott, S. Juggins, A. M. Kreiser, C. Curtis, and R. Harriman, *Ambio*, 1996, **25**, 366–369.
[41] H. Løkke, J. Bak, U. Falkengren-Grerup, R. D. Finley, H. Llvesniemi, P. H. Nygaard, and M. Starr, *Ambio*, 1996, **25**, 510.

11 FOREST DECLINE

Reports of problems in the health of forests in western Europe first appeared in the late 1970s when white fir stands at high altitude in Germany were found to be in an unhealthy state and then in the early 1980s Norway spruce (*Picea abies*), the dominant silviculture tree in central Europe, began to show increasing signs of defoliation and needle discolouration. Extensive death of the fine root system of trees has also been reported. These symptoms have since been observed in the UK, France, the Benelux countries, Scandinavia, the Alps (Switzerland and Austria), the Czech and Slovak Republics, Poland, and the north-eastern USA. The decline in forest health is not confined to Norway spruce; other conifer and broad leaved species have been affected. It has been argued that whilst regional declines of individual species can occur, for example due to disease epidemics, the synchronous decline of several species across a broad area suggests a common cause such as air pollution. In 1987 and 1988 a survey of the severity of damage across the European Community (EC) showed that the forests of southern Europe were in better health than those of central and northern Europe. In the major coniferous forests of northern Europe, 15%, 22% and 28% of *Picea abies, Picea sitchensis*, and *Abies alba* trees were unhealthy. Among the hardwoods of northern and central Europe, 16%, 15% and 12% of *Quercus robur, Quercus petrea*, and *Fagus sylvatica* showed symptoms of damage.[33]

Several reports have been published throughout Europe on the subject of forest decline. However, definitive cause–effect relationships of forest decline have yet to be established. In the early stages it was attributed to acid deposition, through its effect on leaching of nutrients from foliage, loss of nutrients from soils, and increasing bioavailable forms of toxic metals such as aluminium in soils. It is believed that the actual explanation probably involves a combination of factors which include natural and pollution stresses. A number of hypotheses have been put forward. They include multiple stress, soil acidification and aluminium toxicity, interaction between ozone and acid mist, magnesium deficiency, and excess nitrogen deposition. There are different regional types of damage and decline and in Norway spruce five different types have been identified. The widely recognized needle yellowing at higher altitudes appears to be associated with Mg deficiency brought on by a combination of factors such as acid deposition, management practices, and abiotic factors. It has been suggested that excess nitrogen deposition could also play a part by triggering increased growth, resulting in Mg deficiency and/or rendering trees more susceptible to damage by frosts and pests.[42]

[42] L. W. Blank, T. M. Roberts, and R. A. Skeffington, *Nature*, 1988, **336**, 27–30.

It is difficult to reconcile pollution patterns with the distribution of forest decline on a regional scale such as Europe. The distribution of air pollution which is typically a complex mixture of gases (SO_2, NO_x, O_3) and acidic aerosols and ammonium sulfate varies markedly from region to region and over short distances within regions. Equally, forest condition varies enormously across the region and between the major forest species. Field observations on Norway spruce have shown that damage was associated with areas of Europe where air pollution is greatest, in particular deposition patterns of S and N pollutants and summer concentrations of O_3.[43]

Forest decline is frequently associated with foliar deficiency of certain nutrients, and in particular magnesium deficiency, *e.g.* at elevated sites in western Germany, although in other areas potassium deficiency has been found. However, the cause of the nutrient deficiency is unknown. There are reports of increased foliar leaching of mineral nutrients due to acidic deposition or other pollutants but this is thought to be of secondary importance and the supply of mineral nutrients from the soil is probably more critical. There is growing evidence that the deposition of sulfur and nitrogen pollutants have significantly modified soil chemistry and plant nutrition. Increased nitrates and sulfates in the soil solution can increase leaching of cations such as calcium and magnesium from soils and enhance soil acidification. Soil solution chemistry, notably decreases in the ratio of calcium and/or magnesium to aluminium, is known to affect root development and hence water and nutrient uptake.[33,43,44] Recent work has shown that defoliation in *Picea abies* in Norway is associated with reduced Mg and Ca in soil and it has occurred in areas with the highest acidic deposition. It was concluded that acidic pollution in excess of estimated critical loads was consistent with the pattern of defoliation.[45]

12   EFFECTS OF POLLUTANTS ON REPRODUCTION AND DEVELOPMENT: EVIDENCE OF ENDOCRINE DISRUPTION

There are a number of instances of pollution-related events in wildlife populations where changes in reproduction and development have occurred which have led to population declines and on occasions to extinctions. There is growing evidence that reproductive and developmental abnormalities observed are a consequence of effects on the

[43] E.-D. Schulze and P. H. Freer-Smith, 'Acid Deposition—Its Nature and Impacts', ed. F. T. Last and B. Watling, *Proc. Roy. Soc. Edinburgh*, 1991, Section B, **97**, 155.
[44] E.-D. Schulze, *Science*, 1989, **244**, 776–783.
[45] C. Nellemann and T. Frogner, *Ambio*, 1994, **23**, 255–259.

endocrine system, although the fundamental mechanism has yet to be established.

12.1 Eggshell Thinning

A wealth of literature exists on the relationship between DDT (and its metabolite DDE) and the phenomenon of eggshell thinning in various bird populations.[9,46] It was first discovered in the peregrine falcon (*Falco peregrinus*) in the UK. This particular falcon is widespread throughout Eurasia and North America, feeding almost entirely on live birds caught in flight. From about 1955 onwards, for no obvious reason, the numbers of falcons rapidly declined in southern England and subsequently this decline spread northwards to parts of the Scottish highlands. By 1962 in the UK, some 51% of all known pre-war territories had been deserted and this figure was as high as 93% for southern England. At the time biologists in several countries were also investigating populations of birds that had slowly declined to critical levels. It emerged that they all had residues of DDT, its metabolites, other chlorinated hydrocarbon insecticides, polychlorinated biphenyls (PCBs), and other chemicals in their tissues.

Ratcliffe[47] examined the eggshells of peregrine falcons and sparrow hawks in the UK and using an index of eggshell thickness (eggshell weight/(egg length × egg breadth)) found that during 1947 and 1948 this index decreased significantly by, on average, 19%. Eggshell thinning has been found in several American species of birds, notably the bald eagle, osprey, and peregrine falcon. In a study of over 23 000 eggshells of 25 bird species, shell thinning was found in 22 of the species and a correlation was found between the degree of thinning and DDE residues in the eggs. Experimental work with mallards and American kestrels showed that DDE caused eggshell thinning. Lincer[48] compared the degree of thinning with the concentration of DDE in eggs of the American kestrel (*Falco sparverius*) in field populations and in eggs produced by captive birds fed food dosed with DDE. Shell thickness decreased with increasing DDE residues in the eggs of both populations (Figure 13).

It is well accepted that DDE is the primary cause of eggshell thinning in many kinds of birds and that some bird species are more susceptible than others to DDE-induced shell thinning. Eggshell thinning was definitely a major cause of low reproductive success and population

[46] T. J. Peterle, 'Wildlife Toxicology', Van Nostrand Reinhold, New York, 1991.
[47] D. A. Ratcliffe, *J. Appl. Ecol.*, 1970, **7**, 67.
[48] J. L. Lincer, *J. Appl. Ecol.*, 1975, **12**, 781.

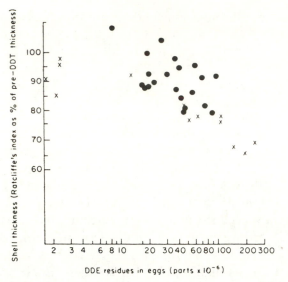

Figure 13   *The relationship between mean clutch shell thickness and DDE residue of American kestrel eggs (*Falco sparverius*) collected from wild populations in Ithaca, New York, during 1970 (●) and the same relationship experimentally induced with dietary DDE (×)*
(Reproduced with permission from *Appl. Ecol.*, 1975, **12**, 781)

decline in some species, but chlorinated hydrocarbons probably contributed to declines in other ways. In general terms, reproductive trouble tends to increase as shells become thinner and thinning of over 20% is likely to result in reproductive failure and population decline. In the case of peregrine falcon populations in the UK, the incidence of broken eggs within clutches rose from the normal 4% to 39% during the period 1951–66. Although thinner eggshells resulted in an increased breakage of eggs and, in turn, a reduction in breeding success of a particular pair of falcons, it was realized that the overall size of the UK population was unaffected by this effect at the time. The actual population decline took place at least five years after the onset of eggshell thinning. This coincided more closely with upsurge in the use of dieldrin as an anti-fungicidal seed dressing. Although it has never been substantiated, it is widely believed that the abrupt decline was due to a combination of factors, involving eggshell thinning and the later ingestion of toxic doses of dieldrin derived from a diet of contaminated pigeons.

One of the earliest examples in which PCBs were implicated in bird mortalities was an incident in the autumn of 1969 when 15 000 dead and dying seabirds were washed up on the coasts around the Irish Sea. Nearly all the birds were guillemots (*Uria aalge*) and almost all the victims were adults just after the annual moult. The bulk of the birds were washed up after storms in late September, but even during the previous fine calm weather some dead and dying birds were spotted. Chemical analyses of dead guillemots for a variety of pollutants revealed the presence of high concentrations of DDE and in particularly very high levels of PCBs in their livers. It has been suggested that moulting, followed by the stress of withstanding stormy conditions resulted in mobilization of fat reserves and in turn release of toxic doses of PCBs into the blood stream.[1]

Experimental studies with mink and ferret have established that PCBs are highly disruptive of reproductive processes. In the American mink (*Mustela vison*) tissue concentrations of over $50 \, \text{mg kg}^{-1}$ PCB are associated with reproductive failure. The otter, which is a closely related species to the mink, has experienced population declines over wide areas of Europe since the 1950s and it has been found that animals from populations showing the greatest decline commonly had PCB concentrations in their tissues of over $50 \, \text{mg kg}^{-1}$.

12.2 GLEMEDS

Organochlorine compounds such as PCBs, DDT, and TCDD have been associated with physiological, reproductive, developmental, behavioural, and population level problems in fish-eating birds of the Great Lakes for over 30 years.[10] Common features have been mortalities and deformities of eggs and chicks which led to the condition being called GLEMEDS (Great Lake Embryo Mortality, Edema, and Deformities Syndrome). The symptoms showed a number of similarities to chick-edema disease in poultry caused by exposure to dioxin contaminated food which was first recognized in the 1950s. The syndrome first came to light in common terns and herring gulls in the 1970s with observations of high incidences of egg and chick mortalities and deformities, particularly in colonies from Lake Ontario. By the mid-1970s organochlorine concentrations had declined and this corresponded with the disappearance of the gross effects in Lake Ontario herring gulls. During the same period there were declines in bald eagle populations. Subsequent studies reported stronger associations with PCBs and dioxins, *e.g.* deformities and egg mortality in colonies of double crested cormorants and Caspian terns were correlated with dioxin contamination and with specific congeners of PCBs in Foster's terns from Lake Michigan. Reproductive and developmental effects have been found in other wildlife populations, *e.g.* egg mortalities and deformities in snapping turtles from the Great Lakes basin, and increased tumour incidence in the Beluga whale in the St Lawrence have been associated with levels of organochlorines and in particular PCBs. Recently developmental problems in infants from regions of Lake Michigan and Lake Ontario have been related to maternal exposure to organochlorines as a result of consumption of contaminated fish.

Between 1992 and 1994 immunological responses and related variables were measured in pre-fledgling herring gulls and Caspian terns at colonies across a broad range of organochlorine contamination (mainly PCBs), as measured in the eggs. There was a strong exposure–response relationship in both species between organochlorines and suppressed T-cell-mediated immunity. Suppression was most severe (30–45%) in colonies in Lake Ontario (1992) and Saginaw Bay (1992–94) for both species and in western Lake Erie (1992) for herring gulls.[49]

Although most of the evidence is through correlation, taken as a whole there is good reason to believe that the reproductive and developmental problems highlighted are due to organochlorines with PCBs and dioxins as the major problem agents. Whilst inputs of organochlorines to the Great Lakes have largely been controlled and concentrations have declined there are still measurable levels in wildlife and concentrations in fish remain a significant hazard. The residual levels are due to the persistence of these substances, leachates from landfill sites, and inputs from the atmosphere.

12.3 Marine Mammals

There is a history of reproductive problems in seal populations in the Baltic and in the Dutch Wadden Sea. Between 1950 and 1970 the population in the Wadden Sea dropped from more than 3000 to less than 500 animals. The reduction in numbers is associated with low breeding success. The populations from both the Baltic and Wadden were found to contain high levels of organochlorine compounds and in particular PCBs.

Case Study 4. Epizootic disease and impaired immune response in marine mammals. There have been a number of major disease outbreaks among seals and dolphins which have been attributed to infection with known or newly recognized morbilliviruses. In 1988 the previously unrecognized phocine distemper virus caused the death of 20 000 harbour seals in north-western Europe. Other morbilliviruses have been shown to be the sources of infections in porpoises and dolphins, and mass mortalities due to this type of virus occurred among striped dolphins in the Mediterranean from 1990 to 1992.[50] The severity and

[49] K. A. Crassman, G. A. Fox, P. F. Scanlon, and J. P. Ludwig, *Environ. Health Perspect.*, 1996, **104**, (suppl. 4), 829–842.
[50] R. L. De Swart, T. C. Harder, P. S. Ross, H. W. Vos, and A. D. M. E. Osterhaus, *Infect. Agent. Dis.*, 1995, **4**, 125–130.

extent of these virus related diseases has led to speculation that pollution and in particular the bioaccumulation of organochlorines could have had damaging effects on the immune systems of the animals and as a consequence made them more vulnerable to infection.

Semi-field experiments in which groups of harbour seals were fed fish contaminated with PCBs have shown a strong link with reproductive failure. Experiments were over two years or more and during this time one group of harbour seals were fed a diet of fish from the contaminated Baltic Sea and another group were fed relatively uncontaminated fish from the Atlantic. The reproductive success of the group receiving the higher dose of PCBs was significantly lower than the other group with the lower dose of PCBs.[51] Follow up studies have shown more conclusively that PCBs are the main group of pollutants associated with these effects and immunotoxicological studies under the same semi-field conditions have demonstrated impaired immune responses in animals with elevated body burdens of PCBs.[52]

12.4 Imposex in Gastropods

Case Study 5. Tributyltin and imposex in gastropods. Sex abnormalities in neogastropod molluscs were first recorded in 1970. Many females in dogwhelk populations (*Nucella lapillus*) from Plymouth Sound, UK were found to have penis-like structures, this was followed by the reporting of male characteristics in female American mud snails from the Connecticut coast, USA. This development of male characters in females is referred to as imposex. In the 1970s reproductive failure and major population declines of the English native oyster occurred and the introduced Pacific oyster, *Crassostrea gigas*, at certain locations along the British east coast and the French west coast, started to show poor growth and unusual thickening of their shells. Populations of both gastropods and bivalve molluscs showing the severest abnormalities were invariably in the vicinity of yachting marinas and it was later shown that their tissues had a high tin content. In the 1980s it was firmly established that antifouling paints containing tributyltin (TBT) were responsible for these effects. This was confirmed by laboratory tests which reproduced imposex in gastropods and shell thickening in the bivalve molluscs at the same concentrations of TBT found near to yachting marinas and it was also shown that the tissue concentrations

[51] P. J. H. Reijnders, *Nature*, 1986, **324**, 456.
[52] R. L. De Swart, P. S. Ross, J. G. Vos, and A. D. M. E. Osterhaus, *Environ. Health Perspect.*, 1996, **104**, (suppl 4), 823–827.

of TBT associated with these effects were the same in field surveys and laboratory experiments.[53]

Many other species of gastropods have been found to exhibit TBT-induced masculinization of the female and the phenomenon is worldwide in occurrence with over 100 species showing varying degrees of imposex. The consequences of imposex vary according to species; in some the abnormality does not appear to affect reproduction, whilst in others such as the dogwhelk, abnormality can be so severe that breeding is prevented and as a result populations have declined drastically and even become extinct. In the dogwhelk the degree of imposex can be related to the level of TBT exposure and the onset of symptoms is associated with exposures of less than 1 ng/l TBT. Concentrations of 1–3 ng/l result in more pronounced masculinization with the penis size approaching that of males, although at this stage breeding is unaffected. However at 5 ng/l the development of the male sex organs begin to cause blockages in the female tissues and the snail in this condition is unable to breed. The no-effect exposure with respect to imposex is very much lower than that estimated from initial screening studies directed at other end-points and mortality. The use of TBT as an antifouling agent on small vessels was banned in the UK in 1987 and affected populations have started to recover. However it is still permitted for use on large vessels and imposex has been found in the edible whelk from the North Sea close to busy shipping routes.

Imposex is associated with elevated titres of testosterone and it has been shown that administration of testosterone to females produces masculine characters. One mechanism, for which there is good evidence, is that TBT inhibits an aromatase enzyme involved in the metabolism of testosterone to 17β-oestradiol.[54–56]

12.5 Endocrine Disruptors

The possible effects of chemicals on endocrine function in humans and wildlife has become one of the major issues. The functions of all organ systems are regulated by the endocrine signalling system and disturbances of this system resulting in hormone imbalances can lead to a long lasting damage especially during the early stages of the life cycle. Chemicals that can disturb normal endocrine homeostasis are referred to as endocrine disruptors.

[53] 'Tributyl Tin: Environmental Fate and Effects', ed. M. A. Champ and P. F. Seligman, Elsevier Applied Science, 1990.
[54] J. M. Ruiz, G. Bachelet, P. Caumette, and O. F. X. Donard, *Environ. Pollut.*, 1996, **93**, 195–203.
[55] G. W. Bryan and P. E. Gibbs, 'Metal Ecotoxicology: Concepts and Applications', ed. M. C. Newman and A. W. McIntosh, Lewis, Ann Arbor, 1991, p. 323.
[56] P. Matthiessen and P. E. Gibbs, *Environ. Toxicol. Chem.*, 1998, **17**, 37–43.

Studies of human populations indicate a deterioration in male reproductive health in recent decades. These include reports from Belgium,
Denmark, France, and the UK of declines in semen quality in the 1990s,
although it has to be said that there is criticism of the methods of analysis
and therefore interpretation of the results. Over the same period there
has been an increased incidence in testicular cancer, and certain disorders
of the male reproductive system, hypospadias and cryptorchidism,
appear to be increasing. The synthetic oestrogen, diethylstilbestrol
(DES) has been widely prescribed to women during pregnancy as a
means of combating abortion and other complications of pregnancy.
There is now strong evidence that this has resulted in a higher incidence
of reproductive disorders in the sons of these women. Diethylstilbestrol
has also been found to produce the same reproductive abnormalities in
animal studies.[57,58]

Several pollutants have been shown to be capable of disrupting
endocrine systems and in particular of mimicking the activity of natural
oestrogens. It is believed that these oestrogenic substances produce
effects through interaction with the oestrogen receptor. A variety of
substances, including the organochlorines, DDT, PCBs, and dioxins
have been shown to have oestrogen disrupting activity, although their
affinity for the oestrogen receptor is much lower than that of natural
oestrogens. There is therefore considerable concern that the widespread
reproductive and developmental problems reported in humans and
wildlife are related and that they are caused by substances which affect
the endocrine system in animals.

As long ago as 1968, DDT was reported to have oestrogenic activity[59,60] and recently its metabolite *pp'*-DDE was reported to have potent
antiandrogen properties (demasculinization).[61] Injecting DDT into eggs
of California gulls (*Larus californicus*) at similar levels to those found in
wild populations in the late 1960s produced feminization in male birds. It
has been observed in large gull populations that the female to male ratio
is skewed to females, suggesting a possible causal link.

Lake Apopka in Florida was polluted by DDT and related compounds following an accidental spill by a chemical company in 1980. In
the following years the alligator population declined whereas at other
nearby locations populations increased. The decline was associated with
reproductive disorders and in particular male alligators showed a
number of reproductive abnormalities. Sex differentiation in the alligator

[57] R. J. Kavlock *et al.*, *Environ. Health Perspect.*, 1996, **104**, (suppl. 4), 715–740.
[58] Medical Research Council, 'IEH Assessment on Environmental Oestrogens: Consequences to Human Health and Wildlife', University of Leicester, Leicester, 1995.
[59] D. M. Fry and C. K. Toone, *Science*, 1981, **213**, 922.
[60] J. C. Bitman *et al.*, *Science*, 1968, **162**, 371–372.
[61] W. R. Kelce *et al.*, *Nature*, 1995, **375**, 581–585.

is temperature dependent and can be altered by oestrogen treatment, so it is suggested that the organochlorine compounds were acting as oestrogenic agents impairing normal sexual development of male alligators.[62] Metabolites of PCBs have been shown to have strong oestrogenic effects in turtles.[63] These animals exhibit temperature-dependent sex differentiation and eggs incubated at 26 °C normally produce 100% males but when they were exposed to PCB metabolites under the same conditions the offspring were female.

Case Study 6. Vitellogenin as a biomarker for oestrogenic disruption in fish. Vitellogenin is a yolk protein in egglaying vertebrates and invertebrates. In fish it is produced in the liver of females and transported in the bloodstream to the oocytes. The synthesis of this protein is regulated by oestrogens and during egg production it is normally elevated in blood serum of females. However in males and immature females vitellogenin production can be induced by exposure to oestrogens and oestrogenic agents. In the UK elevated levels of vitellogenin have been reported in male freshwater fish exposed to sewage effluents and similar results have been reported in the USA and in marine fish exposed to sewage effluent. The oestrogenic agent responsible for these raised levels is unclear and may vary between locations depending on the composition of the contaminating effluent. It has been suggested that natural and synthetic oestrogens could be involved. Also another group of substances, the alkylphenols, which are commonly present in such effluents, have been shown to induce vitellogenin and impair reproductive development in male trout. These substances are degradation products of non-ionic surfactants produced during sewage treatment.[64-68]

The number of instances of reproductive and developmental impairment in wildlife and human populations is clearly of concern and perhaps indicative of a common cause. However the link between environmental oestrogens and human disorders has yet to be established and in wildlife the causal link is known in only a few cases. The endocrine system in all animals is complex and not fully understood and so the consequences of inhibiting or enhancing it at one point or another are largely unknown. The system also interacts with both the nervous and immunological

[62] L. J. Guilette, T. S. Gross, G. R. Masson, J. M. Matter, H. F. Percival, and A. R. Woodward, *Environ. Health Perspect.*, 1994, **102**, 680–687.

[63] J. M. Bergeron *et al.*, *Environ. Health Perspect.*, 1994, **102**, 780–781.

[64] J. P. Sumpter and S. Jobling, *Environ. Health Perspect.*, 1995, **103**, (suppl. 7), 173–178.

[65] C. M. Lye, C. L. J. Frid, M. E. Gill, and D. McCormick, *Mar. Pollut. Bull.*, 1997, **34**, 34–41.

[66] J. E. Harries, D. A. Sheahan, S. Jobling, P. Matthiessen, P. Neall, E. J. Routledge, R. Ryecroft, J. P. Sumpter, and T. Taylor, *Environ. Toxicol. Chem.*, 1996, **15**, 1993–2002.

[67] J. E. Harries, D. A. Sheahan, S. Jobling, P. Matthiessen, P. Neall, J. P. Sumpter, T. Taylor, and N. Zaman, *Environ. Toxicol. Chem.*, 1997, **16**, 534–542.

[68] S. Jobling and J. P. Sumpter, *Aquatic Toxicol.*, 1993, **27**, 361–372.

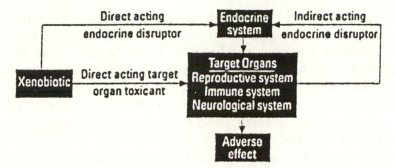

Figure 14 *Summary of the effect of xenobiotic substances on the endocrine system in animals, highlighting the possible interactions between the endocrine, nervous and immunological systems*
(Reproduced with permission from *Environ. Health Perspect.*, 1996, **104** (suppl. 4), 715)

systems and disruption to one of these by pollutants could have repercussions on the others. Figure 14 provides a summary of the possible interactions and it is interesting to note that some of the observations in wildlife, particularly in marine mammals, points towards possible interactions between these systems.

13 HYDROCARBONS IN THE MARINE ENVIRONMENT

Crude oil is a complex mixture of a great variety of organic substances with many different physical and chemical properties. The physical and chemical composition of oils of different origins also vary widely. In March 1978, the supertanker *Amoco Cadiz* released most of its 223 000 tonnes cargo of Iranian and Arabian crude oil into the coastal waters of Brittany. The oil was light, it had a low viscosity, and it contained 30–35% of highly toxic aromatic hydrocarbons, 39% saturated hydrocarbons, 24% polar material, and 3% residuals and there were more than 300 compounds in the fresh crude oil. Gullfaks oil which was involved in the *Braer* spill in the Shetlands in January 1993 is an unusual one in that it is a light biodegraded oil with a low concentration of such components as *n*-alkanes and asphaltenes and a high aromatic and naphthenic content.[69]

Following a spillage, oil is subject to various environmental processes that affect its chemical and physical composition, including evaporation, dissolution, and microbial and photochemical degradation. The lower molecular mass fractions are generally more volatile, more soluble, and more easily degraded and so an ageing or weathered oil has lost these components and it is made up of the more viscous higher molecular

[69] G. A. Wolff, M. R. Preston, G. Harriman, and S. J. Rowland, *Mar. Pollut. Bull.*, 1993, **26**, 567–571.

weight compounds. The more persistent and larger polynuclear aromatic hydrocarbons (PAHs) and their transformation products at this stage assume greater ecological significance. Despite sensitive and modern sophisticated instrumentation only a small spectrum of the compounds in crude oil are routinely determined. Consequently, it is not easy to identify the specific toxic components in oil. In the immediate aftermath of a spill, the lighter aromatic fractions such as monocyclic aromatic hydrocarbons, naphthalenes, and phenanthrenes are present and these are acutely toxic and partially water soluble. The effects on marine organisms at this stage are largely due to short term toxic effects and narcotization. The more persistent PAHs are chronically toxic and 5–6-ring PAHs have carcinogenic and mutagenic properties. Because of their relative insolubility in water but solubility in lipids they accumulate in organisms and in organic matter of suspended particles and sediments. They can remain bound to sediments for several years. The bioavailability in this form is therefore an important factor in the toxicity of PAHs.

A major oil spillage in the marine environment has immediate catastrophic results. However, such events constitute a relatively small percentage of the total quantity of petroleum compounds discharged each year. Chronic inputs from shipping, offshore wells, industry, rivers, and the atmosphere together make a very significant contribution. The fate and effects of these discharges are particularly difficult to monitor and assess, as the more toxic fractions which tend to be more volatile, soluble, and easily degraded in the environment are continually being removed from the oil. In consequence, much of our knowledge and understanding comes from investigations undertaken following a major accident.[70,71]

Accidents of this nature in the marine environment have considerable impact on marine birds and the adjoining coastal ecosystems, bearing in mind that changes in the composition of the oil occur fairly rapidly, and much of the acutely toxic components disappear in a matter of days. Deleterious effects are evident at sea, but because of the added influence of dispersal, they tend to be short lived covering a time course of days or at most weeks. The effect of oil reaching the coastal environment depends on the distance of travel and therefore the time taken for the oil to reach the coast. In this respect it is interesting to compare the *Torrey Canyon* incident (a spillage of 100 000 tonnes some 200 miles off the coast of Cornwall) with that of the *Amoco Cadiz*. By the time that oil was stranded on the Cornish beaches from the *Torrey*

[70] Royal Commission on Environmental Pollution, Eighth Report, Cmnd. 8358, HMSO, London, 1981.
[71] R. B. Clarke, *Phil. Trans. Roy. Soc. London*, B, 1982, **297**.

Canyon it was almost biologically inert and the damage that ensued was largely caused by dispersants used to clean the beaches. In the case of the *Amoco Cadiz* the spillage was only 1.5 nautical miles from the shore and a major part of the acutely toxic aromatic fraction was still present when the oil reached the coast causing considerable fatalities among marine life.

Oil washed onshore interacts with a variety of coastal features, extending from high energy eroding rocky promontories to low energy accumulating environments. The processes that degrade oil in the open water operate in the coastal environment but to differing degrees. In high energy and medium energy coastal systems the hydraulic action of breaking waves mechanically disperses and erodes the oil rapidly. During the *Braer* spill conditions were particularly stormy, so the oil was rapidly dispersed and acute fatalities were confined to the immediate vicinity of the wreck. By contrast, oil may persist for many years in the low energy environments that include lagoons, estuaries, and marshes, since in these areas wave energy is low and degradation is dependent on the slower processes of microbial degradation and dissolution. With time, attention shifts from the immediate short term acute effects of oil and focuses on the more persistent long term effects of PAHs in the low energy coastal systems.

Case Study 7. The impact of oil spills in coastal waters.   About 360 km of the Brittany coastal environment was polluted by oil following the wrecking of the *Amoco Cadiz*. This included rocky and sandy shores, salt marshes, and estuaries. During the first few weeks after the disaster a very heavy mortality or 'acute mortality crisis' affected the intertidal and subtidal fauna. Populations of bivalves, periwinkles, limpets, peracarid crustaceans, heart urchins, and seabirds were most severely affected. Populations of polychaete worms, large crustaceans, and coastal fishes were less affected. Highest mortalities were found in a 5 km radius around the wreck and in locations further afield where the oil was blown and accumulated ashore by the wind. At one of these locations (St Efflam) for example, mortalities included 10^6 heart urchins (*Echinocardium cordatum*), 7.5×10^6 cockles (*Cardium edule*), and 7×10^6 other bivalves (Solenidae, Mactridae, Veneridae). Within a radius of 10 km of the wreck about 10^4 fish were found dead, mainly wrasses (Labridae), sand eels (*Ammodytes* sp.), and pipe fishes (Syngnathidae), as well as large crustaceans (*Cancer crangon, Leander seratus*). There was no evidence of a severe impact of oil pollution on any intertidal or subtidal species of algae. Delayed effects on mortality, growth, and recruitment were still observed up to three years after the spill. For example, estuarine flat fish and mullets exhibited reduced growth, fecundity, and recruitment and many were affected by varying degrees of fish-rot disease. Populations of clams and nematodes in the meiofauna declined one year after the spill. Weathered oil remained for a number of years in low energy areas although the biological consequences of this were unknown.[72]

The longer term biological effects of oil are mainly concerned with PAH residues. The level of exposure can be assessed by measurement of residues accumulated in tissues of fish, but because fish are able to metabolize PAH rapidly other indicators of exposure are used. In fact activation of monooxygenase enzymes (P4501A) involved in PAH metabolism is the primary biological response to PAH contamination in fish and several studies have demonstrated raised levels of this enzyme system in relation to exposure to oil contamination, *e.g. Exxon Valdez* in Alaska, and the Gulf. In the process the metabolites produced, notably *trans*-diols, cause DNA damage and there is growing evidence correlating P4501A induction to PAH levels, DNA damage, hepatic carcinogenesis, and other pathological conditions. The life time of these activated metabolites in fish depends very much on a further group of enzymes (Phase II enzymes) to conjugate and detoxify them.

The *Braer* oil spill occurred close to several salmon farms situated along the coast of Scotland, and caged fish at these farms provided captive sentinels for determining the impact, spatially and temporally, of the oil. A study of these fish demonstrated that, during the initial phase of the spill, fish were exposed to oil in water, resulting in contamination of tissues with PAH and the induction of both monooxygenase and conjugating enzymes. The exposure and biochemical responses declined rapidly with time, such that they had returned to those at control sites after six months. In common dab populations exposed to oil in the water and in sediments, pathological analysis of livers showed changes or lesions in fish from the most contaminated sites symptomatic of the progression to neoplasia. These lesions were only observed in fish a year after the spill but it was not clear whether they were a consequence of the initial short term exposure to oil in the water or to longer term exposure to oil in sediments.[73]

Seabird mortality due to oil pollution generates considerable public outcry and such incidents are mainly associated with major or acute oil pollution episodes. In addition, various surveys have indicated that large numbers of seabirds are killed by oil in the North-West Atlantic throughout the winter, peaking in January–March. These winter deaths

[72] L. Laubier, *Ambio*, 1980, **9**, 268.
[73] R. M. Stagg, A. M. McIntosh, C. F. Moffat, C. Robinson, S. Smith, and D. W. Brun, Proceedings of the Royal Society of Edinburgh Conference on 'The Impact of an Oil Spill in Turbulent Waters: The *Braer*', HMSO, London, XXXX.

are associated more with the ongoing and diffuse chronic oil pollution problem, although clearly various stresses due to the weather are likely to influence mortalities. However, the number of birds killed annually by oil (tens of thousands per annum) is small compared with losses due to natural causes (hundreds of thousands). There has been considerable concern that local populations of certain species are particularly at risk from oil pollution. However, of the many species of seabirds, relatively few, notably the auks (Razorbills, Guillemots, Puffins, *etc.*) and the diving sea ducks (Eiders, Scoters, *etc.*), have suffered severe mortalities. These species are particularly susceptible to oil pollution since they spend almost their entire lives at sea, collect their food by diving and, in most cases, have a low breeding rate. Furthermore, these birds are highly gregarious, particularly in their breeding and wintering areas and so a localized spillage can inflict large numbers of casualties.

The impact of two contrasting refinery effluent discharges on a rocky shore (Milford Haven) and a saltmarsh/seabed community (Southampton Water) have been followed for a number of years. Monitoring has continued at both locations more or less since the commissioning of the refineries. In 1974, in the vicinity of the effluent of the refinery at Milford Haven, the density of barnacle and limpet (*Patella vulgata*) populations were considerably reduced. Laboratory and field experiments showed that this was due to inhibition of larval settlement. When the refinery at Milford Haven closed in 1983, the barnacle and limpet populations showed a steady recovery over the next few years. The saltmarsh communities in Southampton Water, which were extensively damaged during the 1950s and 1960s, have shown a gradual recovery over a number of years, in response to marked improvements in the quality of the effluent.[74]

A widely held view is that oil spills are unlikely to cause long lasting damage to the marine environment and that oil pollution generally does not constitute a chronic threat to either marine ecosystems or indirectly to humans. Recolonization after a spill favours those species with a high fecundity, a short life cycle, and/or a planktonic stage in their life cycle. However, recruitment can be very unstable and full recovery may take several generations; species of such organisms as clams and fish with life expectancies of 5–10 years would not be expected to attain stable populations for up to 30 years, and seabird populations are likely to take even longer.

It is still the case that surprisingly little is known about the long term effects of oil and in particular about the long term bioavailability of

[74] 'Ecological Impacts of the Oil Industry', Institute of Petroleum, John Wiley & Sons, Chichester, 1989.

PAHs to communities foraging among contaminated sediments. There is also little known about the vulnerability of polar and tropical ecosystems to oil pollution.[75] Jackson and co-workers investigated the effects of a spillage of over 8 million litres of crude oil into a complex region of mangroves, seagrasses, and coral reefs just east of the Caribbean entrance to the Panama Canal. At the time it was the largest recorded spillage into coastal habitats in the tropical Americas. The study is significant because many of the habitats damaged by the oil had been studied since 1968, following an earlier oil spill in the region, and also because observations of the effects of the spill began as the oil was coming ashore. In each of the oiled habitats the cover of the major groups was greatly reduced. In the oiled habitats most of the roots sampled for epibiota were dead broken or rotting and so the habitat will not be restored until new trees grow. Seagrass, intertidal reef flat, and subtidal reef habitats were similarly damaged. Of particular note was the extensive mortality of subtidal corals and infauna of seagrasses, as this contradicts the view that these habitats are not affected by oil spills. Also sub-lethal effects of the corals including bleaching or swelling of tissues, conspicuous production of mucus, and dead areas devoid of coral tissue may affect the long term wellbeing of the habitat.[76]

14 HEALTH EFFECTS OF METAL POLLUTION

14.1 Mercury

Several major episodes of mercury poisoning in the general population have been caused by consumption of methyl- and ethylmercury compounds. Methylmercury is a neurotoxin and the main clinical symptoms of poisoning reflect damage to the nervous system. The sensory, visual, and auditory functions, together with those of the brain areas, especially the cerebellum, concerned with co-ordination, are the most commonly affected. Symptoms of poisoning increase in severity in line with increased exposure, as follows: (1) initial effects are non-specific symptoms including paraesthesia, malaise, and blurred vision; (2) in more severe cases, concentric constriction of the visual field, ataxia, dysarthria, and deafness appear more frequently; and (3) in the worst affected cases, patients may go into a coma and die. The effects in severe cases are irreversible due to destruction of neuronal cells. There is a latent period,

[75] E. D. Da Silva, M. C. Peso-Aguiar, M. T. Navarro, and C. Chastinet, *Environ. Toxicol. Chem.*, 1997, **16**, 112–118.
[76] J. B. C. Jackson, J. D. Cubit, B. D. Keller, V. Batista, K. Burns, H. M. Caffey, R. L. Caldwell, S. D. Garrity, C. D. Getter, C. Gonzalez, H. M. Guzman, K. W. Kaufman, A. H. Knap, S. C. Levings, M. J. Marshall, R. Steger, R. C. Thompson, and E. Weil, *Science*, 1989, **243**, 37.

usually of several months between the onset of exposure and the development of symptoms.[6,77]

Case Study 8. Epidemics of mercury poisoning in the general population; exposure–response relationships.

The Iraqi outbreak. This epidemic of methylmercury poisoning occurred in agricultural communities in Iraq in the winter of 1971–72. Over 6000 people were admitted to hospitals in provinces throughout the country and over 400 people died in hospital with methylmercury poisoning. The poisonings arose from misuse of imported high grade seed grain treated with alkylmercury fungicide. The imported seed was intended for sowing but in many areas the grain was ground directly into flour and used in the daily baking of homemade bread. Depending on the number of loaves consumed, individual exposure ranged from a low non-toxic intake to a prolonged toxic intake over 1–2 months. Because of the brief exposure period, epidemiological investigations began 2–3 months after it had ceased and in most cases after the onset of poisoning. This made calculation of ingested dose, and body burden of mercury at the time of exposure to methylmercury difficult to estimate.

Figure 15 shows both dose–effect and dose–response relationships that have been established between symptoms of poisoning and the estimated body burden at the time of cessation of ingestion of methylmercury in bread. An increased body burden of mercury is associated with an increase in the severity of symptoms experienced by patients. The dose–response curve for each sign or symptom shows the same characteristic shape, a horizontal and a sloped line which is referred to as a 'hockey stick' line. The horizontal line represents the background or general frequency of each symptom and the sloped line shows that an increased body burden is associated with an increasing frequency of each symptom. The intersection of the two lines has been taken as the 'practical threshold' of the mercury-related response and this increases with increasing severity of the effects; for paraesthesia it is at a body burden of about 25 mg Hg, for ataxia it is at 50 mg, for dysarthria it is at about 90 mg, for hearing loss it is at about 180 mg, and for death it is over 200 mg. These relationships do not demonstrate cause and effect and the only proof that methylmercury produced the above effects is that the effects followed a known high exposure to methylmercury, the frequency and severity of these effects increased with increasing exposure to methylmercury, the effects are similar to those seen in other outbreaks of methylmercury poisoning, and the major signs have been reproduced in animal models.

---

[77] 'Environmental Health Criteria, 101, Methylmercury', World Health Organization, Geneva, 1990.

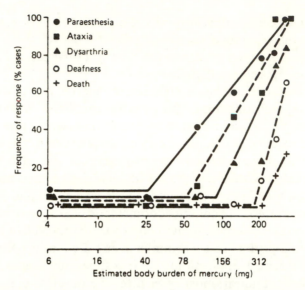

Figure 15 *The relationship between frequency of signs and symptoms of methylmercury poisoning and the estimated body burden of methylmercury (the two scales of the abscissa result from different methods of calculating body burden of methylmercury)*
(Reproduced with permission from *Science*, 1973, **181**, 230)

The Minamata and Niigata outbreaks.   In Japan, two major epidemics of methylmercury poisoning have occurred, one in the Minamata Bay area and the other in Niigata. In each case the problem arose as a result of the local population consuming seafood contaminated with methylmercury. Mercury compounds, including methylmercury were released from industrial sources into the aquatic environment and this resulted in the accumulation of methylmercury in seafood. These outbreaks of poisoning were first discovered during the 1950s and early 1960s and, by the mid-1970s, about 1000 cases (with 3000 suspects) in the Minamata area and over 600 in the Niigata area had been recorded. By and large, symptoms of poisoning followed a very similar pattern to the Iraqi epidemic, although an important difference was in the nature of the exposure which was lower but more prolonged. Moreover, the latent period was longer and in some isolated cases it was as long as 10 years between initial exposure and the onset of symptoms. In other cases it was observed that clinical symptoms worsened with time, despite reduced or discontinued exposure. All this would appear to be related to long term accumulation of mercury in the brain.

Because mercury even from natural sources in fish is predominantly in the methylmercury form, there is concern that certain groups of people who depend largely on a fish diet will be exposed to slightly elevated intakes throughout their lives and hence possibly accumulate mercury to toxic levels. One such group is the Canadian Indians but because of several confounding factors, not least of which are high incidences of malnutrition and alcoholism, the assessment of health risk due to methylmercury is difficult. One study, involving 35 000 samples obtained from 350 communities, found that over two-thirds had mercury blood concentrations within normal limits (<20 μg l^{-1}) but 2.5% (over 900 individuals) had levels in excess of 100 μg l^{-1} and which therefore could be considered as a group 'at risk' and in need of close surveillance.

Mercury analysis of segments of hair provides a meaningful index of past exposures and hence body burden (see Figure 3). As a guideline, blood mercury concentrations of 200–500 μg l^{-1}, hair concentrations of 50–125 μg g^{-1}, and a long term intake of 3–7 μg kg^{-1} body weight are likely to be associated with the onset of the initial symptoms of methylmercury poisoning, such as paraesthesia.

Clinical and epidemiological evidence indicates that prenatal life is more sensitive than adult life to the toxic effects of methylmercury. The first indications came from Minamata in the early stages of the outbreak, where it was found that mothers who were only slightly poisoned gave birth to infants with severe cerebral palsy. A similar situation has been reported in the Iraqi outbreak of 1971–72 with infants which had been prenatally exposed showing severe damage to the central nervous system. Using the maximum maternal hair concentration during pregnancy as an index, the lowest level at which severe effects have been observed was 404 μg g^{-1}. More recent follow-up studies with infants in Iraq have found evidence of psychomotor retardation (delayed achievement of development milestones, a history of seizures, abnormal reflexes) at maternal hair levels well below those associated with severe effects.

14.2 Lead

Lead is a neurotoxin and the overt toxic effects of lead have been known for many centuries. Probably the first reported cases of lead poisoning due to environmental sources was a group of children diagnosed as having lead palsy by clinicians at the Brisbane Hospital in Queensland, Australia, at the turn of the century. A total of ten cases of lead poisoning were found by Health Officials and it was later shown that the source of the lead was lead-based paint which was turning to powder on the walls of homes and railings.

Lead poisoning due to exposure to lead-based paints has affected a large number of children over the years. Most of the epidemiological studies have been centred on the USA. The disease is confined to children, especially those living in inner city areas in dilapidated buildings with surfaces of flaking and peeling lead-based paint. Children playing in the vicinity can take in particles and flakes of paint by inhalation and hand-to-mouth activities. Certain children have a craving for eating non-food items such as flaking paint. This habit called 'pica' and the more normal hand-to-mouth activities of children are capable of introducing excessive amounts of lead into the body: *e.g.* a square centimetre of paint may contain over one milligram of lead.

The disease in the USA was neglected for a long time and the full number of children who have suffered from lead poisoning and the number at risk from the disease emerged over several decades. The probable reasons for this time lag include the poor socioeconomic status of the children in the high risk groups and the difficulties in recognizing the disease. In 1971, the disease became fully recognized when the US Government introduced 'The Lead-Based Paint Poisoning Prevention Act'.

It is important to recognize that 40 years or so ago, a child was diagnosed as suffering from lead poisoning only when acute encephalopathy was evident. Typically this was characterized by a progression to intellectual dullness and reduced consciousness and eventually to seizures, coma, and, in very severe cases, death. Although unknown at the time, acute encephalopathy would only develop when lead concentrations had exceeded 80–100 μg dl^{-1} of blood. At about 80 μg dl^{-1} severe but not life-threatening effects of the CNS can be expected. Lead encephalopathy in children is usually accompanied by peripheral neuropathy, especially foot drop, and general weakness. Acute renal damage is common in severe cases.

Case Study 9. Effects of lead on neurobehavioural development in children. The main focus of attention in recent years has been the effects associated with blood lead concentrations (PbB) less than 40 μg dl^{-1}, which include effects of prenatal and early childhood exposures on physical and neurobehavioural development of children. Behavioural and attentional deficits as rated by teachers (*e.g.* disordered classroom activity, restlessness, easily distracted, not persistent, inability to follow directions, low overall functioning) have been significantly associated with children's tooth and PbB levels. On the basis of these studies, it has been concluded that neurobehavioural deficits may occur at PbB levels at and below 30 μg dl^{-1}.

Less severe symptoms associated with lead poisoning which have been recognized in recent years are deficits in neurobehavioural development and effects on the synthesis of haem in the blood. These effects can be expected to occur at blood lead concentrations of 20–30 μg dl^{-1} and even less.[78,79] It is now well established that levels below 25 μg dl^{-1} have effects on cognitive (reduction in IQ scores) and behavioural development in children. For blood Pb levels below 10–15 μg dl^{-1} range, effects are difficult to detect because of confounding variables and lack of precision of analytical and psychological measurements. Animal studies support a causal relationship between Pb and nervous system effects and there are reports of intellectual deficits in monkeys and rats with blood Pb levels in the 11–15 μg dl^{-1} range. However, attempts to attribute subtle deficits in child development to lead exposure is controversial as many other factors (genetical, nutritional, medical, educational, and parental and social influences) can strongly influence the development of a child. The developing nervous system is particularly sensitive to the toxic effects of Pb and experimental studies have shown that a large number of the effects in the nervous system are due to interference by Pb with biochemical functions dependent on calcium ions and impairment of neuronal connections dependent on dendritic pruning.[80]

Prenatal exposure to lead has been associated with a reduction in the mental development of infants. The effect of environmental exposure to lead on children's abilities at the age of 4 years was studied in a cohort of 537 children born between 1979 and 1982 to women living in the vicinity of a lead smelter at Port Pirie in Australia. The study indicated that elevated PbB concentrations in early childhood had deleterious effects on mental development up to the age of 4 years.[81]

In the US and Europe measures have been taken to reduce Pb exposure. These have included reductions of Pb in gasoline, in food due to reduced use of Pb solder, removal of Pb from paint, and abatement of housing containing lead-based paint. The US Food and Drug Administration (FDA) Market Basket Surveys indicate that the typical daily intake of Pb for 2-year-old children has dropped from 30 μg day^{-1} in 1982 to about 2 μg day^{-1} in 1991. The National Health and Nutrition Examination Surveys (NHANES) of blood Pb levels in children of 1–5 years of age showed that in the 10 years between one survey (1976–80) and the next (1988–91) mean concentrations had decreased to 78% from 15 μg dl^{-1} to 3.6 μg dl^{-1}. Continuing factors that enhance risk to Pb

[78] 'Low Level Lead Exposure: The Clinical Implications of Current Research', ed. H. L. Needleman, Raven, New York, 1980.
[79] Royal Commission on Environmental Pollution, Ninth Report, Cmnd. 8852, HMSO, London, 1983.
[80] R. A. Goyer, *Environ. Health Perspect.*, 1996, **104**, 1050–1054.
[81] A. J. McMichael, P. A. Baghurst, N. R. Wigg, G. V. Vimpani, E. F. Robertson, and R. J. Roberts, *New Eng. J. Med.*, 1988, **319**, 468.

exposure, particularly during fetal life are low socioeconomic status, old housing with Pb-containing paint, and poor nutrition, particularly low dietary intake of Ca, Fe, and Zn. Prenatal exposure may result from endogenous sources such as Pb in the maternal skeletal system or maternal exposures from diet and the environment.[80]

15   CONCLUSION

Pollutants are not of a distinct type and all manner of physical and chemical properties are represented in this broad group of substances. Cause–effect or more specifically exposure–effect relationships form an important basis for the perception and evaluation of risk or damage from pollution. For many of the important pollutants, it is perhaps surprising to learn that these basic relationships have yet to be firmly established.

The problems stem from the complexity of the interaction of pollution with the environment and with systems within organisms. Cause–effect relationships cannot be derived from a single approach, but they are developed from field and experimental investigations and observations. In this regard it is important to establish a consistent relationship between the measured effect and the suspected cause. The observed association should have a reasonable biological explanation. It should be possible to isolate the causal agent and to reproduce the effect under controlled conditions. The cause should normally precede the effect.

In the past, pollution studies were generally initiated in response to an obvious problem and whilst most of the acute problems have been identified and tackled, even today new chronic problems such as endocrine disruption are still being uncovered. This is at a time when there is increased obligation to prevent such events by implementing hazard and risk assessment procedures. As preventive pollution control strategies are implemented for existing forms of pollution and new chemicals it is important that risk assessment procedures take account of all possible risks. There is some doubt whether or not these are sufficiently rigorous to tackle the relatively newly uncovered problem of endocrine disruption. The concluding remarks of the Tenth Report of the Royal Commission on Environmental Pollution are still relevant today, 'an important feature of the type of long-term environmental protection policy ... is that it should guard against creating situations which, though they may initially appear innocuous, have the potential for erupting disastrously—in other words, it should ensure that no new "time bombs" are set'.

Questions

1. Explain the importance of exposure–response relationships in pollution studies. Discuss how information derived from such relationships can be used in the process of risk assessment.
2. Describe the main ecological changes that take place as freshwater systems undergo acidification.
3. What is understood by the term critical loads or concentrations? Demonstrate how the concept is used in the management and control of air pollution.
4. Review the evidence which suggests that endocrine disruption is a fundamental mode of action of organochlorine pollutants.
5. Using specific examples demonstrate how biomarkers can be used to assess biological effects in wildlife.
6. Discuss the way in which epidemics of mercury poisoning in the general population have been important in formulating dose–response relationships.
7. Critically evaluate the effects of low-level lead exposures on neurobehavioural development in children. What are the main sources of lead contributing to lead exposure and how can they be reduced?
8. Assess the fate of hydrocarbons following an oil spill into the sea and evaluate the short and long term effects on wildlife.
9. With reference to experimental and field evidence formulate exposure–response relationships between tributyltin and imposex in gastropods.
10. Critically examine the effects of air pollution on respiratory health in the general population.

CHAPTER 9

Managing Environmental Quality

ANDREW SKINNER

1 INTRODUCTION

The objective of this chapter is to link together the various facets of
environmental chemistry and pollution of the environment and relate
these to the legislative, economic, and social practices adopted by
society to improve and maintain the quality of the natural environment.
It will cover the assessment, both subjective and objective, of environmental quality, the setting of standards, legislative controls on pollution, and the way that these are enforced by government agencies. It
will also consider the way in which society is increasingly seeking
methods, beyond pure legislation, to improve environmental protection
by education and incentive. The topic is a large one and the approach
taken will be to discuss principles and issues rather than delving into
specific details.

What do we mean by environmental quality? How do we go about
measuring it and setting standards and criteria by which we may judge
whether our management is effective and meeting the minimum standards perceived necessary? The setting of standards goes well beyond
measures of the chemical quality of land, air, and water. Standards might
include measures based upon the health, size, and diversity of biological
populations or criteria based on aesthetics and collective personal
judgements about what is acceptable. Someone living near to an urban
watercourse might well consider that the number of supermarket trolleys
to be seen in and around the river was a more relevant measure of the
quality of their environment than the chemical quality of the water in the
river. People living near industrial sites may well feel that noise and smell
were the most important characteristics describing the impact of the
process on their environment, rather than the chemical quality of the
emissions. These considerations lead easily to the conclusion that it is

most unlikely that any one method of assessment will satisfactorily characterize environmental quality.

A good example of an attempt to establish multiple measures is the General Quality Assessment used by the Environment Agency to characterize the quality of surface waters.[1] For surface fresh waters the concept of separate 'windows' measuring chemical, biological, nutrient, and aesthetic quality has been developed and is summarized in Table 1.

A demonstration of how combining different quality assessments can improve the insight into the total environmental quality and help focus remedial action can be seen from a comparison between the chemical and biological quality of urban rivers. It is commonly observed that urban rivers are ranked higher in chemical quality than if the same watercourses are classified by biological means. The reason for this is that, as sewage treatment standards are raised, the prime source of contamination in urban rivers comes from episodic events. A common example is pollution caused during a period of intense rainfall in an urban catchment when accumulated organic pollutants from drains and storm sewers is mobilized, causing a sudden and dramatic reduction in dissolved oxygen and elevation of BOD and ammonia concentrations. In chemical terms the quality of the river can quickly recover after such an event but the fingerprint of pollution is retained by the longer sustained impact on the biota, giving a lower biological score than would be expected from the average chemical quality.

It is also necessary to remember that issues of quality can be strongly linked to issues of availability. This is best demonstrated by the example of water resources. Diminution of flow in watercourses, whether a natural consequence of drought or whether artificially induced by over-abstraction, may often be the most serious impact upon the environmental quality of the watercourse and upon the natural ecosystems that depend upon it. The link to chemical quality is an indirect one, but may be readily apparent, for example, where an authorization for an effluent discharge has been set assuming a particular flow regime which is not being achieved. The reduced dilution beyond that planned will lead to poorer downstream chemical quality.

These examples demonstrate the complexity of the problem and the need for a comprehensive view of environmental media and the links between them if informed judgements and ones that society will accept, are to be made about the quality of our environment and the way in which it is managed.

---

[1] 'The State of the Freshwater Environment in England and Wales', Environment Agency, HMSO, London, May 1998.

Table 1 *Environment Agency General Quality Assessment—windows on water quality*

Window	Characteristics
Chemical	The chemical scheme is a six-grade assessment against chemical standards for dissolved oxygen, biochemical oxygen demand (BOD), and ammonia. Ammonia and BOD are indicators of pollution, which apply to all rivers because of the ubiquitous nature of the risk of pollution from sewage or farms. Dissolved oxygen is essential to aquatic life. High BOD and ammonia concentrations can lead to low DO concentrations so these concentrations are also important in assessing water quality. Some forms of ammonia are also toxic to fish
Biological	The biological scheme is based on the groups (known as taxa) of macro-invertebrates that are found on the riverbed. Macroinvertebrates are used because they do not move far, have reasonably long life cycles and respond to the physical and chemical characteristics of the river. They respond to pollutants which occur only infrequently and which are not measured by the spot sampling procedure used in the chemical GQA scheme. For GQA assessment, species of macroinvertebrates are linked together into 85 taxa. These are given scores of 1 (for pollution-tolerant taxa) to 10 (for pollution-sensitive taxa). The groups are purely taxonomic and assume that the members of each group have similar pollution tolerance. By comparing taxa found in the sample with those you would expect to find if the river were pristine, rivers are classified into one of six grades
Nutrient	The Environment Agency is testing a pilot scheme for phosphorus based on average concentrations of orthophosphate, measured as phosphorus. The six-grade classification is based upon standards related to differing levels of phosphate concentrations and is subject to review in the light of experience. The grade boundaries have been set so that significant changes in ortho-phosphate inputs to rivers will be reflected in a change of grade and to contribute to the definition of eutrophic waters required to implementing the EC Directive on Urban Waste Water Treatment (91/271/EEC)
Aesthetic	The aesthetic quality of a river is determined by a mix of perceptions including the clarity of the water, odour, stagnation, colour, and the presence of oil, litter, or foam. The Agency scheme makes assessments by surveying the river and its bank. A standard method has been devised. Sites are assessed on one or both banks, depending on public access. The number of items of litter are counted, a visual inspection is made of the cover by oil, foam, fungus, and ochre, and the colour and odour of the water is noted. Each type of measurement (litter, odour, and colour) is graded from 1 to 4 and each is given a weighted score for each class according to its accept-ability based on the findings of a public perception study. Sewage litter is weighted as the most unacceptable. The site is then graded from 1 to 4, described as Aesthetically Good Quality, Fair Quality, Poor Quality, and Bad Quality respectively

Table 2   *Objectives, standards, and limits*

Objective	A general environmental objective, often described in terms of more than one environmental standard and possibly involving staged targets
Standard	A measure of quality to be achieved in the environment, generally measured as a concentration in one or more environmental media
Limit	A measure of the quality of an emission into the environment, generally expressed as concentrations of pollutant and volume of effluent or as their load per unit of production

2   OBJECTIVES, STANDARDS, AND LIMITS

The setting of criteria to manage environment quality can be best envisaged as a hierarchy of objectives, standards, and limits. The specific terminology used has varied widely depending upon methods and traditions of pollution control in different countries and for different media. Integration of pollution control procedures and the influence of EU Directives are now driving convergence of terminology, which is summarized in Table 2.

A good description of how this has been tackled in one particular situation can be found in the United Kingdom Government's National Air Quality Strategy.[2]

2.1   Environmental Objectives

An environmental objective may be very general, for example 'to render polluting emissions harmless', but to be helpful objectives have to be more specific and, in general, be set within a time frame for achievement. The objective will usually contain reference to a numeric standard or set of standards that provide a benchmark against which the achievements or objectives can be monitored. Examples of objectives used in the United Kingdom for air and water objectives are given later in the chapter.

Environmental objectives are achieved in practice by regulatory pressures working in combination with changes driven by various types of economic or social measures, for example tax incentives, social changes, and corporate commercial policies. Different organizations, including Government itself, will have different roles to play if objectives are to be achieved. In order to give these objectives force in the policies and priorities of the various organizations involved, it is usual for them to be accorded some kind of statutory or quasi-statutory status. One of

[2] 'The United Kingdom National Air Quality Strategy', HMSO, London, March 1997.

the prime purposes of the United Kingdom National Air Quality Strategy[2] and the United Kingdom National Waste Strategy[3] is to provide a formal basis for setting, monitoring, and achieving standards. National and Local Government policies and the activities of the regulatory agencies are required at all times to support and encourage the achievement of the objectives set.

In the case of water quality objectives in the United Kingdom, the Water Resources Act 1991 gave the Government powers to set statutory objectives independent of a national water strategy. The Surface Waters (River Ecosystem) (Classification) Regulations 1994 introduced a scheme whereby river stretches could be assigned one of five classes relating to the type of ecosystem that should be maintained in that stretch. The ecosystem approach was used to provide a generic status which characterizes the 'use' to which a particular stretch of river might be put and recognized the inherent differences in quality expectations between, say, an upland stream and a lowland river with an urbanized catchment. These classes would become Statutory Water Quality Objectives set by the Secretary of State after consultation. Once set, it would be a duty on the regulator to ensure that the objectives were achieved at all times (after the date set for compliance), as far as this lay within its water pollution control powers. The scheme is intended to establish a formal and open mechanism, involving public consultation, for taking decisions to protect and improve river water quality. It is the intention that following the introduction of the scheme to rivers, further schemes will be developed for other types of controlled waters, such as lakes, estuaries, groundwaters, and for more specifically defined uses, such as abstractions for industry and agriculture, water sports, and specialized and vulnerable ecosystems. The approach has presented both political and practical difficulties and is yet to be implemented. It seems likely that the current system of Non-Statutory Water Quality Objectives first established in 1979, which has provided a reliable basis for planning water quality improvements, will continue for some time yet.

2.2 Environmental Standards

An environmental standard will normally be specified as the concentration of a pollutant in one or more environmental media. It is set with regard to scientific and medical evidence in relation to impacts upon public health or upon natural ecosystems at a level of minimum or zero risk. Where, as is often possible, a no-effect level of pollution for a

[3] 'Making Waste Work: A Strategy for Sustainable Waste Management in England and Wales', HMSO, London, 1995. Command paper 3040.

particular substance can be identified, this will guide the setting of the environmental standard. A large number of standards in relation to public health derive directly from standards created by the World Health Organization, although these may be, and often are, modified when adopted into, say, European legislation. In the absence of adopted European standards, as is the current case for some air quality standards in the United Kingdom, then national standards will be adopted. These standards, often called Environmental Quality Standards (EQS), are those concentrations of pollutant considered to be acceptable in the wider environment.

One way in which Environmental Quality Standards can be met is to ensure that individual permits control all point source discharges into the medium. These permits will contain emission limits appropriate to the achievement of the standard, having regard to the dispersion and dilution that will take place. The assessment of these emission limits will demand a high level of information about the level and variability of the ambient concentrations of the relevant pollutant as well as the level and variability of the expected discharge if emission limits are to be reliably set. For example, in the setting of consents to discharge to water in the United Kingdom, a Monte Carlo simulation technique is used that combines information on the statistical distributions of upstream river and discharge flow and quality, and calculates the appropriate emission limits to meet the downstream Environmental Quality Standard. The standard is therefore set in terms of the environmental capacity of the receiving medium rather than on the basis of what may be achieved by the process producing the emission.

2.3 Emission Limits

An alternative approach is to base the regulatory strategy around emission limits, using criteria based upon the capability of the technology. This leads to uniform emission standard for a process or a site, based upon assessments of process and pollution abatement techniques. This approach has the advantage of setting uniform environmental standards for any particular industry or process and is therefore favoured by industry in providing the much sought after 'level playing field'. It is also an approach that can be used without a wide knowledge of the quality and variability of the receiving medium. However, where such information is available, a sophisticated application of the emission limit approach allows local variation. In Germany, for example, uniform sectoral (that is, specific industry-wide) standards applying to discharges to all media have been established in consultation with industrial representatives and independent technical associations. Uniform emission standards are derived from assessments of the state of the art in the process and pollution abatement technology for dangerous substances and non-biodegradable substances. These have become binding standards that translate directly into process authorizations.

Uniform emission standards are relatively easy to apply. An environmental capacity based system is harder to administer particularly if, as may be the case for air pollution control or for water pollution control to large river catchments such as the Rhine, the issue comes under the jurisdiction of a number of national regulatory agencies. However where environmental capacity is limited, for example in the case of small rivers with little dilution capacity, a uniform emission standard approach could easily lead to problems if account was not taken of the limited assimilative capacity of the receiving medium.

Technology-based emission standards offer a practical way of comparing alternative options for emissions to different environmental media. They are a key element of the Integrated Pollution Control (IPC) regime operated in the United Kingdom. The concept of IPC was developed by the Royal Commission on Environmental Pollution in its 12th (1988) report.[4] They recommended that an IPC inspectorate be established, empowered to 'impose technology-based controls designed to achieve a Best Practicable Environmental Option (BPEO) for the specific process or waste' having consideration of all environmental media. The definition that the Royal Commission gave for BPEO was 'the outcome of a systematic consultative and decision-making procedure, which emphasises the protection and conservation of the environment across land, air and water'. The BPEO procedure establishes, for a given set of objectives, the option that provides the most benefit or least damage to the environment as a whole, at acceptable cost, in the long term as well as the short term. The system of technology-based controls, which defines this ambitious, although difficult to achieve goal, is known in the United Kingdom as BATNEEC (best available technique not entailing excessive cost). The approach still depends upon knowledge of the quality standard in the environment and one difficulty in operating this regime has been to establish emission standards for substances for which no environmental quality standards exist. One approach to solving this problem has been to base standards on established exposure limits used to protect employees in the workplace (workplace exposure limits) under health and safety legislation. Because the impacts in the wider environment may involve longer exposure time to more vulnerable individuals it has been considered appropriate to apply safety factors of up to 100-fold to the workplace exposure limits. Standards formulated in

---

[4] 'Best Practicable Environmental Option', 12th Report of the Royal Commission on Environmental Pollution, HMSO, London, 1988.

this way have been widely criticized by plant operators as being excessively restrictive.

The IPC regime applied in United Kingdom under the 1990 Environmental Protection Act has been influential in the development of a similar but more wide ranging system of controls now to be introduced throughout Europe. An integrated pollution control and permitting regime is to be adopted under the Integrated Pollution Prevention and Control (IPPC) Directive adopted in 1996 (96/61/EC) and due for implementation progressively from 1999. This directive is built around the concept of best available technology (BAT) which in practice is very similar in concept to BATNEEC. Controls on emissions under the directive will be based upon BAT, which describes processes or operations that are 'the most effective and advanced' of those available and which are selected 'bearing in mind the likely costs and benefits and the principles of precaution and prevention'. They should be capable of emission limit values designed to prevent impact on the environment, and if that is not practicable, minimize emissions and the impact on the environment as a whole. If BAT standards do not meet environmental quality standards then standards tighter than BAT must be set to ensure that the environmental quality standard is not breached. In practice BAT will be established on a Europe-wide basis in the form of standard documents drawn up by groups of process specialists. The European adoption of the BAT concept is of great significance for environmental regulation and will extend beyond the IPPC directive to other directives and is expected to drive common European environmental standards.

2.4 Integrating Limit Values and Quality Standards

The IPPC Directive, as described above, brings together the approach of limit values based on best available techniques (BAT) and strict environmental quality standards and will apply to a wide range of processes and industries, considerably wider than the IPC regime in the United Kingdom. This recognition that two approaches can coexist has been long coming and has been most evident in the development of European water quality regulation where three strands are now about to merge.

2.4.1 Use-related Approach. Most of the early EC Directives on water laid down environmental quality standards for individual parameters designed to protect specific water uses such as abstraction for drinking water supply (75/440/EEC), bathing (76/160/EEC), fish life (78/659/EEC), and shellfish life (79/923/EEC). Member States are required to designate waters under the Directives. For designated water the standards in the Directive are legally binding.

2.4.2   Uniform Emission Standards.   The European Directive on dangerous substances (76/464/EEC) first recognized the dual approach and allowed Member States to apply either uniform emission standards or environmental quality objectives for the control of listed substances, which are toxic, persistent in the environment, and bioaccumulative.

2.4.3   Sectoral Approach.   A sectoral approach has been applied more recently in the Directives on titanium dioxide (78/176/EEC and others), nitrate pollution from agriculture (91/676/EEC), and urban waste water treatment (91/271/EEC). These directives require specific control measures appropriate to the special need in each circumstance, including, in the latter two cases, the designation of zones where special preventative measures must be taken.

It\ is the objective of the European Commission to simplify the many individual directives covering water into a Water Resources Framework Directive, which, together with the IPPC Directive, will provide a more consistent framework within which increasingly integrated environmental management can operate. Table 3 gives a list of the large number of current EC directives directly relating to water quality, many of which will be modified or replaced by the proposed framework directive and by the introduction of the IPPC regime.

2.5   Specifying Standards

Quality standards may be specified in a number of different ways, for example:

- as absolute standards to be achieved at all times;
- as standards to be met most of the time, in this case usually expressed as a percentile compliance; or
- as standards to be met as an average over a period.

Depending on circumstance, in particular the frequency of change and the vulnerability of the environmental target, the period over which an average may be taken can vary greatly. Some air quality standards specify 15 min averages, whereas others may be set as an annual average. Examples are shown in Table 4, which lists some of the United Kingdom air quality objects for specific pollutants.

Absolute standards, if translated directly into emission limits, place high demands on process designers and their operators and also on monitoring systems. There are some circumstances, in terms of protecting human health from toxic substances, where absolute standards are necessary, but in many cases properly specified and controlled percentile

Table 3 *List of current EC Directives relevant to water pollution*

Directive Number	EC Directive title
75/440/EEC 79/869/EEC	Quality of surface water abstracted for drinking
76/464/EEC	Pollution caused by the discharge of certain dangerous substances into the aquatic environment
76/160/EEC	Quality of bathing water
78/176/EEC	Waste from the titanium dioxide industry
78/659/EEC	Quality of fresh waters needing protection or improvement in order to support fish life
79/409/EEC	Directive on the conservation of wild birds
80/68/EEC	Protection of groundwater against pollution caused by certain dangerous substances
82/883/EEC	Procedures for the surveillance and monitoring of environments concerned by waste from the titanium dioxide industry
82/176/EEC	Limit values and quality objectives for mercury discharges by the chlor-alkali electrolysis industry
85/513/EEC	Limit values and quality objectives for cadmium discharges
84/491/EEC	Limit values and quality objectives for discharges of hexachloro-cyclohexane
84/156/EEC	Limit values and quality objectives for mercury discharges by sectors other than the chlor-alkali electrolysis industry
86/280/EEC	Limit values and quality objectives for discharges of certain dangerous substances included in List I of the Annex to Directive 76/464/EEC (DDT, PCP)
88/347/EEC	Amending Annex II to Directive 86/280/EEC on limit values and quality objectives for discharges of certain dangerous substances included in List I of the Annex to Directive 76/464/EEC (aldrin, dieldrin, endrin, hexachlorobenzene, hexachlorobutadiene, chloroform)
90/415/EEC	Amending Annex II to Directive 86/280/EEC on limit values and quality objectives for discharges of certain dangerous substances included in List I of the Annex to Directive 76/464/EEC (dichloroethane, trichloroethane, perchloroethane, trichlorobenzene)
91/676/EEC	Protection of waters against pollution caused by nitrates from agricultural sources
91/271/EEC	Urban waste water treatment

Table 4   *Examples of UK air quality objectives*

Pollutant	Objective
Nitrogen oxides	To achieve the standards of 150 ppb hourly mean, and 21 ppb annual mean, absolute compliance, by 2005
Ozone	To achieve the standard of 50 ppb, measured as a running hourly mean, 97th percentile compliance, by 2005
Particulate matter	To achieve the standard of 50 μg m^{-3}, measured as a running 24 h mean, 99th percentile compliance, by 2005
Sulfur Dioxide	To achieve the standard of 100 ppb, measured as a 15 min mean, 99.9th percentile compliance, by 2005

standards provide an adequate level of environmental protection at less cost, both to operator and regulator. It is possible to set emission limits incorporating both concepts. Discharge consents for sewage works in the United Kingdom are set on a percentile compliance basis, but this can be supported for critical determinands by a so-called upper tier, set as a multiple of, say, three times the percentile limit value, which acts as an absolute standard.

Because of the wide variety of standards and methods of expressing them it is not a trivial matter to understand their impact and even to know, when a standard is reformulated in a different way, whether it represents a tightening or a relaxation. A good example in the United Kingdom has recently emerged with the changes made by the United Kingdom Government to its Nitrogen Dioxide Air Quality Standard. In December 1996 the environmental magazine ENDS carried an article 'UK Goes It Alone on Weaker NO$_2$ Standard', claiming that a revised standard to be set in the National Air Quality Strategy was 'considerably less stringent' than that set in the draft strategy. The new standard, now published[2] is an hourly mean NO$_2$ concentration of 150 ppb with one hundred per cent compliance. This revision was based upon advice from the Government's Expert Panel on Air Quality Standards. The previous draft standard had been an hourly mean concentration of 104.6 ppb but with 99.9 per cent compliance and was based upon the WHO recommended standard. It is not a matter of simple assertion, based on a comparison of the numerical values, that revised United Kingdom 1997 is less stringent because of the need to take account of the tighter percentile compliance.

It is a matter of principle in setting standards that the criteria are those of protecting public health and the environment. Depending on the availability or certainty of toxicological or epidemiological evidence, the approach to standard setting will be different. In those circumstances where it is possible to identify a no-effect threshold, at or below which effects are most unlikely, even for sensitive population groups, then this concentration becomes the standard. In other circumstances it will not be possible to establish a zero risk level, in which case it is necessary to follow a risk-based toxicological approach to derive acceptable risk exposure limits. This approach will apply particularly with carcinogenic substances but, as with the case in setting the United Kingdom air quality standard for particulates, the same approach can be applied for a non-carcinogenic substance, if the available data shows that a no-effect threshold cannot be identified.

Issues of costs and benefits and technical feasibility are clearly relevant to standard setting but should not, in principle, affect the setting of the numerical standard. They may influence the political judgements about the way objectives are set and the time-scale over which they need to be achieved. Inevitably questions of cost and feasibility frequently get bound up in political debates arising from environmental standards.

An example is the standard for nitrate in water. A standard of $50 \, \text{mg} \, l^{-1}$ nitrate as NO_3 was set as part of the EC Drinking Water Directive (80/778/EEC). This has now become the environmental standard for nitrate in water because the same numerical standard has been translated into the European Directive for the Control of Nitrates from Agriculture (91/676/EEC). The standard of $50 \, \text{mg} \, l^{-1}$ is based upon epidemiological evidence relating to human health. It has faced many challenges but it has not been changed. When the standard was first set in the early 1980s there was not a great deal of information about the level and variability of nitrate in rivers and groundwaters, although most countries had examples of extreme values, not uncommonly exceeding $200 \, \text{mg} \, l^{-1} \, NO_3$, most often from shallow groundwater sources. The intensive monitoring that has been done over the last fifteen years has shown that, while nitrate concentrations in surface and groundwaters regularly and widely exceed $50 \, \text{mg} \, l^{-1} \, NO_3$ and are still increasing, exceedance of $100 \, \text{mg} \, l^{-1} \, NO_3$, which was the original World Health Organization standard, is rare. The impact of setting a standard of $50 \, \text{mg} \, l^{-1} \, NO_3$ rather than $100 \, \text{mg} \, l^{-1} \, NO_3$ has been to impose significant costs on the water industry in Europe to treat or blend many water supplies. As member states of the European Union progressively implement the Nitrate Directive, significant changes in agricultural practice in vulnerable areas will become necessary. The best medical evidence is that the standard is correct at $50 \, \text{mg} \, l^{-1}$ but there is no doubt that had the earlier $100 \, \text{mg} \, l^{-1} \, NO_3$ standard been preserved the compliance costs would have been dramatically less. The European Drinking Water Directive sets a guideline value of $25 \, \text{mg} \, l^{-1} \, NO_3$, which influenced a number of member states to set objectives of less than 50 mg l^{-1} NO$_3$ but the controversy about the standard has now put this objective in question. For example, the Austrian Drinking Water Decree of 1989, made before Austria was a member of the EU, set an objective of achieving a nitrate standard of 30 mg l^{-1} NO$_3$ by 1999. This has now been abandoned on grounds of high cost of achievement and a lack of epidemiological evidence to support the tighter standard.

Case Study 1: Urban waste water treatment (Germany). This case study illustrates the interrelation between environmental and technology-based standards and how, in practice, new discharge limits can be imposed.

A sewage treatment plant in Germany serving the 60 000 population of the town of Bensheim (Hessen), on the east side of the Rhine Valley needed to be upgraded to include a two-stage activated sludge process. The receiving stream was very small and provided little dilution for the effluent, even in wet weather. During the dry season the effluent from the treatment plant is the source of the stream.

The municipal authorities were notified by the regulatory agency that additional tertiary treatment would be needed, and that the following standards would be set as annual means in the revised discharge consent:

Chemical oxygen demand	20 mg l^{-1}
Biochemical oxygen demand	5 mg l^{-1}
NH$_4$ as N	2 mg l^{-1}
NO$_3$ as N	8 mg l^{-1}

The above values are significantly more stringent than the legally binding minimum (Best Technical Means) requirements for sewage treatment plants the German Federal Government has established. They were imposed because of the need to meet the minimum quality standards relating to the target biological class II of the German system shown below in a situation of low or zero dilution.

Chemical oxygen demand	>6 mg l^{-1}
Biochemical oxygen demand	2–6 mg l^{-1}
NH$_4$ as N	<0.3 mg l^{-1}

The municipality of Bensheim, which owned and operated the plant, appealed against this notification, arguing that the requirements were too demanding for their treatment plant and exceeded the minimum technology-based requirements they were obliged to meet. In particular the standard for chemical oxygen demand of 20 mg l^{-1}, which compared with 90 mg l^{-1} Federal minimum requirement, would impose substantial additional costs for the construction and operation of a filtration plant. The appeal delayed progress during which time the existing plant continued to affect water quality in the receiving stream.

Following the enactment of the European Urban Waste Water Treatment Directive (91/271/EEC) it became necessary to upgrade two other sewage treatment plants serving nearby communities of 10 000 and 20 000 inhabitants. A study of the situation showed that the economically and environmentally optimal solution would be to close down the two smaller plants, serving the nearby municipalities of Lorsch and Einhausen, and extend and upgrade only the Bensheim sewage works.

Lorsch did not agree to the connection but it was decided to enlarge the Bensheim plant and to connect Einhausen to it. In accordance with the Best Technical Means in sewage treatment for this size of plant, the following standards were set in the discharge consent:

Chemical oxygen demand	30 mg l^{-1}
Biochemical oxygen demand	6 mg l^{-1}
NH_4 as N	2 mg l^{-1}
NO_3 as N	10 mg l^{-1}
P_{total}	0.2 mg l^{-1}

The plant will now operate under these consent conditions before an evaluation takes place, possibly leading to further measures being imposed in order to achieve the biological water quality class II.

Case Study 2: Discharge of dangerous substances (United Kingdom). This case study concerns emissions of the dangerous substance pentachlorophenol (PCP) from textile finishing industries in the Irwell catchment in North-West England. PCP is controlled as a dangerous substance under European Directive 86/280/EEC, within the framework of Directive 76/464/EEC (see Table 3). Member States are required to control emissions of PCP through an authorization process.

At the time Directive 86/280/EEC came into force, monitoring of the Irwell catchment showed a number of rivers to contain levels of PCP exceeding environmental quality standards. There were consequential effects upon river biota. Monitoring of PCP levels downstream from sewage treatment plants gave PCP values in the range 3 to 21 $\mu g\,l^{-1}$. Investigations revealed 16 textile finishing companies in the River Irwell catchment with effluents containing PCP, all discharging their trade effluents to sewers.

Discharge consent standards from the sewage works to the river were required to ensure that the environmental quality objective of 2.0 $\mu g\,l^{-1}$ as an annual mean value was met in the receiving water. Because of the limited dilution in the River Irwell catchment, conditions for sewage works effluents were set with maximum PCP concentrations between 2 and 6 $\mu g \, l^{-1}$. This in turn led to water company trade effluent control on the discharges from the individual works to the sewer being set at levels of between 6 and 10 $\mu g \, l^{-1}$ in order to meet the environmental quality standard in the river. It is notable that these levels are much more stringent than the limit value of 2000 $\mu g \, l^{-1}$ specified in the Directive.

As a result of measures taken by the textile industry to reduce PCP emissions, annual mean levels of PCP in the River Irwell catchment are now between 0.06 and 0.15 $\mu g \, l^{-1}$. The reductions in PCP in the discharge were achieved by better waste water treatment facilities in the textile factories but, in particular, by better quality control on the raw materials being taken in to the factories. Imported material in particular was found to have varying and often high levels of PCP.

A very similar set of circumstances existed in the Worcestershire Stour in the English Midlands, as a result of discharges from the carpet industry centred on Kidderminster. Again, quality control on imports, plus better effluent treatment, encouraged by the setting of appropriate discharge controls, has improved the discharge and significantly improved water quality which now meets the environmental quality standard and has improved the diversity of aquatic life in the Stour.

2.6 Remediation Targets

The discussion so far has related to the setting of environmental standards to prevent pollution. A key question is whether the same standards necessarily apply when pollution has already occurred and a judgement is required on whether, to what level, and with what priority remediation should be carried out. These issues come into sharpest focus in relation to the remediation of contaminated land, which also impacts on the protection of water quality. Risk-based approaches to the setting of targets are increasingly being applied in this sector. In remediating contaminated land the most important factors relate to the protection of the human health of those using the land, for the protection of water quality draining from the land and for the protection of biota and habitat. The earliest experience of contaminated land remediation at any scale were those carried out in the United States in the 1980s under the impetus of the 'Superfund' legislation, where federal finance was allocated for clean-up of what were perceived to be the major environmental problems. Massive remediation projects were commissioned using both landowner and federal funds to try and achieve environmental standards often equivalent to natural ambient quality. These were, in many cases, unachievable because of lack of available technology and also of questionable necessity at the high cost specified. The same philosophies began to be applied also in Europe. An example is the 'multi-function-ality' approach adopted by the Dutch Government in its Soil Protection Act of 1987, which established a remediation regime based upon clean-up standards intended to make polluted soil fit for any use.

The United Kingdom Government's approach has been very different. This is set out in the provisions of the 1995 Environment Act and the related draft guidance[5] which requires remedial action to land contam-ination only where:

- the contamination poses unacceptable actual or potential risks to health or the environment; and
- there are appropriate and cost-effective means available to do so, taking into account the actual or intended use of the site.

This has been called the 'fitness for use' approach and leads to use-related targets being set for remediation of land only where it can be shown that there is risk of environmental harm to people, water or biota. In other words land which is contaminated but where the contamination is not mobile and placing a human or environmental receptor at risk need not be remediated unless or until circumstances and change and risk arises. This therefore leads to the concept of latent contamination; not desirable but not a priority for action. This approach has led to the development of a concept of guideline values for soils in different environmental settings, which are used to define risk and therefore set criteria for remediation.

The United Kingdom approach has been widely criticized both inside[6] and outside the United Kingdom but it is finding increasing favour. In a major change of policy the Dutch Government announced, in May 1997, that it would be following a 'fitness for use' approach and that its priorities for remediation would be based more on redevelopment needs. The prime reason for this change is the rising cost of the remediation programme as more data become available about the scale of soil and groundwater contamination, now estimated to be 100 000 sites in The Netherlands with a clean up cost of US$50 billion. This is a hundred-fold increase on estimates made in the early 1980s when the policy was set. In addition there is the realization that the problem will take many years to solve, during which time the size of the problem will increase as polluted soils continue to contaminate groundwaters. Significant changes of approach are now being adopted in the USA, for example in California. In late 1966 the state authorities changed its programme of tackling

---

[5] 'Draft Statutory Guidance on Contaminated Land', Department of Environment, London, 1996.
[6] 'Sustainable Use of Soils', 19th Report of the Royal Commission on Environmental Pollution, HMSO, London, 1996.

contamination caused by underground fuel tanks from a comprehensive remediation programme attempting to tackle every known site to one based upon known problems or identified risk to human health or water supply.

Protocols to provide a framework for the application of what has come to be called Risk-based Corrective Action (RBCA) are now under active development in many countries. These techniques use a tiered approach to target setting for remediation, whereby decreasingly precautionary targets are set, as more information is available from survey and monitoring.

2.7 The Principles of No Deterioration and Precaution

Pollution control regulation based around fitness for intended use has been seen by many as philosophically unsatisfactory because it is seen to give preference to the short term exploitation of the environment rather than its protection over the longer term. In the case of use-related standards there will be a threat, it is argued, of a lower level of protection and a decline in quality of the environment, if a use on which an environmental standard was based were to be discontinued. These concerns are met by two underpinning principles. The first is of no deterioration, which means that, where a quality standard has been set, even if on a use-related basis, no new release to the environment would be permitted which would allow the quality standard to fall below the established threshold. The precautionary principle establishes that, where doubt exists as to the environmental consequences of a proposed new action, an approach based upon caution, supported by the available evidence and avoiding risk of irreversible impact, should be adopted. In various formulations these principles are incorporated in pollution control law in most developed countries. Its application in Europe is described further in Section 3.2

3 LEGISLATION TO CONTROL AND PREVENT POLLUTION

Pollution prevention legislation has evolved along a variety of routes in different countries. Even in the USA and Europe, where there is strong unifying legislation, there is still great diversity in the priorities given to implementing various aspects of the legislation. There is even greater diversity in the number and relationships of the agencies that carry out these tasks. This section will deal with the general features of pollution control legislation and the following section with the agencies charged with enforcing them.

3.1 Origins of Pollution Control Legislation

Pollution control legislation in all countries emerged from concerns about the impacts of both industrial processes and the disposal of human wastes on human health. These gradually evolved to increasingly embrace a specific environmental agenda distinct from health and safety legislation. The duties were vested in a number of agencies, such that there was no integrated control linking pollution of air, water, and soil and land but most countries, in various ways, have now achieved a significant degree of integrated multimedia environmental management. The linkage between control of environmental pollution and health and safety has recently come into closer focus, particularly in relation to the regulation of major industrial plant. Incidents like the major industrial plant disaster at Seveso in Italy led the European Community to adopt the Control of Industrial Major Hazards and Accidents Directive (CIMAH) (82/501/EEC). This is being replaced in 1999 by the Control of Major Accidents and Hazards Directive (COMAH), the so-called Seveso II Directive (96/82/EC), which is aimed at ensuring that operators of industrial plant take all necessary measures to prevent major accident hazards to both people and the environment. It will require joint competency of health and safety and environmental regulators for its implementation in European Member States. In some countries the emphasis on protection of human health has led to environmental regulation being closely linked to public health legislation. In an international context much of the impetus for pollution control initiatives are driven by international and national health agencies. The World Health Organization, as a source of health standards, is a key agency in the setting and reviewing of environmental standards that depend upon human health criteria.

In the United Kingdom the first significant step in the management of industrial pollution came with the passing of the Alkali Act in 1863. This created the Alkali Inspectorate, a direct forerunner, via Her Majesty's Inspectorate of Pollution, of the Environment Agency and the Scottish Environmental Protection Agency (SEPA). In 1874 a subsequent Act introduced the concept of 'Best Practicable Means' to prevent noxious emissions and render them harmless, a direct forerunner of the technology-based direct emission controls and BAT. The two Acts were consolidated into the Alkali Act 1906 which was finally and completely repealed only in 1996, when all remaining industrial process control covering emissions to air came under the Environmental Protection Act 1990.

The first significant controls on water pollution in the United Kingdom came with the Rivers Pollution Prevention Act of 1876 which required sewage discharges to 'be rendered inoffensive' prior to discharge to an inland watercourse. The Act was poorly enforced and did nothing to address water pollution from industrial sources. A Royal Commission on Sewage Disposal investigated the continuing problem of water pollution from domestic sewage and reported in 1912. It recommended emission standards for BOD and suspended solids related to the dilution available in the receiving watercourse. Its recommendations, although never given statutory force, had a very significant impact on standards. A system of discharge controls to rivers was not established until the Rivers (Prevention of Pollution) Act 1951, which was the first time industrial pollution was addressed. These developments led, through successive legislation and agencies, via the National Rivers Authority and the Scottish River Purification Boards, to the water pollution control duties of the Environment Agency and the Scottish Environmental Protection Agency.

Control of pollution to land came very much later in the United Kingdom in 1972 with the introduction of the Deposit of Poisonous Wastes Act, which led quickly to more comprehensive legislation for control of waste to land administered by Local Authorities (Control of Pollution Act 1974, Part I). These provisions were extended and incorporated in the Environmental Protection Act 1990, Part II and responsibilities for waste regulation passed from Local Authorities to the Environment Agency and SEPA in 1996.

Legislation to control and remediate land contamination not specifically covered by waste management or water pollution legislation, although provided for in the Environment Act 1995, is not expected to be fully implemented in the United Kingdom until 1999.

3.2 Trends in European Environmental Legislation

Within the European Community it is possible to see a significant evolution in the style and scope of environmental legislation. The Treaty of Rome, which established the European Community, saw standardization of legislation on the environment as an aspect of standardization of trading across the community. The Single European Act of 1987 made various amendments to the Treaty, in particular providing a new Chapter on the Environment in its own right (Articles 130r, 130s and 130t). Article 130r establishes that European action relating to the environment shall take place to:

- Preserve, protect, and improve the quality of the environment
- Contribute towards protecting human health
- Ensure a prudent and rational utilization of natural resources.

Table 5   *Treaty of the European Union (Article 130r): Key principles in*
environmental regulation

Sustainable	Policies should contribute to the goal of sustainable development
High level of protection at source	Environmental damage should as a priority be rectified at source rather than by treatment at the point of use
Precautionary	Where outcomes are uncertain, particularly if they are likely to be irreversible, then there should be a presumption in favour of a cautious approach
Integrated	Decisions on environmental impact should have regard to impacts and options for all media and be taken in an integrated and holistic way
Subsidiarity	Decisions should be taken at the lowest level and at the most local scale reasonable in the circumstances and recognize that, although principles of environmental protection are general, the diversity of situations applying in different regions will lead to different practices being applied

It also establishes guiding principles for environmental management shown in Table 5.

The early directives primarily focused on the control of toxic substances in the environment, usually in relation to one medium only and in general focused on point source pollution. The regulations were based around specific limits on emissions linked to an environmental quality standard set for a range of listed substances. As already described in Section 2 of this chapter, this approach provoked a strong debate with those countries whose legislation, particularly in the field of water pollution control, had evolved along a route of environmental capacity, that is the ability of the environment to assimilate polluting loads. More recently there has become increasing international acceptance that the two approaches are not mutually exclusive and can work in parallel, with the emission control concept being more specifically relevant for toxic and persistent substances, particularly those for which there is an international or global agenda to seek reduction. One feature of European environmental legislation that has precipitated this convergence of thinking is the move away from specific process or substance controls to broadly integrated environmental directives. The most significant in this regard is the Integrated Pollution Prevention and Control Directive (96/61/EC) which provides for a wide framework of integrated and environmental management of a range of industrial, commercial, and agricultural activities.

The Commission has announced the intention of incorporating a wide range of existing directives relating to water management and water pollution into a Water Framework Directive. This directive proposes to combine, for the first time in any European legislation, the ecological and chemical quality of surface waters and the chemical quality and quantitative status of groundwaters. It is proposed to develop a series of criteria to define 'good' status for each of these, with a target of achieving this by 2010. This draft directive specifically embraces both emission control and environmental quality objectives and will incorporate most of the earlier dangerous substances and daughter directives. A further feature of the proposed directive is to require monitoring, reporting, and action plans to use the river basin rather than political boundaries as the reporting unit.

In the area of waste management the European Commission adopted Waste Framework Directives (75/442/EEC and 91/156/EEC). These directives, together with the directives on defining hazardous wastes and their status (91/689/EEC and 94/31/EC), define waste, encourage its reuse, and control its movement and disposal. The packaging directive (94/62/EC) seeks to further encourage waste minimization and resource conservation by constraining waste production. A specific directive relating to the landfilling of waste has been under consideration for a number of years but has yet to be adopted. There is increasing awareness of the complexities of the environmental and economic arguments, which drive wastes to be recycled, deposited in landfills or incinerated. The legislative framework for each of these, currently included within three different directives or proposed directives, needs to be closely harmonized if the right outcomes for the environment are to be achieved.

3.3   Reporting Environmental Performance

One characteristic feature of most early pollution control legislation, including the early European Directives, was that they were weak on requirements for monitoring, reporting, and verification. It was often left to the discretion of the pollution control agency to determine the extent to which the implementation of legislation was enforced. Obligations to undertake environmental monitoring were also few, thus limiting the judgements that could be made about the efficacy of the pollution control regime. This lack of a requirement for monitoring both implementation and effectiveness has been seen to be a major stumbling block in both the uniform application of environmental control regimes and the objective assessment of future legislative needs. Although governments have been slow to react to these concerns, public opinion has not. Pressure of public opinion has become a major force in demanding and achieving better environmental monitoring and better reporting. The outcome has been much more rigorous attention being given to monitoring and reporting in both European, and consequently in national legislation.

The European Commission first sought to address this problem by promoting in 1991 the Standardized Reporting Directive (91/692/EEC) to provide better information across Europe on compliance with thirty major environmental directives. More recent directives have included their own specific requirements for reporting. In 1995 a European Environment Agency (EEA) was established with the specific initial objective of reporting on the state of the environment in Europe integrating, and over time forcing, standardization of national environmental databases and collating and reporting on specific environmental topics. One of its first actions was to co-ordinate an extensive report on the state of the environment in Europe, the Dobris Assessment.[7] Such assessments will be repeated from time to time. The EEA has established a network of topic centres in various research and environmental institutes around Europe specializing in different issues, including fresh water, air quality, air emissions, and nature conservation. The Agency is also establishing a European Environment and Conservation Network (EIONET) to provide a database for European environmental information much of which will be assembled from the reporting requirements of European directives together with the wider environmental monitoring carried out by national agencies across Europe. These moves parallel in many aspects the existing databases established by the EPA in the USA, which has provided a model for many of these initiatives on public information on the environment.

3.4 Pollution Control and Land Use Planning

A further aspect of legislation, which closely relates to the prevention of pollution, is the control, which all developed countries have, in one form or another, as part of their urban and rural infrastructure planning. The United Kingdom has not been alone in relying for many years on planning controls rather than pollution control legislation to prevent the creation of future pollution threats. The convergence between these different aspects of environmental management was recognized by the United Kingdom Government in 1994 by publishing guidance on liaison between planning and pollution control agencies.[8] In many cases pollution prevention and the protection of the environment will be significant

[7] 'Europe's Environment: The Dobris Assessment', European Environment Agency, Copenhagen, 1995.
[8] 'Planning and Pollution Control', Planning Policy Guidance PPG 23, HMSO, London, 1994.

issues to be addressed in the Environmental Assessment required as part of an application for planning permission.

An Environmental Assessment needs to be undertaken in support of planning approval for a wide range of development which may have a significant effect upon the environment. The statutory requirement comes from the Environmental Assessment Directive (85/337/EEC), first agreed in 1985 but modified extensively to increase the range of topics coming within the scope of the Directive. An Environmental Assessment requires the applicant to prepare an Environmental Statement and to assess, and make available for public scrutiny and comment, the environmental aspects of a proposed new development. Environmental Statements are required to consider issues such as visual appearance and impact on the landscape as well as issues of environmental quality. They are also required under a revision of the original directive to contain an appraisal of options through an 'outline of the main alternatives studies and an indication of the main reasons for the scheme chosen, taking into account the environmental effects'.

Environmental assessments are in general complementary to and not in duplication of the fulfilment of environmental protection legislation and they provide an objective basis for decisions between alternative strategies for meeting infrastructure development needs. There has been concern expressed in many countries, including the United Kingdom, about the lack of integration of environmental assessments under planning regulations and BPEO assessments under pollution control regulations. The concerns arise on account of the additional burden placed on promoters of new schemes because of the duplication of activity in satisfying the demands of different regulatory regimes (planning and pollution prevention), the potential for conflicting outcomes, and uncertainty because of conflicts in timing between the two processes. The European Commission has reacted to this criticism by making provision within the IPPC directive for members states to run Environmental and IPPC assessments in parallel if they wish.

A further criticism of Environmental Assessments is that they take place too late in the evolution of environmental strategies. Many of the environmental issues and impacts caused by individual projects such as roads or industrial developments stem from earlier decisions taken at a higher decision-making level that cannot be tested at the project level. This leads to the concept of Strategic Environmental Assessments (SEAs) which could be carried out at an earlier stage at this higher level and which would help to integrate environmental considerations into policies from the outset.

4 POLLUTION CONTROL AGENCIES

4.1 Structure and Organization of Pollution Control Agencies

As we have seen the pollution control legislation in the United Kingdom has developed piecemeal, whereby successive legislation has led to new agencies being created, usually, but not always, independent of both national and local government structures. Since 1996 these various strands have now combined in the United Kingdom to form the Environment Agency and the Scottish Environmental Protection Agency, with comprehensive duties for pollution control as free standing public bodies independent of government, known as Non-Departmental Public Bodies (NDPBs). These agencies share some of their powers and duties with Local Authorities particularly on contaminated land and also, in England and Wales, on air pollution control. They carry out environmental protection duties with a wide strategic and operational remit, ranging from national policy development and direct interface with Government and the European Community through to the management of the local field inspectorate. The Agencies operate to environmental boundaries, based on river catchments, rather than local political boundaries and are fully independent of Local Authorities. The Agencies combine in one organization environmental management of the whole water cycle (integrated river basin management) and a wide range of pollution control duties (integrated pollution control). They have direct powers of prosecution through the courts without having to rely on any form of political or Government legal intermediary or prosecution service. The system in the United Kingdom is unique in Europe for its comprehensiveness of technical functions, degree of independence, and method of organization.

Many countries in Europe have created unified inspectorates but few of them link water management and the full range of pollution control duties, and few integrate their operational activities over river basins. Most, especially in federal states, are part of the regional government structure.

The UK approach has its supporters and its detractors. It is seen to maximize the possibility that regulatory decisions will be taken consistently, objectively, and on the basis of sound science, independent of local political influence. As national organizations the Agencies can, if necessary in technical collaboration, ensure that, when required, the highest level of technical expertise can be brought to bear on a problem. In addition, because the Agencies have the bulk of their staff located in their field inspectorates they can be responsive to local needs and can easily collaborate in local partnerships with other agencies.

At a policy level, regional as well as national environment decisions are intimately linked with social and economic decisions. Whereas it is right that regulatory decisions should be taken on the basis of sound science and technology, they should not be taken so remotely that there is no public confidence in them and no public under-standing of the technical rationale. The general public tend not to understand the technical complexity of issues but have strong views about the protection of the environment. If not involved they can feel that their voice is not carrying due weight, compared with the technically informed lobbyists from industry, who they might feel have a greater ability to influence regulatory decisions if they are not open to scrutiny. The issue of balance between sound science and public perception and understanding are a challenge for environmental regulators in every country and are a natural and welcome consequence of greater public knowledge of and concern for the environment.

There is also no consistency from country to country as to the best way of managing the environment in an integrated way. Some countries combine management of natural habitat within the same agency as pollution prevention and they might also combine related regulatory activities like the protection of drinking water and aspects of public health protection, which although not part of the natural environment, have very strong links to it.

Countries also differ in the way in which they manage the water cycle in comparison with other environmental media. The Environment Agencies in the United Kingdom combine both the integrated manage-ment of pollution control and the integrated management of the water catchment including water resources, fisheries, and flood defence within one organization and within common operational units. This approach is not replicated in many other countries in the world where basin management, if it exists, is usually independent of pollution control or is managed in discrete operational units. Achieving the optimum solu-tion to complex environmental problems depends on an integrated management of environmental media and a wide range of technical specialisms. Organizations, which are the subject of regulation because their activities impact on the environment, do not like dealing with a plethora of agencies and officials. Both of these arguments favour the creation of large multi-functional inspectorates but, as the above discus-sion reveals, there is no common view about the best way of organizing these complex organizations so that they are effective, efficient and enjoy public confidence.

4.2 Forestalling Pollution

Pollution control agencies have their origins in the need to enforce pollution control legislation. Historically, pollution control legislation has been reactive in character, setting standards, prosecuting breaches, and thereby forcing corrective action and setting an example to others. The philosophy behind the BAT approach to environmental regulation is that it sets standards not only for emissions but also for waste management and emission abatement technology. Compliance with environmental standards is more about having *processes* that conform to the needs of the environment rather than focusing on *end of pipe* regulation. This leads to the concept of the *enforcement notice* by which a pollution control inspector can issue a legally enforceable demand for a procedure to be modified so that acceptable standards of environmental protection can be achieved and risk of pollution forestalled. As a result of the Environment Act 1995 a wide range of enforcement notice powers are now in force, covering many aspects of industrial, commercial and agricultural practices.

Enforcement notices are general powers that can be used at any location where they are relevant without geographical restriction. There are situations, however, where environmental risks are of a localized and specific character and can be best addressed by special powers which would not be appropriate for general application. A good example of this is the restriction of land use activities within the catchment or gathering grounds of water supply sources. There is a long tradition in legislation in European countries for setting protection zones around points of groundwater abstraction. This statutory protection zone approach has not been applied in the United Kingdom largely because of its inflexibility and a consultative approach using planning as well as pollution control law, based upon a general groundwater protection policy[9] has been preferred. Recently, however, a system of protection zones has been established under the EC Nitrate Directive (91/676/EEC) which requires the creation of Nitrate Vulnerable Zones. These are zones where agricultural activity has caused or is in the process of causing pollution of either surface water or groundwater resources in excess of the EC Nitrate Standard of 50 mg l^{-1}. Different countries in Europe have implemented the directive in different ways. In Denmark, where because of the prevailing geology water resources are particularly vulnerable to agricultural activities and where virtually all of the public water supply is taken from groundwater, there is no value in applying

---

[9] 'Policy and Practice for the Protection of Groundwater', HMSO, London, 1998 (National Rivers Authority, 1991).

localized protection zones. Therefore Denmark has designated the whole country as a Nitrate Vulnerable Zone. In the case of the United Kingdom, and in most other Member States, there is significant local variation in the scale of the problem. Therefore zones have been defined on water quality and hydrological criteria to identify the catchments of both ground and surface water supplies, where land use change is necessary to mitigate the impact of agricultural practices on the concentration of nitrate in water resources. Case study 3 describes the approach adopted in the United Kingdom.

Another example of targeting pollution control to localized problem areas, this time in United Kingdom national legislation, is the setting up under the United Kingdom National Air Quality Strategy[2] of Air Quality Management Areas (AQMAs). These are areas where air quality objectives are unlikely to be achieved by existing regulatory measures and special action, through a local action plan, is necessary to force improvements in air quality through the existing planning and regulatory regime. Case study 4 describes the first steps being taken in the United Kingdom to set up these zones.

Case Study 3: Control of nitrate pollution from agricultural sources. The 1991 European Nitrate Directive is specifically directed towards the control of nitrates derived from agricultural sources. The directive applies to:

(1) *surface waters used for human consumption*, which currently have, or are expected in the future to have, nitrate concentrations in excess of 50 mg l^{-1} NO_3;

(2) *groundwaters used for human consumption*, which currently have, or are expected in the future to have, nitrate concentrations in excess of 50 mg l^{-1} NO_3;

(3) *waters, which are eutrophic* on account of their nitrate content.

There are three steps in the process to meet the requirements of the directive.

— Firstly Member States, using information from special monitoring programmes if sufficient data is not otherwise available, must identify the water bodies—rivers, lakes or groundwaters—to which the above criteria apply.

— Secondly they must identify the areas of land where agricultural practice is considered to be the cause, or potential cause, of the nitrate problems identified. These are called Nitrate Vulnerable Zones (NVZs).

— Thirdly they must set up a series of controls or limits on agricultural practice to apply in the defined NVZ, which will limit nitrate leaching and protect the water bodies identified.

Different Member States have adopted different approaches. Some countries, such as Denmark, have designated the whole country as a Nitrate Vulnerable Zone, which has removed many of the technical and administrative problems experienced in other Members States in carrying out the first two stages. The demands of the directive are onerous and are strongly contested by farmers who have lobbied strongly in all Member States to influence the way the directive has been implemented. In 1997 the European Commission published[10] an assessment of the implementation of the Directive and, to varying degrees, found fault with procedures adopted in every Member State.

In the United Kingdom NVZs were defined by identifying the catchment areas of all river reaches or points of groundwater abstraction where nitrate concentrations exceeded $50 \text{ mg l}^{-1} \text{ NO}_3$. In addition, for groundwater sources, assessments were made of those showing rising trends in nitrate concentration to allow for the long lag times when nitrate leaches into underground strata. Catchments for surface waters were defined from the topography; catchments for groundwater sources were defined by using groundwater models to work out flow directions. Where there were a number of adjacent groundwater sources with coalescing zones of pumping, these were combined and the NVZ boundaries where possible followed geological boundaries. No waters were identified which were considered eutrophic under the restricted definition of the directive.

This exercise led to the definition of 68 NVZs in the first phase in England and Wales. These are shown in Figure 1. The directive requires a four year revision which is likely to lead to further designations in 1998 as a result of more comprehensive data which has been collected in the intervening period. Also in 1998, the restrictions on agricultural practice came into force, requiring limitation on the amount and timing of fertilizer use and also limiting or controlling other activities which have been shown to cause excessive nitrate leaching.

Case Study 4: Introducing Air Quality Management Areas.   The United Kingdom Air Quality Strategy[2] set air quality targets to be achieved by the year 2005. The general expectation of Government is that these targets will be met by national programmes of emission controls and enforcement, planning controls, and a transport policy which reduces

---

[10] 'The Implementation of Council Directive 91/676/EEC concerning the protection of waters against pollution caused by nitrates from agricultural sources', European Commission COM(97)473, Luxembourg, 1997.

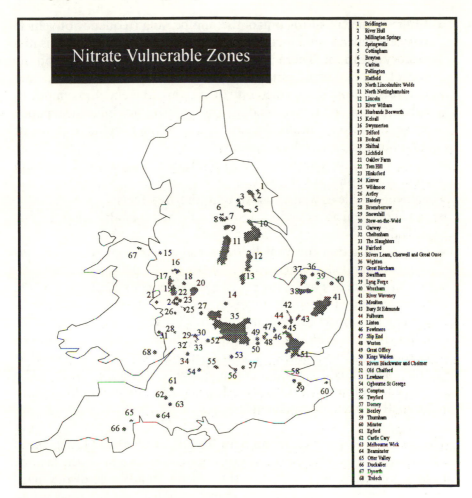

Figure 1 *Nitrate vulnerable zones*

reliance on road transport for both social and commercial transport. There remains considerable debate about both the standards and the time-scale in which they are expected to be achieved but there is no disagreement that there are some areas where national measures alone will not be sufficient to ensure that the targets are met. Local authorities are to be required to review and assess air quality in their areas. Where they believe that air quality objectives will not be met by 2005 by national measures alone, they must declare an Air Quality Management Area (AQMA) and draw up a specific and locally relevant action plan designed to achieve compliance.

The sorts of measures that might be put in place, some of which would require the grant of additional powers to local authorities, are:

- traffic management to control the number and frequency of vehicle movements and to control the movement of vehicles with unsatis-factory emissions. There is a wealth of experience worldwide on the use of flow control, restricted access, pedestrianization, parking controls, automatic traffic control systems, the encouragement of other modes of travel, and giving priority to high occupancy vehicles to reduce traffic density, although the benefits in terms of air quality are not well documented.
- more restrictive controls of point source emissions and problem industries within the AQMA.
- control on sale as well as use of unauthorized fuels.
- roadside vehicle emission tests with fixed penalty charges for offenders.
- access restriction for vehicles to sensitive areas, perhaps perma-nently or limited to certain times of the day or the year.
- use of planning powers to restrict developments likely to impede achievement of air quality objectives. Except in special circum-stances the impact of planning controls is likely to be felt only over a long time-scale although there is already an increasing number of cases where development plans are being opposed on grounds that traffic generated by the development would affect air quality. The existence of an Air Quality Management Area would greatly strengthen such an argument.

The ability to specify an action plan for an Air Quality Management Area and to choose between the various options is limited by the lack of data and predictive capability. For this reason, in advance of the main legislation, there has been a programme to establish 14 trial areas in the United Kingdom covering a variety of geographical locations and air quality problems. The emphasis of the work in these areas has been on data collection, validation of new monitoring techniques, and assessment of models, in each case focused on the particular air quality problems characterizing that area. Many of the areas have covered urban air quality issues but problems in more rural areas, for example north-east Derbyshire and Cornwall where particulates from quarrying are a particular concern, have been included.

4.3   Other Regulatory Action

Even with effective legislation and enforcement it is inevitable that, from time to time, pollution incidents will occur. They may arise because of a breach, accidental or malicious, of pollution control law or they may be due to unplanned circumstances, such as a road traffic accident involving a vehicle with a chemical load causing spillage that pollutes a river or aquifer. Regardless of the question of breach of environmental law the first requirement is to remedy the immediate situation. For this reason the Environment Agencies maintain an emergency response capability to attend accidents and emergencies where the environment might be placed at risk. Emergencies vary in scale and require collaborative action with other agencies but, in all cases, the objectives are to protect human life and health, protect property, and protect the environment. This action will be combined with investigation of the causes of the pollution.

An increasing role for environmental agencies is to encourage and promote environmental best practice, not as a substitute for the enforcement of pollution control regulations but to extend the influencing role beyond the negative and minimalist approach of avoiding prosecution. To this end the Environment Agencies have worked extensively with others to promote better environmental practice in industry, commerce, and agriculture, using the benefits of their extensive technical expertise and, on the ground, contact through environmental protection staff. These initiatives have not only concerned themselves with minimizing emissions but also with conservation of raw materials and environmental resources and the minimization of waste. These sorts of activities are increasingly coming within the scope of environmental legislation, in particular through the EC directive on packaging (94/62/EC) implemented in the United Kingdom through the producer responsibility legislation. These wider objectives of environmental best practice in business are fundamental to the development of Environmental Management Systems, which are discussed in Section 6.

5   ECONOMIC INSTRUMENTS FOR MANAGING POLLUTION

5.1   Alternatives to Pollution Regulation by Permit

In most countries, including the United Kingdom, operations (abstraction of water, discharge of gaseous or liquid effluents, movements and disposals of waste) which have a potential to impact on the environment require a permit (variously called an authorization, licence, or consent) from the pollution control agency. At least part of the cost of operating the pollution control permitting system is recovered by a system of charges for issuing and maintaining the permits. A criticism of this approach is that these permits can be regarded as licences to exploit the environment rather than a means of protecting it. The charges for permits are generally low compared with the costs of investing in environmental management of pollution control systems that would minimize or remove the need for the permit. The costs also often bear only a limited relationship to the impact of the regulated activity on the environment. In general the charging schemes will only offer a limited incentive to responsible operators and no encouragement to the discharger to go beyond the minimum requirements of their consents even if there was financial advantage to them doing so. Economic signals to business about an improved environment are therefore rather weak.

As a result, commentators[11] have suggested that the goal of sustainable development might be advanced if there were a change of emphasis in favour of economic instruments as an additional means of achieving environmental goals. Various forms of economic charge and incentive have been shown to reduce the total impact on the environment and to ensure that there is a more optimal take up of the capacity of the environment to absorb pollution. Examples of these economic instruments are various forms of environmental or green taxes, which levy a charge on particular products or processes to discourage their use or encourage recycling. Closely allied to this is the concept of tradeable permits whereby authorizations to use environmental resources, either to discharge pollutants to the environment or to abstract water, are tradeable assets, which can be bought and sold like any other commodity. This approach is intended to ensure that such environmental capacity as is available, over and above limits set by environmental quality objectives, is used to the greatest economic benefit, thereby giving scarce environmental resources a value, which is lacking under conventional regulatory systems.

These ideas have been widely canvassed in academic and policy circles and the most surprising thing is how little they have actually found their way into existing regulatory regimes, especially in the United Kingdom. Two reasons are quoted for the lack of readiness to introduce green taxes.

- The values of environmental benefits are hard to assess and therefore it is argued that there is considerable uncertainty as to how well any given tax will work, or even if it will have unexpected detrimental effects.
- It is a taxation principle in the United Kingdom that taxes should not be 'hypothecated', that is the decisions about tax raising and tax expenditure should not be linked. It has been argued that environmental taxes not spent to the benefit of the environment would not command public acceptance. Green taxes have their critics who

---

[11] 'Report of the House of Lords Committee on Sustainable Development', HMSO, London, 1995, Vol. 1.

question whether the basic motivation is to improve the environment or to be a convenient source of additional revenue for national
treasuries.

A further incentive to development in green taxation is that economic
theorists have begun to speculate that taxation to penalize pollution and
force investment into environmental improvement will not have a
negative impact on industry, which has been the previous assumption,
but will be a stimulus to the economy. This is the so-called 'double
dividend' of green taxation, both generating revenue and stimulating
growth and jobs in the environmental industries.[12] These arguments
remain controversial and demonstrate the difficulty there has been in
getting environmental issues to a central position in policy thinking.
However during 1997 the UK Government made a number of statements
indicating an increased commitment to green taxation and in November
1997 published a consultation paper on economic instruments for water
pollution.[13] Case study 5 below gives examples of situations around the
world where green taxes are in use.

In the United Kingdom various modest taxation measures with green
credentials have been established (VAT on domestic fuel, differentials in
duty between leaded and unleaded petrol). In October 1996 a landfill tax
was introduced which places a levy on the tonnage of waste deposited to
land. A further innovation of this tax is that arrangements can be made
for a proportion of the income to be returned, through a charitable trust
format, to finance environmental improvement projects, thus in some
way addressing the issue of returning the benefit of environmental taxes
to the environment. Early indications of the impact of this tax are that it
is sending important price signals, which over time are likely to shift
thinking away from waste production to various forms of waste
conservation. In some parts of the country there is evidence that it has
also increased the amount of illegal or opportunistic waste disposal
although this is likely to be a transitional impact as the industry comes to
terms with the new financial regime and is not expected to persist. In the
light of this experience it is now being suggested that an even larger tax
would send a stronger conservation signal.

One of the inherent objectives of a landfill tax is to increase costs of
landfilling and so direct wastes to various forms of recovery or reuse,
either as the product itself or through incineration as waste-to-energy
schemes. This is driven by the 1991 European Waste Framework
Directive (91/156/EEC) which required Member States to 'take appro-

---

[12] 'Environmental Policies and Employment', OECD, Paris, 1997.
[13] 'Economic Instruments for Water Pollution', Department of Environment, Transport and the
Regions, London, 1997.

priate measures to encourage firstly the prevention or reduction of waste
... and secondly the recovery of waste ... or the use of waste as a source
of energy'. This waste management hierarchy also features strongly in the
UK strategy.[3] Preventing the production of waste is the main objective of
producer responsibility legislation in its various forms. A scheme
managed by the Environment Agencies to control and to meet reduction
targets in packaging materials was introduced in the United Kingdom in
1997, and many similar schemes, extending to a wider variety of
materials, are being developed in other European Member States.

Many recycling schemes fail to achieve their potential of preventing
wastes going to landfill because the economic incentive is missing, that is
the environmental waste hierarchy is not matched by a parallel economic
waste hierarchy. The likely way forward is that waste management
decisions will be based upon a hierarchy that is influenced by both
economic and environmental costs. Sustainability can be built into this
equation by using fiscal measures to affect the economic costs. Life Cycle
Analysis, which quantifies the environmental impacts 'from cradle to
grave' of products or processes, is one of the tools being used to help
assess waste management options and the influence of economic incentives upon them.

5.2   Tradeable Permits

Attempts to establish systems of trading in environmental licences have
moved very slowly despite the fact that this approach is strongly
favoured by governments who favour market orientated solutions
rather than conventional regulatory solutions. One of the problems of
instituting a system of tradeable permits is that it has to operate within a
defined environmental unit so that, as emission or abstraction authorizations are bought and sold, the net environmental impact can be
controlled without regulatory intervention. Trading of water rights
within a catchment or within an estuary fits this concept well. In many
cases the overall net impact on the environmental system of moving the
rights to abstract or discharge from one location to another might be
assumed to be small. However a way needs to be found to prevent
unacceptable impact in the situation, for example, that a right to abstract
or discharge at the bottom of a catchment were sold for use nearer the
headwaters where the environmental impact might be unacceptable. It is
harder to apply the same ideas to the trading of air emission permits
because the same concept of a bounded environmental pool either does
not exist or cannot easily be identified or may not be under a common
jurisdiction. In practice any scheme involving tradeable environmental
permits is only likely to be robust in the heterogeneous environmental

conditions which exist in most of Europe over small geographic areas where the number of potential traders is likely to be small. Likely exceptions are water abstraction in areas of intense agricultural use for irrigation and discharges to estuaries.

Case Study 5: Examples of economic instruments with environmental benefits. There are three categories of economic instrument, which can be identified:

- taxation schemes which are targeted at products or services and which are designed to reduce use of potentially harmful substances or otherwise reduce environmental impacts but where the revenue generated is added to general taxation.

    There are a number of well established models, in particular taxes on acid gas wastes in Sweden, on toxic waste in Germany and on various aspects of water pollution in The Netherlands. Nordic countries have led the way in this regard and, for example, Sweden is said to raise in total the equivalent of 6% of its GDP from green taxes. Other taxes which have been, or are currently, under consideration either at a European or Member State level are taxes on VOCs, sulfur, and carbon, the last often called an energy tax.

- taxation schemes which may have a desirable environmental outcome in their own right but where some or all of the revenue derived from the taxation is also devoted to environmental improvement through subsidizing or promoting environmental improvement.

    The DETR report[13] quotes examples from Germany and France where water pollution charging schemes are used to provide incentive for environmental improvement. In Germany discounts on the charging tariff are available if the discharger meets or exceeds the standards required under the best technical means standards. In France a similar scheme also provides rebates for meeting best technical means standards and generates revenue for investment in wastewater treatment and pollution abatement projects. Revenue from environmental taxation in Norway is funding a scheme to separate carbon dioxide from the Norwegian offshore gas fields and to reinject it back into the reservoir formation to minimize the atmospheric impacts.

- tradeable permits; systems of licensing and charging for environmental capacity (to discharge pollutants or to use renewable natural resources) which can be traded and which, by creating a market, encourage the optimal use of the environmental capacity.

    Permit trading is well established in the field of water resources in arid areas. Water use permits for crop irrigation depending upon water transfer schemes incorporate various forms of rights trading, annual licence, and even a rights auction to optimize the economic return from the water transfer. Well established examples exist in the south-west USA (California, Colorado River) and in Australia (Snowy Mountain Scheme). Trading of water pollution rights is much less well established. A number of studies have shown theoretical cost savings and environmental benefits but in practice the achievements have been very limited. One scheme for BOD trading on the Fox River in Wisconsin was established in 1981 but only one trade has taken place. In the field of air quality the greatest consideration in the United Kingdom has been given to sulfur trading as a means of controlling SO_2 and meeting international obligations with minimum commercial impact. To a degree this was achieved by the IPC authorizations issued to power stations which allowed a degree of trade between stations but the full application of a trading approach has not been implemented because it could not be guaranteed to be compatible with BATNEEC. The second international sulfur protocol requires that the UK reduce sulfur emissions progressively by 80% from a 1980 baseline by 2010. It is now apparent that this will be achieved within the existing pollution control regime without recourse to other means and the tradeable permit approach will not be further extended. On the global scale, however, one emission trading agreement has been reached. The negotiations at the UN Kyoto 1997 World Climate Conference on the control of greenhouse gas emissions involved a trade in national emission quotas, which will allow the USA to offset its target reductions by acquiring additional quota from other nations, particularly Russia.

A United Kingdom Government report published in 1993 said 'although theory suggests that economic instruments will save resources, the limited use of economic instruments to date means that there is no widespread body of empirical evidence to demonstrate this in practice'. This view remains valid.

6   PUBLIC AND COMMERCIAL PRESSURES TO IMPROVE THE ENVIRONMENT

6.1   Environmental Management Systems

So far we have considered the management of environmental quality in terms of what Governments or agencies of Government can do through legal and financial incentives to encourage companies and individuals to not harm, and where possible, to improve the environment. These are widely held public goals, so why do organizations not set environmental objectives for themselves independent of or in addition to the demands of regulation? The answer is, of course, that increasingly they do, both to be certain that they achieve compliance with regulation and to meet wider goals, particularly if they see competitive advantage or cost savings accruing. The object of this section is to review various aspects of environmental management systems and procedures, which are adopted by organizations to meet this objective. The term Environmental Management System (EMS) is the generic term used to describe such procedures and is the subject of international standard ISO 14001,[14] which from 1997 has replaced the equivalent British Standard BS7750. The key reasons why a company would wish to develop an EMS are:

- the requirement for regulatory compliance
- the need to establish systems to ensure that compliance is maintained
- financial incentives
- market place pressures
- green credentials

An important factor in the development of environmental management systems has been the influence which companies have on their trading partners, the so-called supply chain pressures. Major manufacturers and traders—particularly in the automotive industry, which has been in the forefront of quality management, supermarkets, and other parts of the retail trade—have been a major force, through their own environmental policies, in encouraging and often obliging their suppliers to adopt an approved EMS.

The principal formalized EMS in Europe is the European Communities EMAS (Ecomanagement and Audit Scheme). This can be achieved by meeting the ISO 14001 standard but also requires a verified public environmental policy to meet the requirements.

The take up of EMAS within Europe has been very variable as shown in Table 6. In four countries no sites had been registered by mid-1997, more than two years after the scheme was initiated. The greatest take up is in Germany which is generally believed to be because companies in that country believe that registration under EMAS will reduce the formal regulatory burden upon them. The position of the Environment Agencies in the United Kingdom is that they encourage companies to establish

---

[14] 'Environmental Management Systems—Specifications with Guidelines for Use', EN ISO 14001, CEN, Brussels, 1996.

Table 6   *Growth in number of EMAS Registered Sites in EU Member States*

	February 1996	April 1997
Austria	6	56
Belgium	2	3
Denmark	3	17
Finland	0	4
France	3	7
Germany	113	518
Ireland	1	2
The Netherlands	2	13
Spain	0	4
Sweden	1	45
UK	9	26

Source: EC Official Journal.
None registered in Greece, Italy, Luxembourg and Portugal as at 1 April 1997. The scheme commenced in 1995.

an EMS but that this, in itself, is not seen as a substitute for environmental regulation. It is notable that one of the first companies to achieve BS7750 accreditation in the United Kingdom was prosecuted for causing water pollution shortly afterwards.[15]

There is good evidence from many companies that investment in EMS is repaid in both tangible benefits, such as reduced operating costs, and intangibles, for example improved safety, improved public perception, and less risk of prosecution. However the achievement of EMAS or ISO 14001 accreditation and the ongoing maintenance of the necessary systems is a significant investment in time and resources, which many smaller organizations are unwilling to commit. The benefits, in reduced operating costs and reduced risk of causing pollution, of adopting environmental best practice are still important to business and bring with them an improvement to the environment. Therefore, following the adage that 'prevention is better than cure', pollution control agencies are increasingly devoting effort to education in pollution prevention and good environmental management through special campaigns, on-site advice and promotion of waste minimization and recovery projects.

6.2   Public Opinion and the Environment

There is a high level of general public awareness of environmental issues, which is itself a positive force for environmental improvement. In rare cases the weight of public opinion can be decisive. This was most

[15] 'First BS7750 holder to be fined for pollution', *ENDS Magazine*, Feb. 1997, **265**, 43.

dramatically shown in the case of the proposed deep sea dismantling of the disused *Brent Spar* oil platform by the Shell Company. This so excited public opinion that a boycott of petrol sales by the company throughout Europe, led by the environmental pressure group Greenpeace, forced the company to abandon their proposal. Less spectacular, but with significant effect over the years, has been the close independent monitoring of environmental issues by environmental interest groups, providing reports to the European Commission of failures to implement directives or prosecute breaches. Publicly accessible registers of environmental monitoring data provide an important source of public information on the state of the environment and demonstrate accountability to the public for the way it is regulated.

There is also an increasing interest and demand by the public for a close involvement in decisions affecting the environment. Public consultation takes place on applications for the issue of permits to release waste into the environment. The style of this consultation has changed significantly as public awareness and interest in issues affecting their environment has increased. Pollution control agencies are increasingly being required to engage in debate about the technical issues surrounding the granting of permits and provide greater public documentation of their decisions and the reasons for them. Whereas traditionally newspaper advertisements were the medium to make the public aware of new applications, some pollution control agencies are beginning to make this information available on the Internet. Many issues which attract public concern are those for which the technical assessments are complex and involve judgements of level of risk often involving widely varying public perceptions of what is acceptable. Regulatory decisions must be based upon sound science but the decisions are being made on behalf of the public and for the wider public good. Therefore pollution control agencies cannot ignore issues of public perception of risk and must, through consultation and information, seek to maintain the confidence of the communities they serve while making regulatory decisions that contribute to the goal of environmental improvement.

QUESTIONS

1. Discuss with examples the concepts of *objectives*, *standards*, and *limits* in relation to environmental quality.
2. Compare and contrast the relative merits of pollution control strategies based respectively upon environmental quality objectives and uniform discharge standards, illustrating with examples as appropriate.

3. Discuss in the context of air, freshwaters, and soil the appropriateness of applying a 'fitness for use' approach to environmental quality.

4. Discuss the appropriateness and effectiveness of using taxation measures as an instrument to control pollution.

5. Describe the concept of integrated pollution control. Discuss the factors which would need to be taken into account in evaluating the pollution impact of a lead battery works whose activities involve discharge to air and water, and contamination of soils within the works boundary. If the primary aim is to protect the health of the public, what considerations would be necessary to minimize the effect of the works in the optimal manner?

Subject Index